A TEXTBOOK OF
MANUFACTURING TECHNOLOGY–II

[For the students of B.E. / B.Tech Anna University & other Technical Universities of India]

AF673196

Dr. P.C. SHARMA

B.Sc. Engg. (Mech.) (Hons.)
M.Sc. Engg. (Mech.) (Distinction)
Ph.D. (I.I.T. Delhi)
MISTE, LMISME

Visiting Professor

Department of Mechanical Engineering
S.U.S. College of Engineering and Technology
Tangori (Mohali), Punjab
Formerly of Punjab Engineering College Chandigarh

S. CHAND
PUBLISHING

S Chand And Company Limited
(ISO 9001 Certified Company)

S Chand And Company Limited
(ISO 9001 Certified Company)
Head Office: D-92, Sector–2, Noida – 201301, U.P. (India), Ph. 91-120-4682700
Registered Office: A-27, 2nd Floor, Mohan Co-operative Industrial Estate,
New Delhi – 110 044, Phone: 011-49731800
www.**schandpublishing.com;** e-mail: **info@schandpublishing.com**

S. CHAND
PUBLISHING

Marketing Offices:

Chennai	:	Ph: 23632120; chennai@schandpublishing.com
Guwahati	:	Ph: 2738811, 2735640; guwahati@schandpublishing.com
Hyderabad	:	Ph: 40186018; hyderabad@schandpublishing.com
Jalandhar	:	Ph: 4645630; jalandhar@schandpublishing.com
Kolkata	:	Ph: 23357458, 23353914; kolkata@schandpublishing.com
Lucknow	:	Ph: 4003633; lucknow@schandpublishing.com
Mumbai	:	Ph: 25000297; mumbai@schandpublishing.com
Patna	:	Ph: 2260011; patna@schandpublishing.com

First Edition 2008
Reprints 2010, 2011, 2012, 2013, 2014, 2015, 2017, 2018, 2020, 2021, 2024
Reprint 2025

ISBN : 978-81-219-2846-5 **Product Code :** H3MFT62MAFT10ENAA08O

PRINTED IN INDIA

By Vikas Publishing House Private Limited, Plot 20/4, Site-IV, Industrial Area Sahibabad, Ghaziabad – 201 010 and Published by S Chand And Company Limited, A-27, 2nd Floor, Mohan Co-operative Industrial Estate, New Delhi – 110 044.

PREFACE

This book "A Textbook of Manufacturing Technology–II " is the result of persistent demand from the students and teachers of various Technical Institutes affliated to University of Tamil Nadu. The book fully covers the contents of the syllabus of the course : ME 1203, "Mechanical Technology–II", taught to undergraduate Mechanical Engineering students of the University.

The author sincerely hopes that the book will meet the needs of the readers. All the suggestions to improve the text of the book will always be welcomed.

Pointing out of errors and misprints in the text will be highly obliged and thankfully acknowledged.

P.C. SHARMA

CONTENTS

1

THEORY OF METAL CUTTING

1.1. GENERAL

Metal cutting or "Machining" is the process of producing a workpiece by removing unwanted material from a block of metal, in the form of chips. This process is most important since almost all the products get their final shape and size by metal removal, either directly or indirectly. The major drawback of the process is loss of material in the form of chips. Inspite of these drawbacks, the machining process has the following characteristics :

1. They improve the dimensional accuracy and tolerances of the components produced by other processes.
2. Internal and external surface features which are difficult or not possible to be produced by other processes, can be produced by machining processes.
3. Specified surface characteristics or texture can be achieved on a part or whole of the component.
4. It may be economical to produce a component by machining process.

   In this chapter, we shall have a fundamental understanding of the basic metal cutting process.

1.2. THE MECHANICS OF CHIP FORMATION

A typical metal cutting process can be schematically represented as shown in Fig. 1.1. A wedge-shaped tool is made to move relative to the workpiece. As the tool makes contact with the metal, it exerts a pressure on it resulting in the compression of the metal near the tool tip. This induces shear-type deformation within the metal and it starts moving upward along the top face of the tool. As the tool advances, the material ahead of it is sheared continuously along a plane called the "Shear plane". This shear plane is actually a narrow zone (of the order of about 0.025 mm) and extends from the cutting edge of the tool to the surface of the workpiece. The cutting edge of the tool is formed by two intersecting surfaces. The surface along which the chip moves upwards is called "Rake surface" and the other surface which is relieved to avoid rubbing with the machined surface, is called "Flank". The angle between the rake surface and the normal is known as "Rake angle", α (which may be positive or negative), and the angle between the flank and the horizontal machined surface is known as the "relief or clearance angle", γ. Most cutting processes have the same basic features as shown in Fig. 1.1, where a single point cutting tool is used (a milling cutter, a drill, and a broach can be regarded as several single-point tools joined together and are known as multi-point tools).

1.3. SINGLE POINT CUTTING TOOL

A single point cutting tool consists of a sharpened cutting part called its point and the shank, (Fig. 1.2). The point of the tool is bounded by the face (along which the chips slide as they are cut by the tool), the side flank or major flank the end flank, or minor flank and the base. The side cutting edge, *a-b*, is formed by the intersection of the face and side flank. The end cutting

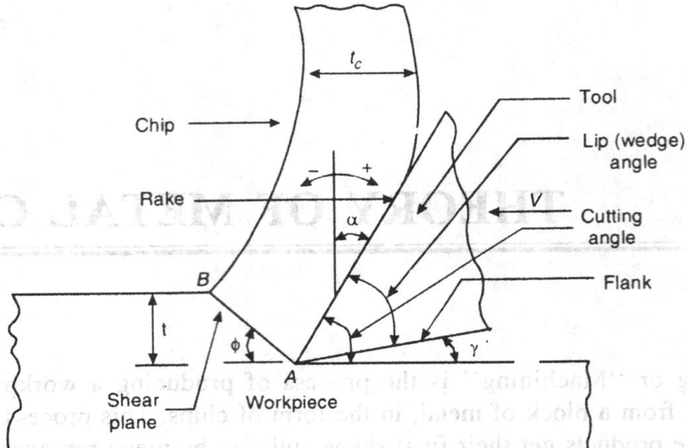

ϕ = shear angle ; t = uncut chip thickness ;
t_c = chip thickness after the metal is cut

Fig. 1.1. Schematic Representation of Machining Process

edge *a-c* is formed by the intersection of the face and the end flank. The chips are cut from the work piece by the side-cutting edge. The point '*a*' where the end and side-cutting edges meet is called the "nose" of the tool. Fig 1.2 is for a right hand tool. Below, we give the definitions of the various tool elements and tool angles :–

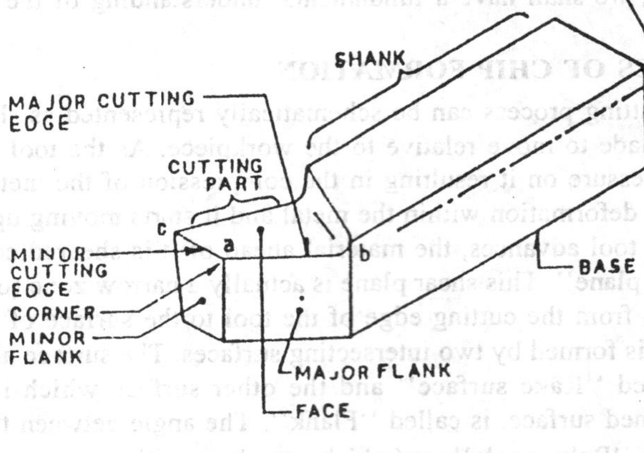

Fig. 1.2. A Single Point Cutting Tool.

Shank. It is the main body of the tool.

Flank. The surface or surfaces below and adjacent to the cutting edge is called flank of the tool.

Face. The surface on which the chip slides is called the face of the tool.

Heel. It is the intersection of the flank and the base of the tool.

Nose. It is the point where the side cutting edge and end cutting edge intersect.

Cutting Edge. It is the edge on the face of the tool which removes the material from the workpiece. The total cutting edge consists of side cutting edge (major cutting edge), end cutting edge (minor cutting edge and the nose).

A single point cutting tool may be either right or left hand cut tool depending on the direction of feed. In a right cut tool, the side cutting edge is on the side of the thumb when the right hand is placed on the tool with the palm downward and the fingers pointed towards the tool nose (Fig. 1.3 *b*). Such a tool will cut when fed from right to left as in a lathe in which the tool moves from tailstock to headstock. A left-cut tool is one in which the side cutting edge is on the thumb side when the left hand is applied (Fig. 1.3 *a*). Such a tool will cut when fed from left to right.

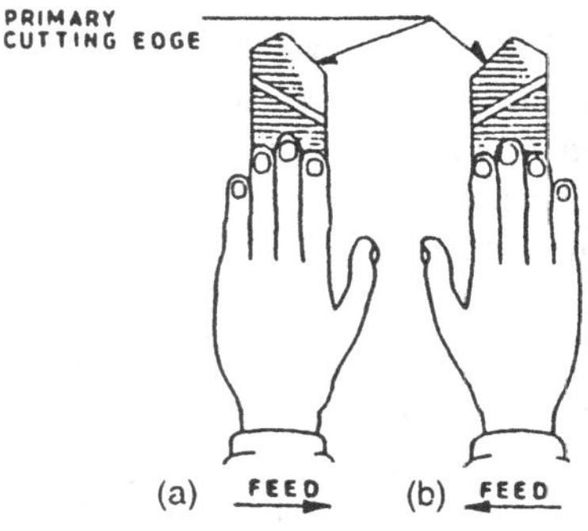

Fig. 1.3. Left and Right Cut Tools.

The various types of surfaces and planes in metal cutting are explained below with the help of Fig. 1.4, in which the basic turning process is shown. The three types of surfaces are :

(1) the work surface, from which the material is cut.

(2) the machined surface which is formed or generated after removing the chip.

(3) the cutting surface which is formed by the side cutting edge of the tool.

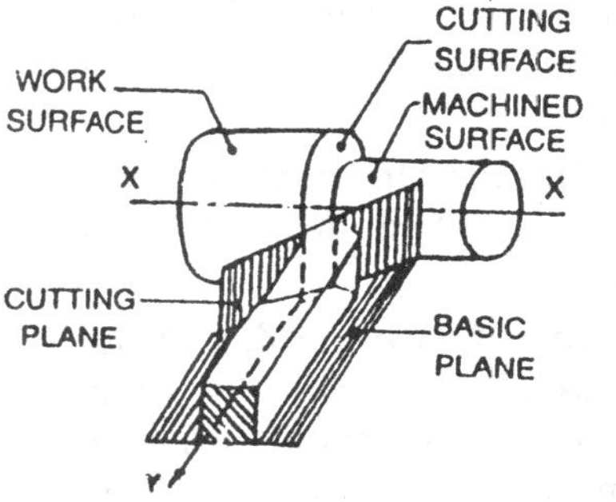

Fig. 1.4. Principal Surfaces and Planes in Metal Cutting.

The references from which the tool angles are specified are the 'cutting plane' and the 'basic plane' or the 'principal plane'. The cutting plane is the plane tangent to the cutting surface and passing through and containing the side cutting edge. The basic plane is the plane parallel to the longitudinal and cross feeds, that is, this plane lies along and normal to the longitudinal axis of the workpiece. In a lathe tool, the basic plane concides with the base of the tool.

1.3.1. Designation of Cutting Tools. By designation or nomenclature of a cutting tool is meant the designation of the shape of the cutting part of the tool. The two systems to designate the tool shape, which are widely used, are :—

1. American Standards Association System (ASA) or American National Standards Institute (ANSI).
2. Orthogonal Rake System (ORS).

ASA System. In the ASA system, the angles of tool face, that, is its slope, are defined in two orthogonal planes, one parallel to and the other perpendicular to, the axis of the cutting tool, both planes being perpendicular to the base of the tool. For simple turning operation, this system is illustrated in Fig. 1.5.

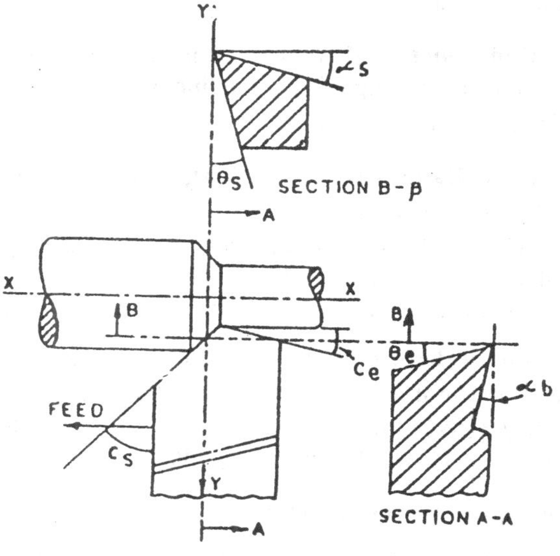

Fig. 1.5. ASA System.

The typical right hand single point cutting tool terminology is given in Fig. 1.6 (*a*). Fig. 1.6(*b*) gives the three views of the single point cutting tool, with all the details marked on it.

The various tool angles are defined and explained below :

Side Cutting Edge Angle (SCEA). Side cutting edge angle, C_s, also known as lead angle, is the angle between the side cutting edge and the side of the tool shank.

The complimentary angle of SCEA is called the "approach angle".

End Cutting Edge Angle (ECEA). This is the angle between the end cutting edge and a line normal to the tool shank C_e.

Side Relief Angle (SRA). It is the angle between the portion of the side flank immediately below the side cutting edge and a line perpendicular to the base of the tool, and measured at right angle to the side flank θ_s.

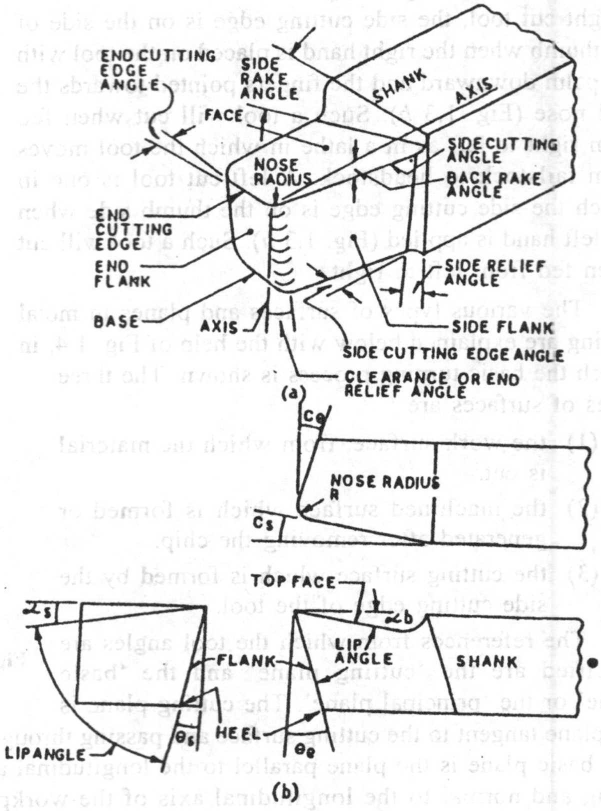

Fig. 1.6. Tool Terminology.

End Relief Angle (ERA). It is the angle between the portion of the end flank immediately below the end cutting edge and a line perpendicular to the base of the tool, and measured at right angle to the end flank θ_e.

Back Rake Angle (BRA), α_b. It is the angle between the face of the tool and a line parallel to the base of the tool and measured in a plane (perpendicular) through the side cutting edge. This angle is positive, if the side cutting edge slopes downwards from the point towards the shank and is negative if the slope of the side cutting edge is reverse. So this angle gives the slope of the face of the tool from the nose towards the shank.

Side Rake Angle (SR), α_s. It is the angle between the tool face and a line parallel to the base of the tool and measured in a plane perpendicular to the base and the side cutting edge. This angle gives the slope of the face of the tool from the cutting edge. The side rake is negative if the slope is towards the cutting edge and is positive if the slope is away from the cutting edge.

Importance of Tool Angles :

1. Side Cutting-Edge Angle, C_s. It is the angle which prevents interference as the tool enters the work material. The tip of the tool is protected at the start of the cut, Fig. 1.7, as it enables the tool to contact the work first behind the tip. This angle affects tool life and surface finish. This angle can vary from 0° to 90°. The side cutting edge at increased value of SCEA will have more of its length in action for a given depth of cut and the edge lasts longer. Also, the chip produced will be thinner and wider which will distribute the cutting and heat produced over more of the cutting edge. On the other hand, the larger this angle, the greater the component of force tending to separate the work and the tool. This promotes chatter. Satisfactory values of SCEA vary from 15° to 30°, for general machining. The shape of the workpiece will also determine the SCEA. To produce a 90° shoulder, zero degree SCEA is needed. No SCEA is desirable when machining castings and forgings with hard and scaly skins, because the least amount of tool edge should be exposed to the destructive action of the skin.

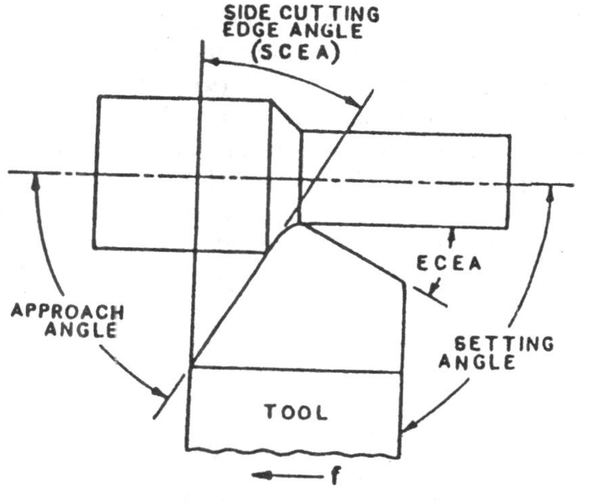

Fig. 1.7. SCEA and ECEA.

2. End Cutting Edge Angle, C_e. The ECEA provides a clearance or relief to the trailing end of the cutting edge to prevent rubbing or drag between the machined surface and the trailing (non-cutting) part of the cutting edge. Only a small angle is sufficient for this purpose. Too large an ECEA takes away material that supports the point and conducts away the heat. An angle of 8° to 15° has been found satisfactory in most cases on side cutting tools, like boring and turning tools. Sometimes, on finishing tools, a small flat (1.6 to 8 mm long) is ground on the front portion of the edge next to the nose radius, to level the irregular surface produced by a roughing tool. End cutting tools, like cut off and necking tools often have no end cutting-edge angle.

3. Side Relief Angle, (SRA) and End Relief Angle (ERA). These angles (denoted as θ_s and θ_e in the figure) are provided so that the flank of the tool clears the workpiece surface and there is no rubbing action between the two. Relief angles range from 5° to 15° for general turning. Small relief angles are necessary to give strength to the cutting edge when machining hard and strong materials. Tools with increased values of relief angles penetrate and cut the workpiece material more efficiently and this reduces the cutting forces. Too large relief angles weaken the cutting edge and there is less mass to absorb and conduct the heat away from the cutting edge.

4. Back and Side Rake Angle (α_b, α_s). The top face of the tool over which the chip flows is known as the rake face. The angle which this face makes with the normal to the machined surface at the cutting edge is known as "Back-rake angle, α_b", and the angle between the face and a plane parallel to the tool base and measured in a plane perpendicular to both the base of the tool holder and the side cutting edge, is known as "Side-rake angle, α_s". The rake angles may be positive, zero, or negative. Cutting angle and the angle of shear are affected by the values for rake angles. Larger the rake angle, smaller the cutting angle (and larger the shear angle) and the lower the cutting force and power. However, since, increasing the rake angle decreases the cutting angle, this leaves less metal at the point of the tool to support the cutting edge and conduct away the heat. A practical rake angle represents a compromise between a large angle for easier cutting and a small angle for tool strength. In general, the rake angle is small for cutting hard materials and large for cutting soft ductile materials. An exception is brass which is machined with a small or negative rake angle to prevent the tool form digging into the work.

The use of negative rake angles started with the employment of carbide cutting tools. When we use positive rake angle, the force on the tool is directed towards the cutting edge, tending to chip or break it, Fig. 1.8 (*a*). Carbide being brittle lacks shock resistance and will fail if positive rake angles are used with it. Using negative rake angles, directs the force back into the body of the tool away from the cutting edge, Fig. 1.8 (*b*), which gives protection to the cutting edge. The use of negative rake angle, increases the cutting force. But at higher cutting speeds, at which carbide cutting tools can be used, this increase in force is less than at normal cutting speeds. High cutting speeds are, therefore, always used with negative rakes, which requires ample power of the machine tool.

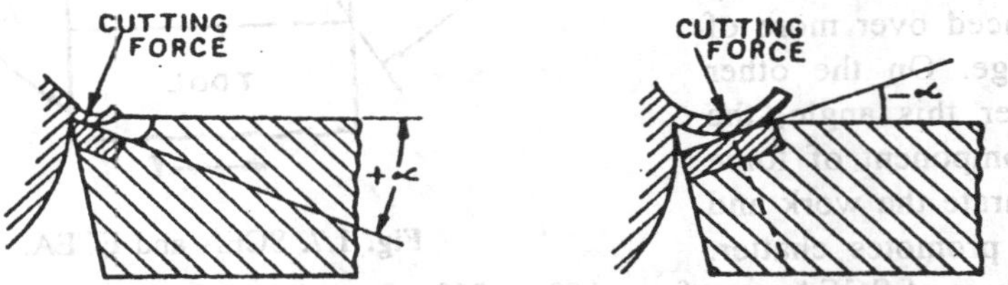

(a) WITH POSITIVE RAKE (b) WITH NEGATIVE RAKE

Fig. 1.8. Cutting with Positive and Negative Rake Tools.

The use of indexable inserts has also promoted the use of negative rake angles. An insert with a negative rake angle has twice as many cutting edges as an equivalent positive rake angle insert (as will be discussed ahead). So, to machine a given number of components, smaller number of negative rake inserts are needed as compared to positive rake inserts.

The use of positive rake angles is recommended under the following conditions :

1. When machining low strength ferrous and non-ferrous materials and work-hardening materials.
2. When using low power machines.
3. When machining long shafts of small diameters.
4. When the set up lacks strength and rigidity.
5. When cutting at low cutting speeds.

The use of negative rake angles is recommended under the following conditions :

1. When machining high strength alloys.
2. When there are heavy impact loads such as in interrupted machining.
3. For rigid set ups and when cutting at high speeds.

Recommended rake angles are given in Table 1.1.

Table 1.1. Recommended Rake Angles

Work Material	*Tool Material*					
	H.S.S. and Cast Alloys		*Cemented Carbide*		+	
			Brazed		*Throw away*	
	Back	*Side*	*Back*	*Side*	*Back*	*Side*
Free Machining Steels	10	12	0	6	−5	−5
Mild Steel	8	10	0	6	−5	−5
Med. Carbon Steels	0	10	0	6	−5	−5

Alloy tool Steels	0	10	– 5	– 5	– 5	– 5
Stainless Steel	0	10	0	6	– 5	– 5
Cast Iron	5	5 to 10	0	6	– 5	– 5
Aluminium Alloys	20	15	3	15	0	5
Copper Alloys	5	10	0	8	0	5
Magnesium Alloys	20	15	3	15	0	5
Titanium Alloys	0	5	0	6	– 5	– 5

5. Nose Radius. Nose radius is favourable to long tool life and good surface finish. A sharp point on the end of a tool is highly stressed, short lived and leaves a groove in the path of cut. There is an improvement in surface finish and permissible cutting speed as nose radius is increased from zero value. Too large a nose radius will induce chatter. The use of following values for nose radius is recommended :

R = 0.4 mm, for delicate components.

$\geq$ 1.5 mm for heavy depths of cut, interrupted cuts and heavy feeds.

= 0.4 mm to 1.2 mm for disposable carbide inserts for common use.

= 1.2 to 1.6 mm for heavy duty inserts.

The rules of thumb for selection of Nose radius are :-

(*i*) For a strong cutting edge, select the largest prossible Nose radius.

(*ii*) A large nose radius permits higher feeds.

(*iii*) Select a smaller nose radius if there is a tendency to vibrate.

For rough machining, the most commonly used nose radii are 1.2 to 1.6 mm.

Tool Designation. The tool designation or tool signature, under ASA system is given in the order given next :

Back rake, Side rake, End relief, Side relief, End cutting edge angle, Side cutting edge angle, and nose radius that is,

$$\alpha_b - \alpha_s - \theta_e - \theta_s - C_e - C_s - R$$

If tool designation is :

$$8 - 14 - 6 - 6 - 6 - 15 - \frac{1}{8}, \text{ it means that,}$$

$$\alpha_b = 8°, \qquad \alpha_s = 14°$$

$$\theta_e = 6°, \qquad \theta_s = 6°$$

$$C_e = 6°, \qquad C_s = 15°$$

and $R = \frac{1''}{8}$.

In ASA system of tool angles, the angles are specified independently of the position of the cutting edge. It, therefore, does not give any indication of the behaviour of the tool in practice. Therefore, in actual cutting operation, we should include the side cutting edge (principal cutting edge) in the scheme of reference planes. Such a system is known as Orthogonal Rake System (ORS).

Orthogonal Rake System (ORS). As mentioned above, in this system the planes for designating tools are the plane containing the principal or side cutting edge and the plane normal to it. In the plane NN which is normal to the principal cutting edge and is known as Orthogonal

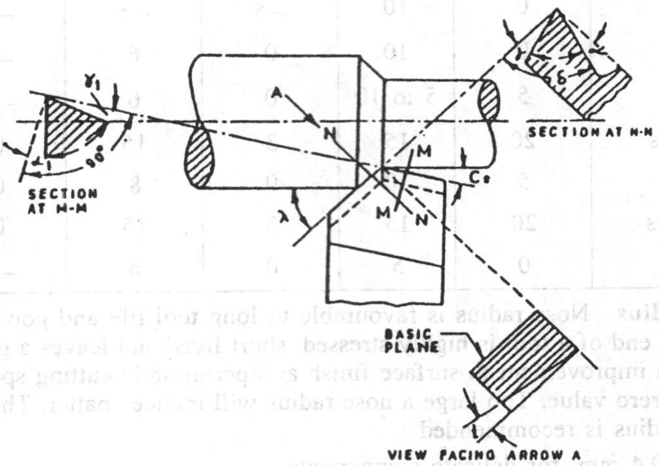

Fig. 1.9. ORS of Tool Angles.

plane or the chief plane, we have the following angles : side relief angle γ, the side rake angle (known as Orthogonal rake angle) α, wedge (lip angle) and the cutting angle (see Fig. 1.9).

The side relief angle is the angle between the side (main) flank and the cutting plane. The side rake angle, α, is the angle between the toolface and a plane normal to the cutting plane and passing through the main cutting edge. This angle is positive when the face slopes downward from the plane perpendicular to the cutting plane (as shown in Fig. 1.9), equal to zero when the face is perpendicular to the cutting plane and negative when the face slopes upwards. The "wedge angle, β" is the angle between the tool face and the main flank. The "cutting angle, δ" is the angle between the tool face and the cutting plane. When α is positive, we have,

$$\alpha + \gamma + \text{wedge angle} = 90°$$

$$\gamma + \beta = \delta$$

The usual values of α and γ are :

$\alpha = -10°$ to $+15°$, $\gamma = 6°$ to $12°$

In the ORS, the back rake angle is the inclination angle (i) between the principal cutting edge and a line passing through the point of the tool parallel to the principal plane. This angle is measured in a plane passing through the main cutting edge and perpendicular to the basic plane. In Fig. 1.9, the angle i is negative with tool nose being the highest point of the cutting edge. It will be zero when the cutting edge is parallel to the basic plane and positive if the cutting edge is towards the right (Fig. 1.9) of the line passing through the point of the tool and parallel to the principal (basic) plane, that is, the tool nose is the lowest point of the cutting edge.

In addition to the angles discussed above, angles are also measured in the plane MM (known as Auxiliary reference plane) which is normal to the projection of the end cutting edge on the basic plane. These angles are the end relief angle γ_1, and the back rake angle α_1 (also called auxiliary rake angle). The plan angles are the Approach angle or entering angle λ which is equal to $(90° - C_s)$ and the end cutting edge angle, C_e.

$\gamma_1 = 8°$ to $10°$, $\lambda = 30°$ to $70°$, $C_e = 10°$ to $15°$

The tool designation under ORS is :

$$i - \alpha - \gamma - \gamma_1 - C_e - \lambda - R$$

A typical tool designation (signature) is :

$$0 - 10 - 6 - 6 - 8 - 90 - 1 \text{ mm}$$

Interconversion between ASA system and ORS

$$\tan\alpha = \tan\alpha_s \sin\lambda + \tan\alpha_b \cos\lambda$$

$$\tan\alpha_b = \cos\lambda \tan\alpha + \sin\lambda \tan i$$

$$\tan\alpha_s = \sin\lambda \tan\alpha - \cos\lambda \tan i$$

$$\tan i = -\tan\alpha_s \cos\lambda + \tan\alpha_b \sin\lambda$$

In the second and third equations above, the values of angles α and i are taken with their signs.

1.4. METHODS OF MACHINING

In the metal cutting operation, Fig. 1.1, the tool is wedge-shaped and has a straight cutting edge. Basically, there are two methods of metal cutting, depending upon the arrangement of the cutting edge with respect to the direction of relative work-tool motion :

1. Orthogonal cutting or two dimensional cutting.
2. Oblique cutting or three dimensioning cutting.

In orthogonal cutting, Fig. 1.10, the cutting edge of the tool is arranged perpendicular to the cutting velocity vector, V, whereas in oblique cutting, it is set at some angle other than 90° to the cutting velocity vector, which gives an "inclination angle i". The analysis of oblique cutting being very complex, the relatively simple arrangement of orthogonal cutting is, therefore, widely used in theoretical and experimental work.

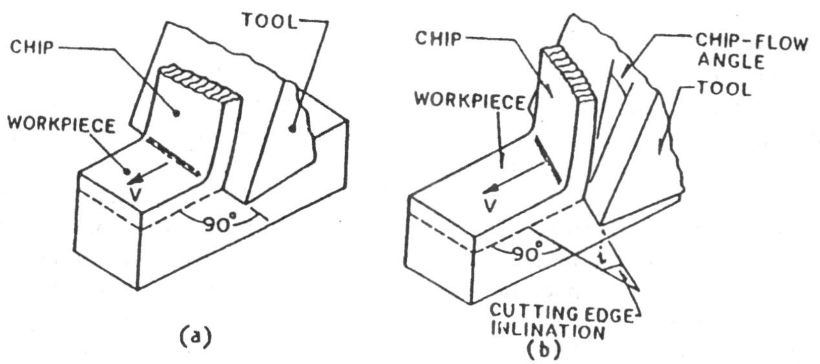

Fig. 1.10. Methods of Machining.

In pure orthogonal cutting, $i = 0°$, $C_e = 0°$, and $\lambda = 90°$. This is also known as orthogonal system of second kind. When $i = 0$, and $0 < \lambda < 90°$, it is called as orthogonal system of first kind. A common example of pure orthogonal cutting process is the turning of a thin pipe with a straight edged tool set normal to the longitudinal axes.

1.5. TYPES OF CHIPS

Whatever the cutting conditions can be, the chips produced may belong to one of the following three types, (Fig. 1.11) :

1. Discontinuous Chips.
2. Continuous Chips.

3. Continuous Chips with build-up-edge (BUE).

Discontinuous Chips. These types of chips are usually produced when cutting more brittle materials like grey cast iron, bronze and hard brass. These materials lack the ductility necessary for appreciable plastic chips formation. The material ahead of the tool edge fails in a brittle fracture manner along the shear zone. This produces small fragments of discontinuous chips. Since the chips break up into small segments, the friction between the tool and the chips reduces, resulting in better surface finish. These chips are convenient to collect, handle and dispose of. Discontinuous chips are also produced when cutting more ductile materials under the following conditions :

(*i*) large chip thickness.

(*ii*) low cutting speed.

(*iii*) small rake angle of the tool.

(*iv*) cutting with the use of a cutting fluid.

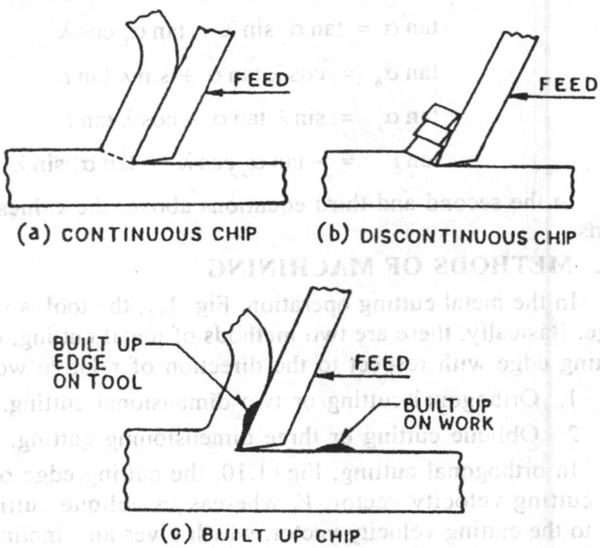

Fig. 1.11. Types of Chips.

Continuous Chips. These types of chips are produced when, machining more ductile materials. Due to large plastic deformations possible with ductile materials, longer continuous chips are produced. This type of chip is the most desirable, since it is stable cutting, resulting in generally good surface finish. On the other hand, these chips are difficult to handle and dispose off. The chips coil in a helix (chip curl) and curl around the work and the tool and may injure the operator when break loose. Also, this type of chip remains in contact with the tool face for a longer period, resulting in more frictional heat. These difficulties are usually avoided by attaching to the tool face or machine on the tool face, a 'chip breaker', (Fig. 1.12). The function of chip breaker is to reduce the radius of curvature of the chip and thus break it. The following cutting conditions also help in the production of continuous chips :

(*i*) small chip thickness.

(*ii*) high cutting speed.

(*iii*) large rake angle of the cutting tool.

(*iv*) reducing the friction of the chip along the tool face, by : imparting high surface finish to the tool face, use of tool material with low co-efficient of friction, and use of a good cutting fluid.

Continuous Chips with Built-up-edge (BUE). When machining ductile materials, conditions of high local temperature and extreme pressure in the cutting zone and also high friction in the tool-chip interface, may cause the work material to adhere or weld to the cutting edge of the tool forming the built-up edge. Successive layers of work material are then added to the built-up edge. When this edge becomes larger and unstable, it breaks up and part of it is carried up the face of the tool alongwith the chip while the remaining is left over the surface being machined, which contributes to the roughness of the surface. The built-up edge changes its size during the cutting operation. It first increases, then decreases, then again increases etc.

This cycle is a source of vibration and poor surface finish. Although, the built-up edge protects the cutting edge of the tool, it changes the geometry of the cutting tool. Low cutting speed also contributes to the formation of the built-up edge. Increasing the cutting speed, increasing the rake angle and using a cutting fluid contribute to the reduction or elimination of the built-up edge.

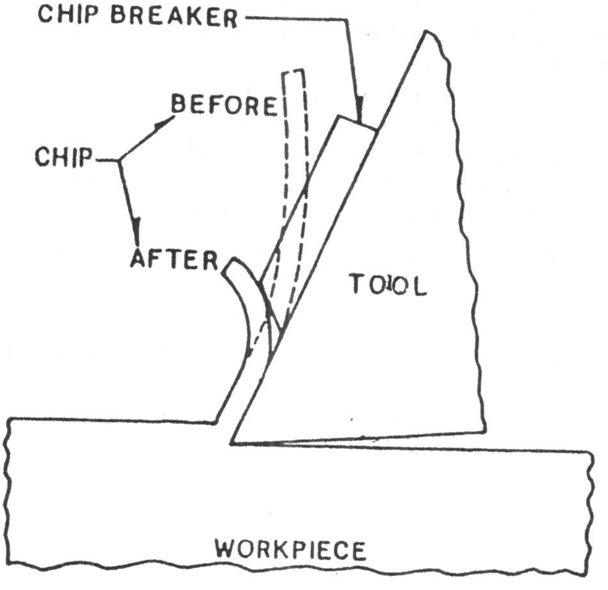

Fig. 1.12. Chip Breaker.

From the above discussion, we can summarize the factors that are likely to influence the formation of various types of chips.

Factors	*Types of Chips*		
	Continuous	*Continuous with B.U.E.*	*Discontinuous*
Material	Ductile	Ductile	Brittle
Tool :-			
Rake angle	Large	Small	Small
Cutting edge	Sharp	Dull	–
Cutting conditions :			
Speed	High	Low	Low
Feed	Low	High	High
Friction	Low	High	–
Cutting Fluid	Efficient	Poor	–

1.6. PRINCIPAL ELEMENTS OF METAL MACHINING

The principal elements of metal machining are :

(*a*) Cutting Speed, (*b*) Feed, (*c*) Depth of Cut

Cutting Speed. The cutting speed can be defined as the relative surface speed between the tool and the job. It is a relative term, since either the tool or the job or both may be moving during cutting. It is expressed in metres per minute (mpm). It is thus the amount of length that will pass the cutting edge of the tool per unit of time.

Feed. It may be defined as the relatively small movement per cycle of the cutting tool, relative to the work piece in a direction which is usually perpendicular to the cutting speed direction. It is expressed in millimetres per revolution (mm/rev) or millimetres per stroke (mm/str). It is more complex element as compared to cutting speed, since it is expressed differently for various operations. For example, in turning and drilling, the feed is the axial advance of the tool along or through the job during each revolution of the tool or job; for the shaper and planer, it is lateral offset between the tool and work for each stroke and for multitooth milling cutters, feed is the advance of the work or cutter between the cutting action of two successive teeth (expressed basically as mm/per tooth).

Depth of Cut : The depth of cut is the thickness of the layer of metal removed in one cut, or pass, measured in a direction perpendicular to the machined surface. The depth of cut is always perpendicular to the direction of feed motion.

Selection of Cutting Speed. The cutting speed to be used will depend upon the following factors :

(*i*) **Work Material.** Hard and strong materials require a lower cutting speed; whereas soft and ductile materials are cut at higher cutting speeds.

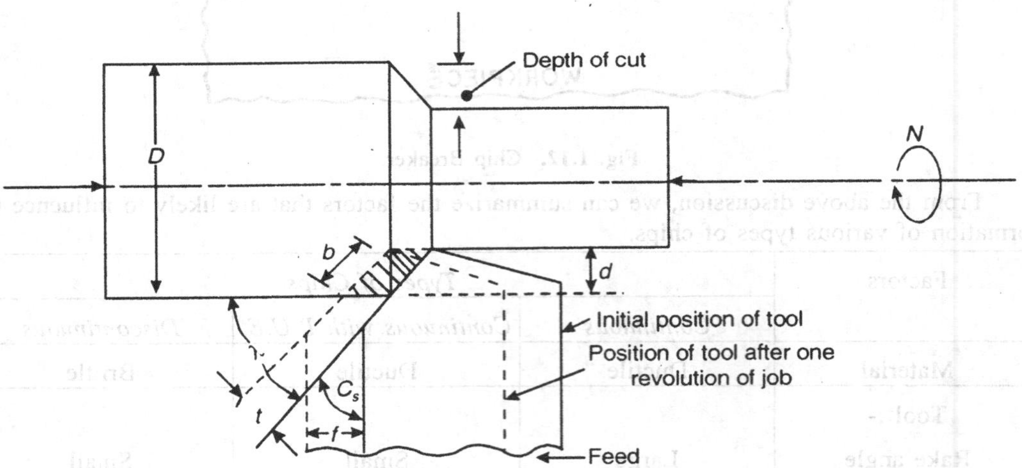

Fig. 1.13. Elements of Machining Process.

(*ii*) **Cutting Tool Material.** Special cutting tool materials, for example, cemented carbides, ceramics, Stellite and H.S.S. will cut at much higher cutting speeds than alloy or carbon steel tools.

(*iii*) **The Depth of Cut and Feed.** A light finishing cut with a fine feed may be run at a higher speed than a heavy roughing cut.

(*iv*) **Desired Cutting Tool Life.** The tool life is a direct function of cutting temperature which increases with increase in cutting speed. Thus as the cutting speed is increased, cutting tool life is decreased.

(*v*) **Rigidity and Conditions of the Machine and Tool and the Rigidity of the Work.** An old, loose machine working with a poorly supported tool on a thin bar, will not cut at such a high speed, as a good machine with rigid tool operating on a well supported bar of reasonable dimensions.

Table 1.2
Cutting Speeds and Feed Rates

Work material	*Cutting Speed V in mpm*								*Feed rate f in mm/rev.*	
	H.S.S. tool				*Carbide tool*		*Stellite tool*			
	Turn		*Ream and Thread*	*Drill*	*Turn*		*Turn*			
	Rough	*Finish*			*Rough*	*Finish*	Rough	*Finish*	*Rough*	*Finish*
Mild Steel	40	60	7.5 to 15	30	90	180	50	75	0.625 to 2.0	0.125 to 0.75
Cast Steel	15	24	3.5	12	45	100	24	33	0.5 to 1.25	0.125 to 0.5
Stainless Steel	15	18	3	12	27	45	22	25	0.5 to 1.0	0.075 to 0.175
Grey C.I.	18	27	3.5	13	60	100	33	45	0.4 to 2.5	0.2 to 1.0
Aluminium	90	150	15	72	240	360	120	180	0.1 to 0.5	0.075 to 0.25
Brass	75	100	18	60	180	270	90	150	0.375 to 2.0	0.2 to 1.25
Phosphor Bronze	18	36	4.5	13	120	180	30	50	0.375 to 0.75	0.125 to 0.5

Selection of Feed. Feeds, to be used, will depend upon the following factors :

(*i*) **Smoothness of the Finish Required.** A coarse feed will give wider and deeper machining marks and an inferior finish to a fine feed. A blunt nosed tool will give a better finish than a sharp tool for the same feed.

(*ii*) **Power Available, Condition of the Machine and its Drive.** The product of the speed, feed and depth of cut gives the amount of metal being removed and hence the power necessary. A coarse feed on a poor or badly driven machine will be harmful both for the machine and the tool. This will also result in slipping of the drive or belt.

(*iii*) **Type of Cut.** As a general rule, give coarsest feed possible for a roughing cut because finishing is unimportant. For a finishing cut, the feed should be fine enough to give the class of finish required.

(*iv*) **Tool Life.** The cutting temperature increases with increase of feed, resulting in decreased tool life.

A simple rule of thumb connecting feed and nose radius is, for rough turning :

$$f\text{ (mm/rev.)} = 0.5 \times \text{Nose Radius}$$

The maximum recommended feed rate is of the order of 2/3 of the nose radius.

The higher feeds apply for inserts :-

(*i*) having a strong cutting edge with at least a 60° cutting edge angle.

(*ii*) that are single sided.

(*iii*) that are used with a smaller entering angle than 90°.

(*iv*) that are used in materials with good machinability and moderate cutting speeds.

Selection of Depth of Cut. The depth of cut to be used will depend upon the following factors :-

(*i*) **Type of Cut.** Use large depths of cut for roughing operations than for finishing operations.

(*ii*) **Tool Life.** The cutting temperature increases with increase of depth of cut, resulting in decreased tool life.

(*iii*) **Power Required**. As discussed above, the cutting speed multiplied by area of cut (feed × depth of cut) gives the metal removal rate, which gives the power requirements. For a given area of cut, a large ratio of depth of cut to feed usually gives the most efficient performance as well as a better surface finish.

The three elements of machining process are shown in Fig. 1.13, for a simple turning operation on a centre lathe.

The cutting speed is given as :

$$V = \frac{\pi D N}{1000}, \text{ m/min.}$$

where D = Diameter in mm of work or cutter

N = rev. / min. of work or cutter

V = cutting speed of work or cutter.

The common values of cutting speed and feed are given in Table 1.2. The depth of cut can be taken as equal to 4 to 5 mm in rough turning, 0.5 to 2 mm for semi finish turning and 0.1 to 0.4 mm for finish turning.

Depth of cut is usually taken 3 to 5 times the feed for rough opeatious. The values for finishing operations are usually small.

1.7. DETERMINATION OF SHEAR ANGLE

In the simplified model of two dimensional cutting operation, the cutting tool is completely defined by the rake angle α and clearance angle γ. In addition, the following assumptions are made :

1. The tool is perfectly sharp and contacts the chip on its front, or rake face.

2. The primary deformation takes place in a very thin zone adjacent to the shear plane *AB*, Fig. 1.1.

3. There is no side flow of the chip, that is, plain strain condition.

Shear angle, φ, is defined as the angle made by the shear plane, with the direction of the tool travel. If,

t = uncut or underformed chip thickness

and t_c = chip thickness after the metal is cut.

Then
$$r = \frac{t}{t_c} \quad \text{...(1.1)}$$
is called the "cutting ratio" or "chip-thickness ratio", or "chip-compression factor".
$$\zeta = \frac{1}{r} = \frac{t_c}{t} \quad \text{...(1.2)}$$
is called the "chip reduction factor".

't' is the distance between two consecutive positions of the cutting edge, measured in a direction normal to the cutting edge along the plane of the tool face. In case of turning with $\lambda = 90°$, t = feed (see Art. 1.8).

The shear angle can be determined in the following ways :

(*a*) From Fig. 1.14, simple geometry gives,

$$r = \frac{t}{t_c} = \frac{AB \sin \varphi}{AB \cos(\varphi - \alpha)}$$

or
$$r = \frac{t}{t_c} = \frac{\sin \varphi}{\cos(\varphi - \alpha)} \quad \text{...(1.3)}$$

Solving this equation for φ, we get,

$$\tan \varphi = \frac{r \cos \alpha}{1 - r \sin \alpha} \quad \text{...(1.4)}$$

After the cut has been taken, the measurements of 't' and 't_c' can be taken, and so the cutting ratio can be determined. Then the shear angle existing during the cut can be obtained with the help of Eqn. 1.4.

(*b*) If the length of cut, l, is known then the continuity equation gives,

$$\rho\, lbt = \rho\, l_c btc$$

where r is the density of the material, assumed incompressible and l_c is the length of chip.

$$\therefore \quad \frac{l_c}{l} = \frac{t}{t_c} = r \quad \text{...(1.5)}$$

Then, with the help of equation (1.4), ϕ can be estimated.

(c) When the length of the cut is not directly known, then it can be estimated by weighing a known length of chip. Then, weight of the chip,

$$W_c = \rho l_c b t_c = \rho l b t$$

or $$l = \frac{W_c}{\rho b t} \quad ...(1.6)$$

Knowing ρ, b and t, the length of cut 'l' can be determined. Then, φ can be estimated from equations (1.5) and (1.4).

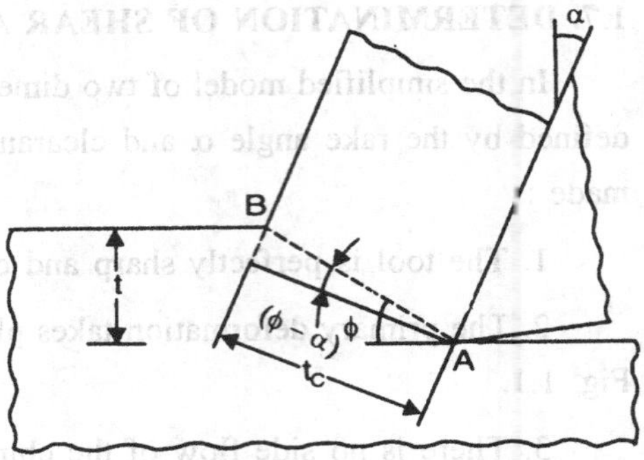

Fig. 1.14

(d) If the chip and workpiece velocities V_c and V are known, then from the continuity equation :

$$V\rho\, bt = V_c \rho b t_c$$

or $$\frac{t}{t_c} = \frac{V_c}{V} = r \quad ...(1.7)$$

Then φ can be determined from eqn. (1.4).

The shearing process in metal cutting can be represented by a simple model shown in Fig. 1.15.

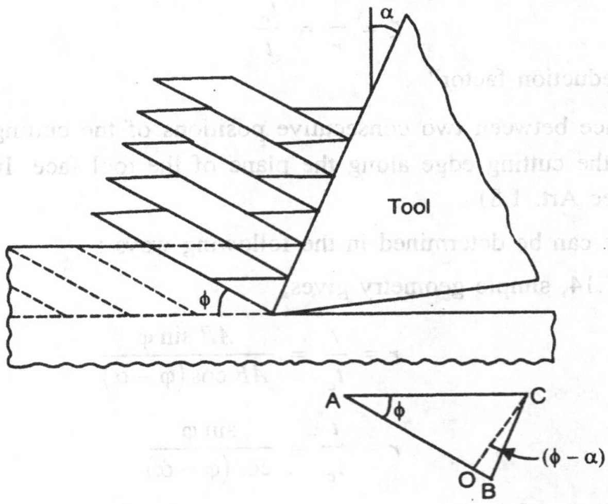

Fig. 1.15. Model of Chip Formation.

The shearing process is analogous to the displacement of a stack of cards, with each card sliding slightly over the adjacent card.

From this model, the shear strain that the material is undergoing can be determined. Based on its definition, the shear strain will be given as, (View at tool tip)

$$s = \frac{\text{distance sheared}}{\text{thickness of zone}}$$

$$s = \frac{AB}{OC} = \frac{OA + OB}{OC} = \frac{AO}{OC} + \frac{OB}{OC}$$

or, $$s = \cot\varphi + \tan(\varphi - \alpha) \quad ...(1.8)$$

Velocities. There are three types of velocities involved in any metal cutting process :

(*i*) The velocity of the tool relative to the workpiece, V, usually called the cutting speed.

(*ii*) The velocity of the chip relative to the work, V_s, called the shear velocity, that is, the velocity at which shearing takes place.

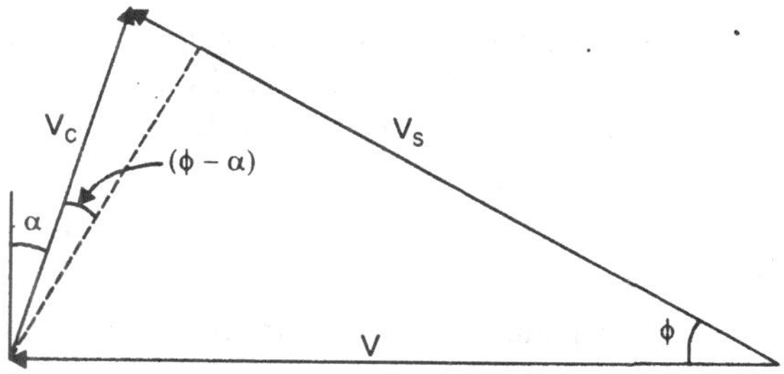

Fig. 1.16

(*iii*) The velocity of the chip up the face of the tool, V_c, called the chip velocity.

We have from constancy of volume,

$$V.t = V_c t_c, \text{ width remaining constant}$$

$$\therefore \qquad V_c/V = \frac{t}{t_c} = r \qquad \text{...(1.9)}$$

The relationship in velocities can be determined from Fig. 1.16. It can be shown that,

$$V_s/V = \frac{\cos\alpha}{\cos(\varphi-\alpha)} \quad \text{and} \quad V_c = \frac{\sin\phi}{\cos(\phi-\alpha)}.V \qquad \text{...(1.10)}$$

From here, V_s, can be determined, and so the shear strain rate during cutting can be estimated, which is given as

$$\dot{s} = \frac{V_s}{\text{thickness of shear zone}} = \frac{V_s}{t_s} \qquad \text{...(1.11)}$$

The maximum value of 't_s' is approximately 25×10^{-3} mm.

1.8. DETERMINATION OF UNDERFORMED CHIP THICKNESS

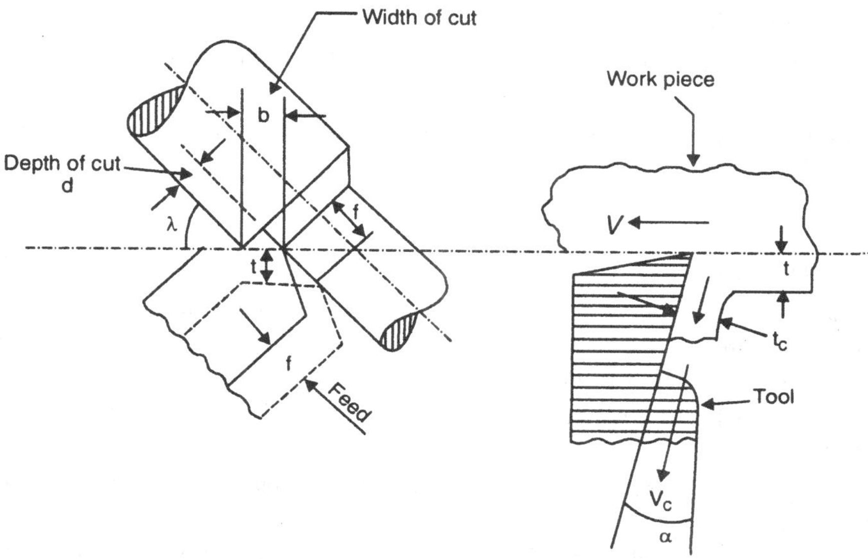

Fig. 1.17

To estimate the underformed or unçut chip thickness, 't', refer to Fig. 1.17, where the two consecutive cuts have been shown and the various parameters, that is, feed f, depth of cut d, width of cut b, t and t_c have been marked. It can be easily seen that the following relations exist :

$$t = f \sin \lambda$$

and $$b = \frac{d}{\sin \lambda} \qquad ...(1.12)$$

It is clear that the uncut chip thickness depends upon the primary cutting edge angle λ, as shown in Fig. 1.18. Uncut chip thickness is measured perpendicular to the side cutting edge of the tool.

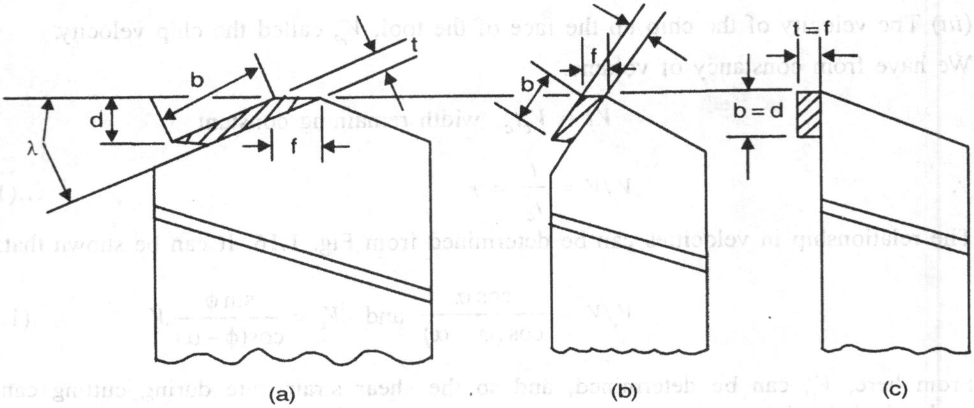

Fig. 1.18

For Fig. 1.18 (*c*) where $\lambda = 90°$ and $C_e \neq 0$.

$$t = f;\ \text{and}\ b = d. \qquad ..(1.13)$$

i.e., uncut chip thickness, t = (feed, mm/rev)

and width of cut = depth of cut

Such a case is called " Restricted Orthogonal cutting".

These relations hold good for metal cutting with single point tools (turning, shaping, planing etc.).

For drilling operation, since, tool has two cutting edges, thus, the feed per tooth will be one-half of the feed. Thus

$$t = \frac{f}{2} \sin \lambda$$

1.9. FORCE RELATIONS

Here the analysis is limited to two dimensional or orthogonal cutting which is simpler to understand as compared to the complicated three dimensional cutting process. When a cut is made, Fig. 1.19 (*a*), the forces acting on the metal chip are :

1. Force F_s, which is the resistance to shear of the metal in forming the chip. It acts along the shear plane.

2. Force F_n normal to the shear plane. This is the "backing up" force on the chip provided by the workpiece.

3. Force N at the tool chip interface acting normal to the cutting face of the tool and is provided by the tool.

4. Force F is the frictional resistance of the tool acting on the chip. It acts downward against the motion of the chip as it glides upwards along the tool face.

Fig. 1.19 (*b*) is a free body diagram showing the forces acting on the chip. R is the resultant force of F and N and R' that of F_s and F_n. Neglecting the couples which curl the chip, considering the equilibrium of the chip, R and R' are equal in magnitudes opposite in direction and collinear as shown.

All these forces can be represented with the help of a circle known as the 'Merchant force circle', Fig. 1.20. Here the two force triangles have been superimposed by placing the two equal forces R and R' together. In the figure, β is the angle of friction. In this diagram, for convenience, the resultant forces have been moved to the point of the tool. Since the forces F_s and F_n are at right angle to each other, their intersection lies on a circle with diameter R'. The forces F and N may be placed in the diagram as shown to form the circle diagram.

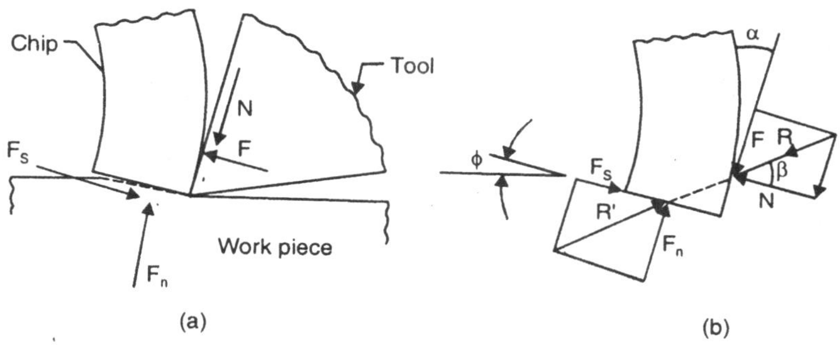

Fig. 1.19. Forces in Metal Cutting.

The two orthogonal components (horizontal and vertical) F_c and F_t of the resultant force R' can be measured by using a dynamometer. The horizontal component is the cutting force F_c and the vertical component is the thrust force, F_t. The power consumed during the process is the product of F_c and V. The thrust force does not contribute to the work done. It holds the tool against the workprice. After F_c and F_t are known, they can be laid as shown. The rake angle α can be laid off and the force F and N determined. The shear angle can be determined as explained earlier. Knowing, F_c, F_t, α and φ, all the component forces acting on the chip can be determined. It is easily shown that

$$F = F_c \sin \alpha + F_t \cos \alpha \qquad ...(1.14)$$

$$N = F_c \cos \alpha - F_t \sin \alpha$$

The co-efficient of friction will then be given as

$$\mu = \frac{F}{N} = \frac{F_c \tan \alpha + F_t}{F_c - F_t \tan \alpha} \qquad ...(1.15)$$

$$\beta = \tan^{-1} \mu$$

On the shear plane,

$$F_s = F_c \cos \varphi - F_t \sin \varphi$$

$$F_n = F_c \sin \varphi + F_t \cos \varphi \qquad ...(1.16)$$

From here,

$$F_t = F_n \cos \phi - F_s \sin f$$

$$F_c = F_n \sin f + F_s \cos f$$

The shear angle, φ, can be determined from equation (1.4).

Now shear plane area,

$$A_s = bt/\sin\varphi \qquad ...(1.17)$$

The average stresses on the shear plane area are then

$$\tau_s = F_s/A_s,$$
$$\sigma_s = F_n/A_s \qquad ...(1.18)$$

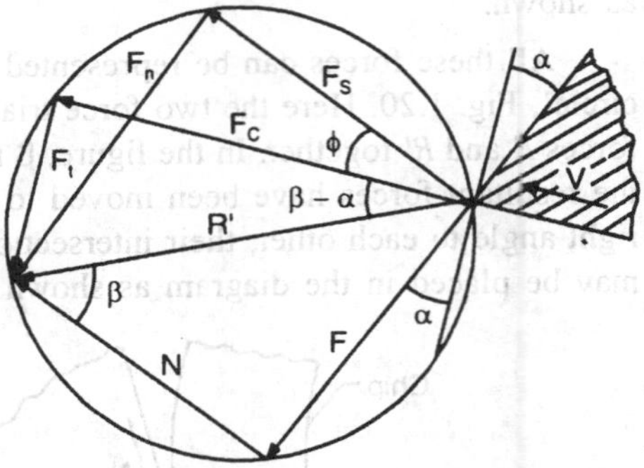

Fig. 1.20. Merchant Force Circle.

The shear angle has great influence on the overall cutting geometry. So, its relationship to material and process variables should be determined so that this angle can be predicted. Shear angle should be such that the cutting force is a minimum or that the maximum shear stress occurs on the shear plane. Now the shear force can be written as

$$F_s = R' \cos(\varphi + \beta - \alpha) \qquad ...(1.18\,a)$$

and

$$F_c = R' \cos(\beta - \alpha)$$
$$F_t = R¢ \sin(b - a)$$

∴ From equation (1.18)

$$\tau_s = \frac{F_c \cdot \sec(\beta - \alpha) \cdot \cos(\varphi + \beta - \alpha) \cdot \sin\varphi}{bt}$$

Assuming that β is independent of φ, therefore, for maximum shear stress,

$$\frac{\partial \tau_s}{\partial \varphi} = 0$$

$$\therefore \quad \cos(\varphi + \beta - \alpha)\cos\varphi - \sin(\varphi + \beta - \alpha)\sin\varphi = 0$$

$$\therefore \quad \tan(\varphi + \beta - \alpha) = \cot\varphi = \tan(90 - \varphi)$$

$$\therefore \quad \varphi = 45° + \alpha/2 - \beta/2 \qquad ...(1.19)$$

For practical purposes, the following approximate values for φ has been suggested,

$$\varphi = \alpha, \quad \text{for } \alpha > 15°$$
$$\varphi = 15° \quad \text{for } \alpha < 15° \qquad ...(1.20)$$

It is clear from equation (1.19), that for the same rake angle, a low friction angle, *i.e.*, less friction between chip and tool results in a higher shear angle, which reduces the cutting force, which in turn results in less friction. A higher shear angle also indicates a thin chip and less severe deformation of the chip, since the shear strain is also lower, equation (1.8). The above results can also be obtained by increasing the rake angle.

In making the shear angle predictions (Eqn. 1.19), Merchant made the following assumptions :

1. The work material behaves like an ideal plastic.
2. The theory involves the minimum energy principle.
3. A_s, τ_s and β are assumed to be constant, independent of φ.

The second assumption is not supported by evidence and also, it is an experimental fact that β (friction angle) varies greatly with α, and also is not independent of φ. On reconsidering these assumptions, Merchant included the following relation into his theory :

$$\tau_s = \tau_{so} + k\sigma_s \quad ...(1.21)$$

where k is a constant and σ_s is the normal stress on the shear plane.

At $\sigma_s = 0$

$$\tau_s = \tau_{so}$$

He, then obtained the following relation :

$$2\varphi + \beta - \alpha = c$$

where $c = \text{arcot } k.$ ...(1.22)

Theory of Lee and Shaffer. Lee and Shaffer applied the theory of plasticity for an ideal-rigid-plastic material, for analysing the problem of orthogonal metal cutting. They also assumed that deformation occurred on a thin-shear plane. They considered that there must be a stress field within the metal chip to transmit the cutting forces from the shear plane to the tool face. For this, they constructed a slip-line field for stress zone and derived the following relation for shear angle,

$$\varphi = \frac{\pi}{4} - (\beta - \alpha) \quad ...(1.23)$$

Many authors have given shear-angle relations and the problem is treated in detail in books on metal cutting.

Cutting Power. The cutting power or the rate of energy consumpting, is the product of the cutting speed, V, and the cutting force, F_C. Thus,

$$E = F_c . V \quad ...(1.24)$$

now $V = \dfrac{\pi DN}{1000 \times 60}$ m/s

where D = Diameter of job/tool in mm

N = Velocity of job or tool, rev/min

Power consumed in cutting, $P_C = \dfrac{F_C \times V}{60 \times 102}$, kW , if F_C is in kgf, and V is in m/min.

If F_C is in newtons and V in m/s, then

$$P_C = \frac{F_C \times V}{1000}, \text{kW}$$

Design power rating of the electric motor of the main drive,

$$P_m = \frac{P_c}{\eta_{mt}}$$

where 'η_{mt}' is the efficiency of the machine tool.

The values for mean efficiency at full load for various machine tools as determined by experimental methods, are given below :

Lathes, Milling machines = 0.8 to 0,90

Drilling machines = 0.85 to 0.90

Shapers, Planers = 0.65 to 0.75

Grinding machines = 0.8 to 0.85

To find the cutting power, the cutting force should be known. The cutting force depends upon depth of cut, feed, cutting velocity and many other factors. The following relations can be used to determined the cutting force.

1. For turning operations with carbide tipped tool on which $\alpha = +10°$, $\lambda = 45°$, $R = 2$ mm, $i = 0°$, $C_e = 10°$, and using no lubrication, the relation is :

$$F_C = Cd^x f^y V^n, kgf$$

where
d = depth of cut, mm
f = feed, mm/rev.
V = cutting speed, m/min

The values of C, x, y and n for work material steel and steel castings with tensile strength of 75 kgf/mm² and for feed less than or equal to 0.75 mm/rev., are :

$$C = 300, x = 1.0, y = 0.75, n = -0.15$$

For the other machining conditions one can refer to hand books.

2. For H.S.S. cutting tools with tool designation (ASA) of 18–14–6–6–6–15–1.8 mm, the following expressions may be used to determine the cutting force (in absence of lubrication) :

(*a*) **Hot Rolled 0.2% *C* – Steel**

$$F_C = 162.4 f^{0.85} d^{0.98} \text{ kgf}$$

(*b*) **18 – 8 Stainless Steel**

$$F_C = 196.7 f^{0.85} d^{0.96} \text{ kgf}$$

(*c*) **Yellow brass**

$$F_C = 123.6 f^{0.81} d^{0.96} \text{ kgf}$$

(*d*) **C.I.** $F_C = Cf^{0.78} d^{0.89}$ kgf

The constant C is given as, C = 60.2 for BHN 126
= 111.6 for BHN 181
= 131.1 for BHN 241

Above, d and f are in mm.

The cutting force may also be determined by the following simple relaton :

$F_C = K_s.b.d$, where
K_s = Specifc cutting resstance, N/mm2
= (2.5 to 4.) × Out for ductile materals or continuous chp
= (0.5 to 1) × HB for brittle materials or segmental chip

and HB = Brinnel Hardness number of work-material.

Note : Power needed to make a particular cut can be approximately estimated from the following formula :

$$\text{Power} = d \times f \times V \times C, \text{ watts}$$

where d and f are in mm and mm/rev respectively and V is in m/min. C is a power constant whose values are given below, for positive rake cutting :

Material →	Low C-Steel	Medium C-steel	Ni-Cr Steel	Grey C.I.	Cu Brass	Bronze	Al Alloys
Values of *C*	24	32	56	12	16	28	16

For negative rake cutting, increase the value of C by 50 per cent.

Metal removal rate. Metal removal rate is expressed as mm^3/min. A process which removes metal at a faster rate may not be the most economical process, since power consumed and cost factors must be taken into account. Due to this, to compare two processes, the amount of metal removed per unit of power consumed is determined. This is called "specific metal removal rate" and is expressed as, mm^3/W/min, if the power is measured in watts. For a single point tool, the rate of metal removal

$$= A_C \times V \qquad ...(1.25)$$

where A_C = cross-sectional area of the uncut chip

$$= b \times t$$

1.10. ENERGY CONSIDERATION IN METAL CUTTING

The total energy required per unit time (power) during metal cutting, E, is given (as already noted),

$$E = F_C \times V, \text{ Nm/s} \qquad (V \text{ being in m/s})$$

This total energy consists of two main parts : the shear energy, E_s, required to produce the plastic deformation in the shear zone, and the friction energy, E_f, used as the chip slides along the tool. Other small contributors to the total energy are : energy required to curl the otherwise straight chip, kinetic energy required to accelerate the chip, and surface energy required to produce the new surface area. These three types of energies are negligible, so, as a first approximation,

$$E \cong E_s + E_f$$

The energy per unit time divided by the volume removed per unit time, is known as the specific energy.

Total specific energy, $$e = \frac{E}{b.t.V} = \frac{F_C}{b.t} \text{ N / mm}^2$$

Specific shear energy, $$e_s = \frac{E_s}{b.t.V} = \frac{F_s.V_s}{b.t.V} \text{ N / mm}^2 \qquad ...(1.26)$$

Specific friction energy, $$e_f = \frac{E_f}{b.t.V} = \frac{F.V_c}{b.t.V} \text{ N/mm}^2$$

Now $$\frac{V_s}{V} = \frac{\cos\alpha}{\cos(\varphi - \alpha)}$$

and $$\frac{V_c}{V} = \frac{t}{t_c} = r$$

$\therefore$ $$e_s = \frac{F_s}{b.t} \cdot \frac{\cos\alpha}{\cos(\varphi - \alpha)}$$

$$e_f = \frac{F}{b.t_c}$$

Because of the many factors involved, it is not possible to exactly determine the cutting force and the cutting power. The reliable predictions are based on the experimental data as given in Table 1.3. The wide range in values is due to the cutting variables, cutting fluids, friction conditions and the variation in the material properties under the same group. While arriving at these values, the efficiency of the drive motor is taken as 0.8. These values should be multiplied by 1.25 for dull tools.

Table 1.3. Approximate Cutting Power Requirements

Material	*Specific energy, W.s/mm³*
Aluminium alloys	0.4 to 1.1
Magnesium alloys	0.4 to 0.6
Copper alloys	1.4 to 3.3
Cast irons	1.6 to 5.5
Steels	2.7 to 9.3
Stainless steels	3.0 to 5.2
High temperature alloys	3.3 to 8.5
Refractory alloys	3.8 to 9.6
Titanium alloys	3.0 to 4.1
Nickel alloys	4.9 to 6.8

1.11. TOOL WEAR AND TOOL LIFE

During any machining process the tool is subjected to three distinct factors : forces, temperature and sliding action due to relative motion between tool and the workpiece. Due to these factors, the cutting tool will start giving unsatisfactory performance after some time. The unsatisfactory performance may involve : loss of dimensional accuracy, increased surface roughness, and increased power requirements etc. The unsatisfactory performance results from tool wear due to its continued use. When the tool wears out, it is either replaced or reconditioned, usually by grinding. This will result in loss of production due to machine downtime, in addition to the cost of replacing or reconditioning the tool. Thus, the study of tool wear is very important from the stand point of performance and economics. Due to a large number of factors over which the tool wear depends (hardness and type of tool material, type and condition of workpiece, dimensions of cut, *i.e.*, feed and depth of cut, tool geometry, tool temperature, which, in turn, is a function of cutting speed, surface finish of tool temperature and cutting fluid), the majority of studies in tool wear are based on experimental observations, since the analytical study will be very difficult.

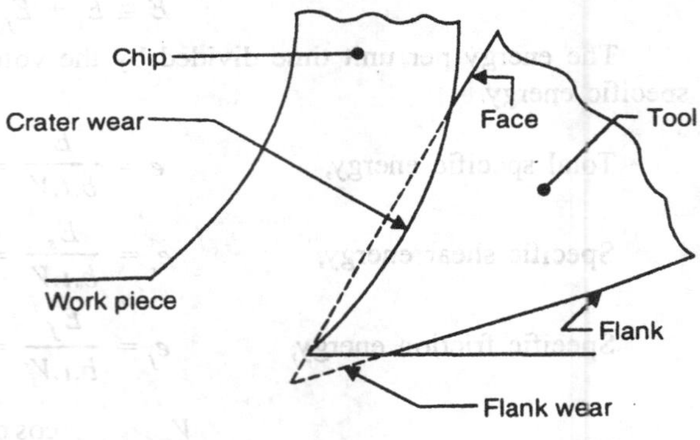

Fig. 1.21. Tool Wear.

Tool wear or tool failure may be classified as follows :

(*a*) Flank wear.

(*b*) Crater wear on tool face.

(*c*) Localized wear such as the rounding of the cutting edge, and

(*d*) Chipping off of the cutting edge.

Flank wear and crater wear are shown in Fig. 1.21. Flank wear is attributed usually to the following reasons :

1. Abrasion by hard particles and inclusions in the workpiece.

2. Shearing of the micro welds between tool and work-material.

3. Abrasion by fragments of built-up edge plowing against the clearance face of the tool.

Crater wear usually occurs due to :

1. Severe abrasion between the chip and tool face.

2. High temperatures in the tool-chip interface reaching the softening or melting temperature of tool resulting in increased rate of wear. The sharp increase in wear rate after the interface temperature reaches a certain temperature is attributed to 'diffusion'. Diffusion is the movement of atoms between tool and chip materials resulting in loss of material from the face of the tool. It depends upon the chemical composition and microstructure of tool and workpiece materials, in addition to temperature. So, unless these conditions are favourable, crater wear due to diffusion may be absent.

Crater wear is more common in cutting ductile materials which produce continuous chips. Also, it is more common in HSS (high speed steel) tools than ceramic or carbide tools which have much higher hot hardness.

The reasons for 'Nose wear' may be one or more of the reasons discussed above. Chipping of the tool may occur due to the following factors :

1. Tool material is too brittle.
2. As a result of crack that is already in the tool.
3. Excessive static or shock loading of the tool.
4. Weak design of the tool, such as a high positive rake angle.

1.11.1. Tool Life. The total cutting time accumulated before tool failure occurs is termed as 'tool life'. There is no exact or simple definition of tool life. However, in general, the tool life can be defined as tool's useful life which has been expended when it can no longer produce satisfactory parts. The two most commonly used criteria for measuring the tool life are :

1. Total destruction of the tool when it ceases to cut.

2. A fixed size of wear land on tool flank. On carbide and ceramic tools where crater wear is almost absent, tool life is taken as corresponding to 0.038 or 0.076 mm of wear land on the flank for finishing respectively.

As discussed above, tool wear and hence tool life depends on many factors. The greatest variation of tool life is with the cutting speed and tool temperature which is closely related to cutting speed. Tool temperature is seldom measured and much study has been done on the effect of cutting speed on tool life. Tool life decreases with increased V, the decrease being parabolic. To draw these curves, the cutting tools are operated to failure at different cutting speeds. In 1907, Taylor gave the following relationship between cutting speed and tool life,

$$VT^n = C \qquad ...(1.27)$$

where V is the cutting speed (m/min), T is the time (min) for the flank wear to reach a certain dimension, *i.e.*, tool life, C is constant and n is an exponent which depends upon the cutting conditions. If cutting speed-tool life curves are plotted on a log-log graph, straight lines are obtained, Fig. 1.22, n is the negative inverse slope of the curve and C is the intercept velocity at $T = 1$. The results are valid only for the particular test conditions employed. Thus 'C' is the cutting speed for tool life of 1 min.

The following values may be taken for 'n' :

$$\begin{aligned} n &= 0.1 \text{ to } 0.15 \text{ for HSS tools} \\ &= 0.2 \text{ to } 0.4 \text{ for carbide tools} \\ &= 0.4 \text{ to } 0.6 \text{ for ceramic tools} \end{aligned}$$

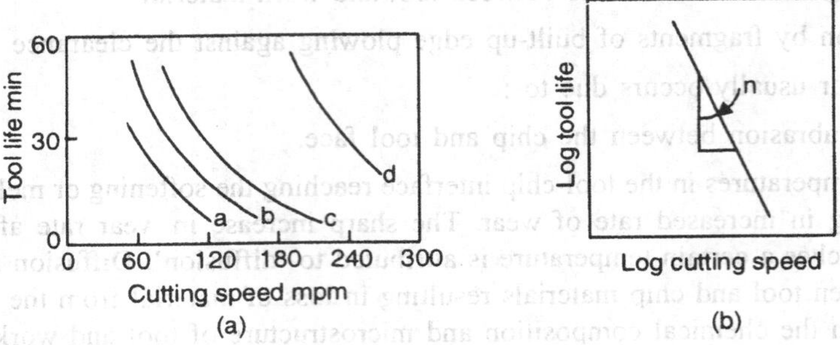

Fig. 1.22. Cutting Speed-Tool Life Curves.

The tool life also depends to a great extent on the depth of cut d and feed rate per revolution, f. Assuming a logarithmic variation of C with d, the equation (1.27) can be written as,

$$VT^n \cdot d^m = C \qquad ...(1.28)$$

It has been seen that decrease of life with increased speed is twice as great (exponentially) as the decrease of life with increased feed.

Considering feed rate also, the general equation can be :

$$VT^n . d^m . f^x = C \qquad ...(1.29)$$

1.11.2. Tool-life Criteria. As pointed out above, there is no exact and simple definition of tool life. The tool life between reconditioning or replacement can be defined in a number of ways, namely

1. Actual cutting time to failure. In the case of interrupted cutting process, such as milling, it will be the total time to failure.
2. Volume of metal removed to failure.
3. Number of parts produced to failure.
4. Cutting speed for a given time to failure.
5. Length of work machined to failure.

Each one of the above methods may be useful at one time or other. A tool fails when it no longer performs its function properly. This will have different meaning under different circumstances. In a roughing operation, where, surface finish and dimensional accuracy are of little importance, a tool failure can mean an excessive rise in cutting forces and power requirements. In the case of finishing operation, where surface finish and dimensional accuracy are most important, a tool failure will mean that the specified conditions of surface finish and dimensional accuracy can no longer be achieved. All these failures are principally related to the wear on the clearance face of the cutting tool. The method of complete tool failure or total tool destruction which occurs due to high cutting forces or shock load, is usually not considered because of the total loss of cutting tool and possible damage to the component. It is clear that the tool life/tool failure is related to tool wear and condition of the finished parts. The various tool life criteria can be listed as given below :

1. Chipping or fine cracks developing at the cutting edge.
2. Total destruction of the cutting tool.
3. Wear land size on the flank of the tool.
4. Crater depth, width or other parameters of the crater wear on the rake face of the tool.
5. A combination of (3) and (4).

6. Volume of weight of material worn off the tool.
7. Limiting value of surface finish produced on the component.
8. Limiting value of change in component size.
9. Fixed increase in cutting forces or power required to perform a function.

'Tool Life' can now be defined as the cutting time required to reach a tool-life criterion. Thus when tool-life values are quoted or compared it is essential to clearly state the tool-life criterion. A 'tool-life criterion' can be defined as a predetermined threshold value of a tool wear measure or the occurrence of a phenomenon.

The first type of failure can occur due to : faults in tool design, poor tool grinding technique, wrong selection of tool material and non-steady cutting conditions. This type of failure can be prevented by improving the cutting tool design and production of the cutting edges. Thus, on a lathe tool, the SCEA is made as large as possible, so that the initial contact between the workpiece and the tool occurs away from the tool tip and the full depth of cut and the cutting forces are obtained gradually. In the case of tools made of cemented carbide and other brittle tool materials, the rake angle is small or even negative to strengthen the tool and allow initial contact to occur on the rake face away from the cutting edge.

The tool wear volume or weight criterion is applicable when radioactive tracer methods are used to study the tool wear. Such tests are expensive and are used only for laboratory experiments. Similarly, the limiting surface finish and component size criteria are expensive and cumbersome. The forces on a worn tool increase and a selected increase in force can be used as a failure criterion. This method is useful when the wear land is the major type of wear. When crater wear is predominant, the tool will act as a restricted contact tool and the cutting forces decrease, with increase in wear. As the wear further increases, the crater merges with the flank, creating a new cutting edge. This results in increased cutting forces followed by total destruction of tool.

The most important tool-life criteria are; wear land size and the crater width or depth. The wear of the face and flank is not uniform along the active cutting edge, therefore, it is necessary to specify the locations and degree of wear, when deciding on the amount of wear allowable before regrinding the tool. Common values of wear-land size, w, Fig. 1.23 (a), are given below, as a guide

Table 1.4. Wear Land Size

w mm	*Tool*	*Remarks*
0.76	Cemented carbide	Roughing passes
0.25 to 0.38	"	Finishing passes
1.50 or total destruction	H.S.S.	Roughing passes
0.25 to 0.38		Finishing passes
0.25 to 0.38	Cemented oxides	Roughing and Finishing passes

The crater depth h, is measured at the deepest point of the crater.

The 'Tool-life' criteria as per I.S.O., are given below :

(a) **For H.S.S. and Ceramic Tools :**

1. Catastropic failure, or

2. $w = 0.3$ mm, if the flank is regularly worn, or

3. $w_{max} = 0.6$ mm, if the flank is irregularly worn.

(*b*) **Sintered Carbide Tools :**

1. $w = 0.3$ mm, or

2. $w = 0.6$ mm if the flank is irregularly worn, or

3. $h = 0.06 + 0.3 f$.

1.11.3. Variables Affecting Tool Life. The various variables that play a role in tool life are :

1. Process variables–speed, feed and depth of cut.
2. Tool material.
3. Tool geometry.
4. Workpiece material, its hardness and micro-structure.
5. Surface condition of the work-piece.
6. Cutting fluid.

Process Variables. The process variables, speed, feed and depth of cut play an important role (Eqn. 1.29) since they control the rate of metal removal and the production rate. Equation 1.29 is reproduced below :

$$VT^n d^m f^x = C$$

or

$$T = \frac{C^{\frac{1}{n}}}{V^{\frac{1}{n}} \cdot d^{\frac{m}{n}} \cdot f^{\frac{x}{n}}}$$

$$= \frac{K}{V^{\frac{1}{n}} \cdot f^{\frac{1}{n_1}} \cdot d^{\frac{1}{n_2}}} \qquad ..(1.30)$$

The values of the exponents $\frac{1}{n}, \frac{1}{n_1}, \frac{1}{n_2}$ and the constant K will depend upon the tool failure criterion. The values of the exponents which describe the effect of the variables on tool life, vary with different tool and workmaterial. It has been found that,

$$\frac{1}{n} > \frac{1}{n_1} > \frac{1}{n_2}$$

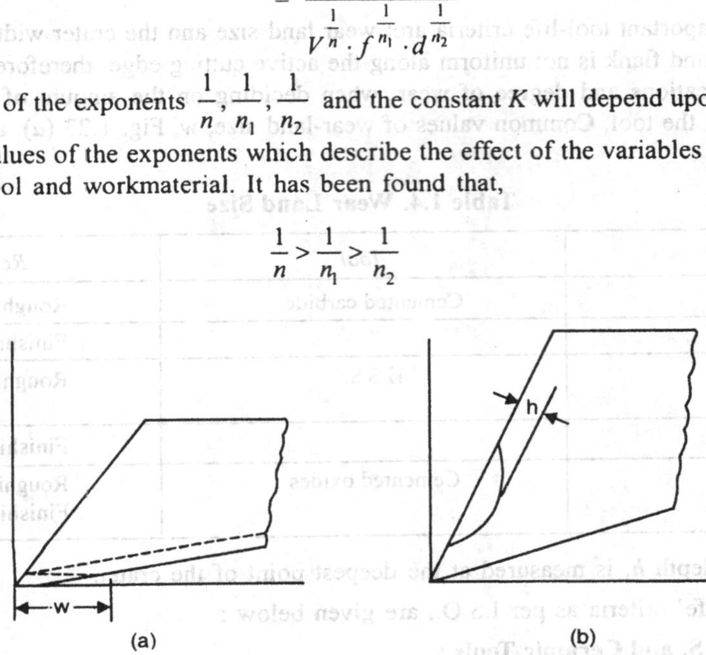

Fig. 1.23. Wear Land and Crater Depth.

It is clear that the cutting speed has the greatest effect on tool life followed by feed and depth of cut, respectively. Their effect on tool life can be explained in terms of tool-workpiece interface temperature. It has been seen that the tool life is a direct function of cutting temperature, irrespective of the cutting condition. The cutting temperature increases with increases in the three process variables.

Tool Material. The major requirements of cutting tool materials (discussed in next chapter) are : Hot hardness, impact toughness, and wear and abrasion resistance. High hardness gives the tool good wear resistance but its toughness decreases. For thermal shock resistance, the tool material should have high thermal conductivity and specific heat, a low co-efficient of thermal expansion and high tensile strength.

The variations in the above tool material requirements with cutting temperature are of considerable importance to tool life. Taking hardness as a guide to wear resistance and hence tool life, the effect of cutting temperature on tool material hardness is of major concern. Carbon steels are very sensitive to temperature and they rapidly lose their hardness at low temperatures. Thus, they are used for slow cutting of soft non-ferrous materials which give low cutting temperatures. The hardness of H.S.S. is affected only slightly till about 600°C, after which its hardness starts falling rapidly with temperature. So, H.S.S. will give good performance below 600°C, even better than cast alloys. Cast alloys are superior to H.S.S. above 600° C. These can be used at slightly higher speeds than H.S.S. and have less tendency to form a B.U.E. Cast alloys are widely used for the machining of C.I., malleable iron and hard bronzes. Cemented carbides can retain their hardness at temperatures as high as 1200°C and can therefore be used at much higher cutting speeds than H.S.S. or cast alloys. However, because of their reduced toughness, they have a greater tendency to chip and fracture under heavy loads and interrupted cutting conditions. Sintered oxides or ceramics can be used at cutting speeds of 2 or 3 times those employed with carbides. But these being very hard are extremely brittle and can only be used where shock and vibrations do not occur.

Tool Geometry. The tool life is greatly affected by tool geometry. It has already been noted that larger the rake angle, smaller will be the cutting angle and larger will be the shear angle. This will reduce the cutting force and power and hence the heat generated during cutting. This means reduced cutting temperature resulting in longer tool life. But increasing the rake angle will reduce the mass of metal behind the cutting edge resulting in poor transfer of heat. This can tend to increase the cutting temperature. Also, the cutting edge becomes mechanically weak resulting as a guide, there is an optimum value of rake angle (about 14°) which gives maximum tool life. From the curves between clearance angle and flank wear, it has been observed that for small values of clearance angle, an increase in clearance angle results in significant reduced wear rake. But the cutting edge will become weaker as the clearance angle is increased. The best compromise is 8° with H.S.S. tools and 5° with carbide tools for most of the work materials.

Work Material. As the hardness of the work material increases, the power consumption and temperatures increase. This increases tool wear resulting in shorter tool life. Impurities or hard constituents in the work material increases tool wear. Micro-structure of the work material is very significant since there is a great variation in hardness of the micro-constituents of an alloy. Out of the various micro-constituents of C.I. and steel, free graphite and ferrite are the softest constituents and martensite is the hardest with spheroidite and pearlite coming in between. Scales and oxide layers on the workpiece being quite abrasive, have an adverse effect on tool life. In such cases, the material should be cut below the oxide layer or below the work hardened zone. Pure metals tend to adhere to the tool surface and give high friction and high wear rates resulting in shorter tool life.

Cutting Fluid. The function of a cutting fluid is to reduce the friction and interface temperature which have an adverse effect on tool life. So, tool life can be increased by the effective use of a cutting fluid.

1.12. SURFACE ROUGHNESS

The quality of surface produced during a machining process will depend upon :

(*i*) the tool geometry and the feed rate, and

(*ii*) the irregularities in the machining process. These irregularities include :

1. Occurrence of B.U.E.
2. Occurrence of vibration and chatter.
3. Deflections of the tool or workpiece.
4. Inaccuracies and irregularities in the movements of spindles, slides and feed mechanism of the machine tools.
5. Formation of discontinuous chips when cutting brittle materials.
6. Tearing of the ductile work material when machining at low cutting speeds.
7. Defects in the structure of the work material.
8. Flow of chips on the machined surface.

If the above contributing factors can be completely eliminated, then the surface roughness produced will depend entirely on tool geometry. Such a surface roughness is called "Ideal surface roughness". However in actual practice, the surface roughness obtained will depend upon the combined effect of tool geometry and the contributing factors. Such a surface roughness is called "natural surface roughness."

Ideal Surface Roughness. As indicated above, the ideal surface roughness will depend entirely on tool geometry. For a sharp cornered single point tool used for turning operation, the surface roughness obtained will be like that shown in Fig. 1.24. The arithmetic mean value is given as,

$$R_a = \frac{\text{Sum of absolute values of all the areas above and below the mean line}}{\text{Sampling length}}$$

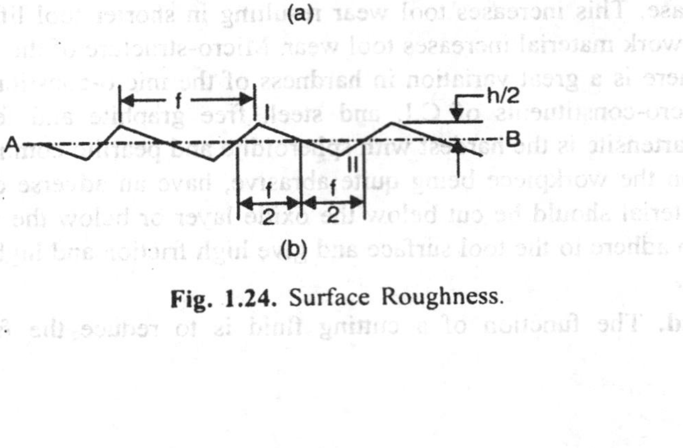

Fig. 1.24. Surface Roughness.

The mean or datum line AB is located such that the sum of the areas above the line is equal to the sum of the areas below the line. For Fig. 1.24 (b), if the sampling length is taken as one feed, f, then the surface roughness will be given as,

$$R_a = \frac{|\text{area I}| + |\text{area II}|}{f}$$

But areas I and II are equal, therefore,

$$R_a = \frac{2}{f}(\text{area I}) = \frac{h}{4}$$

where h is the height of the geometry of the surface roughness.

Now from geometry, $$h = \frac{f}{\tan SCEA + \cot ECEA}$$

$$\therefore \quad R_a = \frac{f}{4(\tan SCEA + \cot ECEA)} \qquad \text{...(1.31)}$$

Practical cutting tools have their corners rounded off and surface geometry produced will be as shown in Fig. 1.25. For such a tool, the height of the geometry will be given as,

For Fig. 1.25 (a),

$$h = (1 - \cos C_e)R + f \sin C_e \cos C_e - \sqrt{(2fR \sin^3 C_e - f^2 \sin^4 C_e)} \qquad \text{...(1.32)}$$

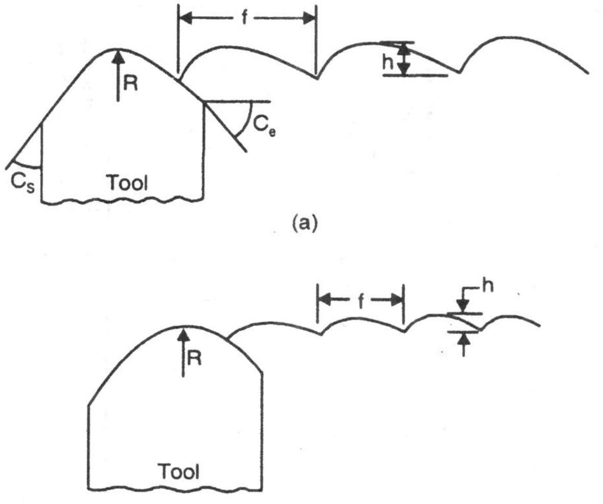

Fig. 1.25. Surface Roughness.

For small values of feed, so that the tool is cutting entirely on its nose radius, Fig. 1.25 (b), that is,

$$f \leq 2R \sin Ce$$

and h gets reduced to

$$h = R - \frac{1}{2}\sqrt{(4R^2 - f^2)}$$

By expanding the second term in binominal series, h, can be approximated as,

$$h \approx \frac{f^2}{8R} \quad ...(1.33)$$

Thus, a change in the rate of feed is more important than a change in the nose radius. The surface roughness will be given as,

$$R_a = \frac{f^2}{18\sqrt{3}.R} \quad ...(1.34)$$

It is clear that the depth of cut has no effect on the surface geometry.

Natural Surface Roughness. In actual practice, it is not possible to entirely eliminate the effects of the contributing factors. To minimise the effect of vibration and chatter, it is recommended that both tool and workpiece should be held rigidly with as little overhang as possible. The damping characteristics of the tool and the machine tool are also important. A major factor which contributes to natural surface roughness is the occurrence of B.U.E. The B.U.E. may be continually building up and breaking down and its broken portions get deposited on the machined surface. Larger the B.U.E. the rougher would be the surface produced. So, the factors which 'eliminate or reduce the formation of B.U.E. will give better surface furnish. A reduction in chip-tool friction will also give better surface finish. Such factors are : an increase in cutting speed, use of cutting tool materials with low co-efficient of friction, use of correct cutting fluid and the use of free machining materials.

Table 1.5

Problems	Remedies and solutions									
	Reduce the cutting speed	Increase the cutting speed	Reduce the feed	Increase the feed	Reduce the depth of cut	Increase the depth of cut	Select a more wear resistant grade	Select a tougher grade	Select a smaller nose radius	Select a positive geometry
Flank Wear	✓						✓			
Notch wear	✓						✓			
Cratering	✓		✓				✓			✓
Plastic Deformation	✓		✓				✓			
Built-up edge (BUE)		✓								✓
Small cracks normal to the cutting edge								✓		
Small cutting edge fracture (frittering)		✓						✓		✓
Insert Breakage			✓		✓			✓		
Curling of Long Chips				✓		✓			✓	
Vibrations	✓			✓					✓	✓

At very low cutting speeds, the chip is usually discontinuous which results in a poor surface finish due to the interrupted cutting action. As the cutting speed increases, the chip becomes continuous, but B.U.E. starts forming. But at higher speeds the B.U.E. begins to disappear resulting in better surface finish. The formation of B.U.E. is greatly dependent upon the temperature and friction at the tool-chip interface. An effective cutting fluid reduces the possibility of formation of B.U.E. and thus improves surface finish. Cemented carbide, ceramics and diamond have low co-efficient of friction as compared to H.S.S. tool material. Their use reduces the tendency of formation of B.U.E. and produces a better surface finish.

Table 1.5 gives the various problems encountered during a machining process and the possible remedies and solution.

1.13. SOLVED EXAMPLES

Example 1. *The useful tool life of a HSS tool machining mild steel at 18 m/min. is 3 hours. Calculate the tool life when the tool operates at 24 m/min.*

Solution.

$$VT^n = C$$

$$V = 18 \text{ m/min}$$

$$T = 3 \times 60 = 180 \text{ min.}$$

$$\therefore \quad C = 18 \times (180)^n$$

Let $n = 0.125$

$$\therefore \quad C = 18 \times (180)^{0.125} = 34.45$$

Now $V = 24$ m/min.

$$\therefore \quad VT^n = C$$

$$T = \left(\frac{34.45}{24}\right)^{\frac{1}{0.125}} = \mathbf{18 \text{ min.}}$$

Example 2. *For a metal machining, the following information is available :*

Tool change time, = 8 min

Tool regrind time = 5 min

Machine running cost, = Rs 5 per hour

Tool depreciation per re-grind, = 30 p

$n = 0.25, \quad C = 150$

Calculating the optimum cutting speed.

Solution. Tooling cost C_t = Tool change cost + tool regrind cost + tool depreciation

$$= \frac{5}{60} \times 8 + \frac{5}{60} \times 5 + 0.30$$

$$= \text{Rs } 1.38$$

$$\text{Machining cost } C_m = \text{Rs } \frac{5}{60}$$

$$\therefore \quad V_T = C\left[\frac{C_m}{C_t} \cdot \frac{n}{1-n}\right]^n$$

$$= 150\left(\frac{5}{60 \times 1.38} \cdot \frac{0.25}{0.75}\right)^{0.25}$$

$$= \mathbf{56.5 \text{ m/min.}}$$

Example 3. *In an orthogonal cutting operation, the following data have been observed :*

Uncut chip thickness,	$t = 0.127$ *mm*
Width of cut,	$b = 6.35$ *mm*
Cutting speed,	$V = 2$ *m/s*
Rake angle,	$\alpha = 10°$
Cutting force,	$F_c = 567$ *N*
Thrust force,	$F_t = 227$ *N*
Chip thickness,	$t_c = 0.228$ *mm*

Determine : Shear angle, the friction angle, shear stress along the shear plane and the power for the cutting operation. Also find the chip velocity, shear strain in chip and shear strain rate.

Solution. (*i*) Shear angle,
$$\varphi = \tan^{-1}\left(\frac{r\cos\alpha}{1 - r\sin\alpha}\right)$$

$$r = \frac{t}{t_c} = \frac{0.127}{0.228} = 0.557$$

$$\alpha = 10°$$

$$\therefore \quad \varphi = \tan^{-1}\left(\frac{0.557 \times 0.985}{1 - 0.557 \times 0.1736}\right)$$

$$= \tan^{-1}(0.607) = \mathbf{31.25°}$$

(*ii*)
$$\mu = \tan\beta = \frac{F_c \sin\alpha + F_t \cos\alpha}{F_c \cos\alpha - F_t \sin\alpha}$$

$$= \frac{567 \times 0.1736 + 227 \times 0.985}{567 \times 0.985 - 227 \times 0.1736}$$

$$= \frac{98.4 + 223.6}{558.5 - 39.4} = \frac{332}{519.1} = 0.64$$

or
$$\beta = \tan^{-1}(0.64) = \mathbf{32.62°}$$

(*iii*) Shear force,
$$F_s = F_c \cos\varphi - F_t \sin\varphi$$

$$= 567 \times 0.855 - 227 \times 0.519$$

$$= 484.8 - 117.8 = 367 \text{ N}$$

$$\therefore \quad \tau_s = \frac{F_s}{A_s} = \frac{367 \times \sin\varphi}{bt}$$

$$= \frac{367 \times 0.519}{6.35 \times 0.127}$$

$$= \frac{190.5}{0.806} = \mathbf{236.5 \ N/mm^2}$$

(*iv*) Cutting power $= \dfrac{F_c V}{1000} = \dfrac{567 \times 2}{1000}$

$= 1.134$ kW

(*v*) Now $\dfrac{V_c}{V} = r$

$\therefore$ Chip velocity, $V_c = 2 \times 0.557 = 1.114$ m/s

(*vi*) Shear strain, $s = \cot \varphi + \tan(\varphi - \alpha)$

$= \cot 31.25 + \tan(31.25 - 10)$

$= \mathbf{2.037}$

(*vii*) Shear plane length $= \dfrac{t}{\sin \varphi} = \dfrac{0.127}{0.519}$

$= 0.245$ mm.

Taking the thickness of deformation zone equal to one-tenth of shear plane length.

t_s, (eq. 1.11) $= 0.0245$ mm

$\therefore$ Shear strain rate, $\dot{s} = \dfrac{V_s}{t_s}$

Now $V_s = \dfrac{V \cos \alpha}{\cos(\varphi - \alpha)}$

$= \dfrac{2 \times 0.985}{\cos 21.25} = 2.11$ m/s

$\therefore$ $\dot{s} = \dfrac{2.11 \times 1000}{0.0245}$

$= 86.12 \times 10^3 \text{ s}^{-1}$

Example 4. *The following equation for tool life is given for a turning operation :*

$$VT^{0.13} f^{0.77} d^{0.37} = C$$

A 60 minute tool life was obtained while cutting at $V = 30$ m/min, $f = 0.3$ mm/rev. and $d = 2.5$ mm.

Determine the change in tool life if the cutting speed, feed and depth of cut are increased by 20% individually and also taken together.

Solution.

$$VT^{0.13} f^{0.77} d^{0.37} = C$$

$\therefore$ $C = 30 \times 60^{0.13} \times 0.3^{0.77} \times 2.5^{0.37}$

$= 30 \times 1.702 \times 0.396 \times 1.403$

$= 28.38$

(*i*) Now $V = 30 \times 1.2 = 36$ m/min.

$$\therefore \quad T^{0.13} = \frac{28.38}{0.396 \times 1.403 \times 36}$$

$$= 1.419$$

$$\therefore \quad T = (1.419)^{\frac{1}{0.13}} = (1.419)^{7.69}$$

$$= \mathbf{14.75\ min.}$$

(*ii*) Now $f = 0.3 \times 1.2 = 0.36$ mm/rev.

$$\therefore \quad T^{0.13} = \frac{28.38}{(0.36)^{0.77} \times 1.403 \times 30} = 1.48$$

$$\therefore \quad T = (1.48)^{7.69} = 20.39 \text{ min}$$

(*iii*) Now $d = 2.5 \times 1.2 = 3$ mm.

$$\therefore \quad T^{0.13} = \frac{28.38}{30 \times 0.396 \times (3)^{0.37}} = 1.591$$

$$\therefore \quad T = (1.591)^{7.69} = \mathbf{35.55\ min}$$

$\therefore$ The maximum effect on tool life is of cutting speed, and the least effect is of depth of cut.

(*iv*) Now $V = 36$ m/min, $f = 0.36$ mm/rev., $d = 3$ mm

$$\therefore \quad T^{0.13} = \frac{28.28}{38 \times 0.455 \times 150} = 1.154$$

$$\therefore \quad T = (1.154)^{7.69} = \mathbf{3.011\ min.}$$

Example 5. *During an orthogonal machining (turning) operation of C–40 steel, the following data were obtained :*

Chip thickness = *0.45 mm*

Width of cut = *2.5 mm*

Feed = *0.25 mm/rev*

Tangential cut force = *1130 N*

Feed thrust force = *295 N*

Cutting speed = *2.5 m/s*

Rake angle = *+ 10°*

Calculate : (a) Force of shear at the shear plane. (b) Kinetic co-efficient of friction at the chip-tool interface.

Solution. Given data are : $t_c = 0.45$ mm, $b = 2.5$ mm, $\alpha = 10°$, $t = f = 0.25$ mm, (Eqn. 1.13) $F_c = 1130$ N, $F_t = 295$ N, $V = 2.5$ m/s.

(*i*) Shear force, F_s, is given as :

$$F_s = F_c \cos \varphi - F_t \sin \varphi, \text{ where } \varphi \text{ is shear angle.}$$

$$\tan \varphi = \frac{r \cos \alpha}{1 - r \sin \alpha}$$

$$r = \frac{t}{t_c} = \frac{0.25}{0.45} = 0.555$$

$$\therefore \quad \tan \varphi = \frac{0.555 \times 0.985}{1 - 0.555 \times 0.1736} = 0.605$$

$$\therefore \quad \varphi = 31.2^\circ$$

$$\therefore \quad F_s = 1130 \times 0.855 - 295 \times 0.518$$

$$= 966.2 - 152.8 = 813.4 \text{ N}$$

(*ii*)

$$\mu = \frac{F_c \tan \alpha + F_t}{F_c - F_t \tan \alpha} \text{ (eqn. 14.9)}$$

$$= \frac{1130 \times 0.1763 + 295}{1130 - 295 \times 0.1763}$$

$$\therefore \quad \mu = \frac{199.25 + 295}{1130 - 52} = 0.458$$

Example 6. *The following data relate to an orthogonal turning process :*

Chip thickness = 0.62 mm

Feed = 0.2 mm/rev.

Rake angle = 15°

(*i*) *Calculate cutting ratio and chip reduction co-efficient.*

(*ii*) *Calculate shear angle.*

(*iii*) *Calculate the dynamic shear strain involved in the deformation process.*

Solution. $t = f = 0.2$ mm (equ. 1.13), $t_c = 0.62$ mm, $\alpha = 15°$

(*i*)

$$\text{Cutting ratio, } r = \frac{t}{t_c} = 0.322$$

$$\text{Chip-reduction co-efficient} = \frac{1}{r} = 3.1$$

(*ii*)

$$\text{Shear angle, } \tan \varphi = \frac{r \cos \alpha}{1 - r \sin \alpha}$$

$$= \frac{0.322 \times 0.966}{1 - 0.322 \times 0.259} = 0.33933$$

$$\phi \approx 18.74^\circ$$

(*iii*)

$$\text{Shear strain, } s = \cot \varphi + \tan(\varphi - \alpha) \text{ (Eqn. 1.8)}$$

$$= 2.947 + 0.065 = \mathbf{3.012}$$

Example 7. *The Taylorian tool-life equation for machining C-40 steel with a 18 : 4 : 1 H.S.S. cutting tool at a feed of 0.2 mm/min and a depth of cut of 2 mm is given by $VT^n = C$, where n and C are constants. The following V and T observations have been noted :*

V, m/min	25	35
T, min	90	20

Calculate (i) n and C.

(ii) Hence recommend the cutting speed for a desired tool life of 60 minutes.

Solution.

$$VT^n = C$$

$\therefore$ $$25 \times 90^n = C \text{ and } 35 \times 20^n = C$$

$\therefore$ $$5 \times 90^n = 35 \times 20^n$$

$\therefore$ $$\left(\frac{90}{20}\right)^n = \frac{35}{25} = 1.4$$

$\therefore$ $$n = 0.225$$

$\therefore$ $$C = 25 \times 90^{0.225} = 68.8$$

(*ii*) $$V \times 60^{0.225} = 68.8$$

$\therefore$ $$V \times 2.512 = 68.8$$

$\therefore$ $$V = \mathbf{27.39\ m/min.}$$

Example 8. *The following data from an orthogonal cutting test is available :*

Rake angle = *15°*
Chip-thickness ratio = *0.383*
Uncut chip thickness, t = *0.5 mm*
Width of cut, b = *3 mm*
Yield stress of material in shear, = *280 N/mm²*
Average co-efficient of friction
on the tool face = *0.7*

Determine the normal and tangential forces on the tool face.

Solution. Now, $$\tan \varphi = \frac{r \cos \alpha}{1 - r \sin \alpha}$$

$$r = 0.383 \text{ and } \alpha = 15°$$

$\therefore$ $$\tan \varphi = \frac{0.383 \times 0.966}{1 - 0.383 \times 0.259} = \frac{0.37}{0.9}$$

$\therefore$ $$\varphi = \text{Shear angle} = 22.35°$$

Now $$\tan \beta = 0.7$$

$\therefore$ $$\text{Friction angle, } \beta = 35°$$

Now $$F_c = \frac{\tau_s bt}{\sec(\beta - \alpha) \cdot \cos(\varphi + \beta - \alpha) \sin \varphi}$$

$$= \frac{280 \times 3 \times 0.5}{1.064 \times 0.739 \times 0.38} = 1405.7 \text{ N}$$

Now $$\mu = \frac{F_c \tan\alpha + F_t}{F_c - F_t \tan\alpha}$$

$$0.7 = \frac{1405.7 \times 0.268 + F_t}{1405.7 - 0.268 \times F_t}$$

$$984 - 0.188\, F_t = 376.6 + F_t$$

$\therefore$ $$F_t = 511.3 \text{ N}$$

$\therefore$ F = Tangential force on tool face

$$= F_c.\sin\alpha + F_t \,.\cos\alpha$$

$\therefore$ $$F = 1405.7 \times 0.259 + 511.3 \times 0.966$$

$$= 364 + 494 = 858 \text{ N}$$

N = Normal force on tool face

$$= F_c.\cos\alpha - F_t \sin\alpha$$

$$= 1405.7 \times 0.966 - 511.3 \times 0.259$$

$$= 1357.9 - 132.4 = \mathbf{1225.4\ N}$$

Example 9. *The following observations were made during orthogonal cutting of steel tube on a lathe :*

Width of cut, $b = 0.5$ *cm*

Cutting speed, $V = 8.2$ *m/min*

Rake angle, $\alpha = 20°$

$t = 0.25$ *mm*

$r = 0.351$

$$\varphi + \beta - \alpha = 35°$$

Find F_c *and* F_t *given tensile property of material as*

$$\bar{\sigma} = 784\,(\bar{\varepsilon})^{0.15}\ N/mm^2$$

Solution. Shear angle, $$\tan\varphi = \frac{r\cos\alpha}{1 - r\sin\alpha} = \frac{0.351 \times \cos 20°}{1 - 0.351 \times \sin 20°}$$

From here, $\varphi = 20.5°$

$\therefore$ $\beta = 34.5°$

Now shear strain $$s = \cot\varphi + \tan(\varphi - \alpha)$$

$$= \cot 20.5° + \tan(20.5° - 20°)$$

$$= 2.774$$

Now from the relations for simple tension test,

$$\sigma = k \cdot (\varepsilon)^n$$

For generalised state of stress, $$\bar{\sigma} = K.(\bar{\varepsilon})^n$$

and $$\bar{\varepsilon} = \frac{s}{\sqrt{3}} = 1.605$$

(Considering Von Mise's yield condition)

$$\therefore \quad \bar{\sigma} = 784\,(1.605)^{0.15}$$

$$= 784 \times 1.0735 = 841.66 \text{ N/mm}^2$$

$$\therefore \quad \text{Yield shear stress, } \tau_s = \frac{\bar{\sigma}}{\sqrt{3}} = \frac{841.66}{\sqrt{3}} = 485.95 \text{ N/mm}^2$$

$$\text{Now shear plane area, } A_s = \frac{bt}{\sin \varphi} = \frac{5 \times 0.25}{\sin 20.5} = 3.65 \text{ mm}^2$$

$$\therefore \quad F_s = \tau_s . A_s = 485.95 \times 3.65 = 1773.7 \text{ N}$$

Now
$$R' = \frac{F_s}{\cos(\varphi + \beta - \alpha)} = \frac{1773.7}{\cos 35°} = 2165.7 \text{ N}$$

Now
$$F_c = R' \cos(\beta - \alpha) = 2165.7 \cos(34.5 - 20)$$

$$= 2096.4 \text{ N}$$

and
$$F_t = R' \sin(\beta - \alpha) = 2165.7 \sin 14.5$$

$$= \mathbf{541.5 \text{ N}}$$

Example 10. *During machining of C-25 steel with 0–10–6–6–8–90–1 mm (ORS) shaped tripple carbide cutting tool, the following observations have been made :*

Depth of cut	*= 2 mm*
Feed	*= 0.2 mm/rev*
Speed	*= 200 m/min*
Tangential cutting force	*= 1600 N*
Feed thrust force	*= 850 N*
Chip thickness	*= 0.39 mm*

Calculate :

(i) Shear force

(ii) Normal force at shear plane

(iii) Friction force

(iv) Kinetic co-efficient of friction

(v) Specific cutting energy.

Solution. From tool designation,

$$\alpha = 10°, \lambda = 90°$$

Other given data are : $d = 2$ mm, $t = f = 0.2$ mm (Since $\lambda = 90°$, Eqn. 1.13),

$V = 200$ m/min, $t_c = 0.39$ mm, $F_c = 1600$ N, $F_t = 850$ N.

(*i*) Shear force,
$$F_s = F_c \cos \varphi - F_t \sin \varphi$$

Now
$$\tan \varphi = \frac{r \cos \alpha}{1 - r \sin \alpha}$$

$$r = t/t_c = 0.2/0.39 = 0.513$$

$$\therefore \quad \tan\varphi = \frac{0.513 \times \cos 10°}{1 - 0.513 \times \sin 10°} = 0.551$$

$$\therefore \quad \varphi = 32°$$

$$\therefore \quad F_s = 1600 \times \cos 32° - 850 \times \sin 32°$$

$$= 1600 \times 0.876 - 850 \times 0.482 = 992 \text{ N}$$

(*ii*) Normal force at shear plane,

$$F_n = F_c \sin\varphi + F_t \cos\varphi$$

$$= 1600 \times 0.482 + 850 \times 0.76 = 1515.8 \text{ N}$$

(*iii*) Friction force, $F = F_c \sin\alpha + F_t \cos\alpha$

$$= 1600 \times \sin 10° + 850 \times \cos 10° = 1089.8 \text{ N}$$

(*iv*)

$$\mu = \frac{F_c \tan\alpha + F_t}{F_c - F_t \tan\alpha}$$

$$= \frac{1600 \times \tan 10° + 850}{1600 - 850 \times \tan 10°} = 0.753$$

(*v*) Specific cutting energy $= \dfrac{F_c}{b.t}$

Now $b = d = 2$ mm (See equ. 1.13)

$$\therefore \quad \text{Specific cutting energy} = \frac{1600}{2 \times 0.2} = \mathbf{4000\ N/mm^2}$$

Example 11. *In 'ORS', the tool angles are :*

Inclination angle $(i) = 0°$

Orthogonal rake $(\alpha) = 10°$

Principal cutting edge angle $(\lambda) = 75°$

Calculate : (i) Back rake (ii) Side rake. *(AMIE 1974 W)*

Solution. We know

$$\tan a_b = \cos\lambda \tan\alpha + \sin\lambda \tan i$$

$$= \cos 75° \tan 10° + \sin 75° \tan 0°$$

$$= 0.259 \times 0.176 = 0.0456$$

$$\therefore \quad a_b = \text{Back rake} = \tan^{-1}(0.0456) = 2°37'$$

Also,

$$\tan\alpha_s = \sin\lambda \tan\alpha - \cos\lambda \tan i$$

$$= \sin 75° \tan 10°$$

$$= 0.966 \times 0.176 = 0.17$$

$$\therefore \quad \text{(side rake) } \alpha_s = \tan^{-1}(0.17) = \mathbf{9°, 40'}$$

Example 12. *In a single point cutting tool used for turning, the geometry as per ASA is :*

Back rake = 8°

Side rake = 4°

Side cutting edge angle = 15°

Find the values of inclination angle and rake angle in ORS of tool nomenclature.

(AMIE 1975 S)

Solution. As per ASA system, $\alpha_b = 8°, \alpha_s = 4°, C_s = 15°$

In ORS of nomenclature, λ = approach angle

$$= 90° - C_s$$

$$= 90° - 15° = 75°$$

Now we know $\tan\alpha = \tan\alpha_s \sin\lambda + \tan\alpha_b \cos\lambda$

$$= \tan 4°.\sin 75° + \tan 8°.\cos 75°$$

$$= 0.07 \times 0.966 + 0.1405 \times 0.259$$

$$= 0.0676 + 0.0364$$

$$= 0.104$$

∴ Orthogonal rake angle, $\alpha = \tan^{-1} 0.104$

$$= \mathbf{5°, 56'}$$

Also, $\tan i = \sin\lambda \tan\alpha_b - \cos\lambda \tan\alpha_s$

$$= 0.966 \times 0.1405 - 0.259 \times 0.07$$

$$= 0.1357 - 0.0181$$

$$= 0.1176$$

∴ Inclination angle, $i = \tan^{-1}(0.1176)$

$$= \mathbf{6.7°}$$

Example 13. *For a turning operation with H.S.S. tool for hot rolled 0.2% C-Steel, the following data is given :*

Cutting speed = *0.2 m/s*

Depth of cut = *3.2 mm*

Feed = *0.5 mm/rev.*

C_s = *15°*

Determine : Cutting power, motor power, specific cutting resistance and unit power.

Solution. The cutting force is, $F_c = 162.4 f^{0.85} d^{0.98}$ kgf

$$= 1593 \times f^{0.85} \times d^{0.98} \text{ newtons}$$

$$= 1593 \times (0.5)^{0.85} \times (3.2)^{0.98}$$

$$= 2763.74 \text{ N}$$

Cutting Power, $P_c = \dfrac{F_c \times V}{1000}$ kW

$$= \frac{2763.74 \times 0.2}{1000} = 0.553 \text{ kW}$$

$$\text{Motor Power, } P_m = P_c / \eta_{mt}$$

$$\text{Let } \eta_{mt} \text{ for lathe} = 0.85$$

$$\therefore \quad P_m = \frac{0.553}{0.85} = 0.65 \text{ kW}$$

$$\text{Area of uncut chip, } A_c = t \times b = f \times d = 0.5 \times 3.2 = 1.6 \text{ mm}^2 \quad \text{(Eqn. 1.13)}$$

$$\therefore \quad \text{Specific cutting resistance} = \frac{F_c}{f \times d} = 1727.34 \text{ N/mm}^2$$

$$\text{Unit power} = \frac{P_c}{A_c V} = \frac{0.553 \times 1000}{1.6 \times 0.2 \times 1000}$$

$$= \mathbf{1.728 \ W/mm^3/s.}$$

Example 14. *Using Taylor equation and using n = 0.5, C = 400. Calculate the percentage increase in tool life when cutting speed is reduced by 50%.*

Solution. $VT^n = C$

$$\therefore \quad V_1 T_1^n = V_2 T_2^n$$

$$= 0.5 V_1 \cdot T_2^n$$

$$\therefore \quad \left(\frac{T_2}{T_1}\right)^n = 2$$

$$\therefore \quad T_2 = 4T_1$$

$$\therefore \quad \text{Percentage increase} = \frac{T_2 - T_1}{T_1} \times 100 = \mathbf{300\%}$$

Example 15. *For an Orthogonal cutting process :*

Uncut chip thickness = 0.127 mm

V = 120 m/min

Rake angle = 10°

Width of cut = 6.35 mm

Chip thickness = 0.229 mm

Cutting force = 556.25 N

Thrust force = 222.50 N

Calculate the percentage of total energy that goes into overcoming friction at the tool-chip interface.

Solution.

$$\frac{\text{Friction energy}}{\text{Total energy}} = \frac{F.V_C}{F_C.V}$$

Now

$$\frac{V_C}{V} = \frac{t}{t_C} = r \quad \text{(Eqn. 1.9)}$$

$$\therefore \quad \frac{\text{Friction energy}}{\text{Total energy}} = \frac{F.r}{F_C}$$

$$r = \frac{t}{t_C} = \frac{0.127}{0.229} = 0.555$$

It is clear from Fig. 1.20,

$$F = R' \sin \beta$$

$$F_c = R'.\cos(\beta - \alpha)$$

and

$$R' = \sqrt{F_c^2 + F_t^2} = 600.725 \text{ N}$$

$$\therefore \quad F_c = 556.25 = 600.725 \times \cos(\beta - 10)$$

From here, $\beta = 32°$

$$\therefore \quad F = 600.725 \times \sin(32°)$$

$$= \mathbf{318.34\ N}$$

$$\therefore \quad \frac{\text{Friction energy}}{\text{Total energy}} = \frac{318.34 \times 0.555}{556.25} \times 100$$

$$= \mathbf{31.75\%}$$

Example 16. *A 300 mm diameter bar is turned at 45 rev/min with depth of cut of 2 mm and feed of 0.3 mm/rev. The forces measured at the cutting tool point are :*

Cutting force = 1850 N

Feed force = 450 N

Calculate :

(i) Power consumption

(ii) Specific cutting energy

(iii) Energy consumed if the total metal removed during the turning operation is 2.5×10^6 mm^3.

Solution. Cutting velocity,

$$V = \frac{\pi \times 300 \times 45}{60 \times 1000} = \mathbf{0.707\ m/s}$$

(*i*) ∴ Cutting power,

$$P_C = \frac{F_C.V}{1000}, \text{ kW}$$

$$= \frac{1850 \times 0.707}{1000}$$

$$= \mathbf{1.308\ kW}$$

$$\text{Now feed velocity} = \frac{0.3 \times 45}{60 \times 1000}, \text{ m/s}$$

$\therefore$ Feed power $= \dfrac{0.3 \times 45}{60 \times 1000} \times 450$

$= \mathbf{0.10\ W}$ (negligible)

(ii) Now, $MRR = A_C V$

$= b \times t \times V$

$= d \times f \times V$ (See Eqn. 1.12)

$= 2 \times 0.3 \times \pi \times 300 \times 45$

$= 2.545 \times 10^4\ mm^3 / min.$

$\therefore$ Specific cutting energy $= \dfrac{1308 \times 60}{2.545 \times 10^4} = \mathbf{3.08\ W.s/mm^3}$

(iii) Energy consumed $= 3.08 \times 2.5 \times 10^6$, W.s

$$= \frac{3.08 \times 2.5 \times 10^6}{1000 \times 3600} = \mathbf{2.139\ kWh}$$

Example 17. *A M.S. bar of 100 mm is being turned with a tool having ASA tool signature as : 6°–10°–5°–7°–10°–30°–0.5 mm. Determine the various components of the machining force and the power consumption. Take : depth of cut = 2.5 mm, feed = 0.125 mm/rev, turning speed of job = 300 rev./min., co-efficient of friction at the tool-work interface = 0.6, ultimate shear stress of the work material = 400 MPa.*

Solution. It is clear from the tool designation, that the side cutting edge angle is 30°. Therefore, the tool approach angle, $\lambda = 90° - 30° = 60°$

Friction angle, $\beta = \tan^{-1} \mu = \tan^{-1} 0.6 = 30.96° \cong 31°$

Now orthogonal rake angle is given as,

$$\tan a = \tan a_s \,.\, \sin \lambda + \tan a_b \,.\, \cos \lambda$$

Now $a_s = 10°$ and $a_b = 6°$, $\lambda = 60°$

From here, $a = 11.6°$

Now, from Merchant's relation, the shear angle is,

$$\phi = 45° + \alpha/2 - \beta/2$$

$$= 45° + 5.8° - 15.5° = 35.3°$$

Merchant's theory is more accurate for plastics but agrees poorly for machining metals. With Lee and Shaffer relation,

$$\phi = \frac{\pi}{4} - (\beta - \alpha)$$

$$= 45° - 31° + 11.6° = 25.6°$$

Now, the cutting force is given as,

$$F_c = \frac{\tau_s .b.t.}{\sec(\beta - \alpha) . \cos(\phi + \beta - \alpha) . \sin \phi}$$

Now b = width of cut $= \dfrac{\text{depth of cut}}{\sin\lambda} = \dfrac{d}{\sin\lambda}$

and uncut chip thickness, $t = f \,.\, \sin\lambda$

$\therefore \quad b \,.\, t = d \,.\, f$

$$\therefore \quad F_c = \frac{400 \times 2.5 \times 0.125}{\sec(19.4°) \times \cos 45° \times \sin 25.6°} = 386 \text{ N.}$$

Now, from equation (1.18a), the thrust component is,

$$F_t = F_c \,.\, \tan(\beta - \alpha)$$

$$\therefore \quad F_t = 386 \times \tan 19.4° = 135.932 \text{ N}$$

The thrust force is normal to the tool-job interface, that is, normal to the principal cutting edge of the tool, See Fig. 1.17.

$\therefore$ Feed force (along the axis of the job),

$$F_f = F_t \,.\, \sin\lambda = 135.932 \times \sin 60° = 117.7 \text{ N}$$

Radial force (Normal to the axis of the job),

$$F_r = F_t \,.\, \cos\lambda = 67.961 \text{ N}$$

Now cutting power $= F_c \,.\, V$ watts

and $V = \dfrac{\pi DN}{1000 \times 60}$, m/s

$$= \frac{\pi \times 100 \times 300}{1000 \times 60} = 1.57 \text{ m/s}$$

$\therefore$ Power $= 386 \times 1.57 =$ **606 watts**

Example 18. *A turning tool with side and end cutting edge of 20° and 30° respectively operates at a feed of 0.1 mm/rev. Calculate the CLA of the surface produced if the tool nose radius is 3.00 mm.*

Solution. Given : $C_s = 20°$, $C_e = 30°$, $f = 0.1$ mm/rev., $R = 3.00$ mm.

Refer to Fig. 1.24 (a), the peak of valley roughness is given as, eqn. 1.32,

$$h = (1 - \cos C_e) \,.\, R + f \,.\, \sin C_e \,.\, \cos C_e - \sqrt{(2 f R \sin^3 C_e - f^2 \sin^4 C_e)}$$

$$= 0.402 + 0.0433 - 0.2727 = 0.1726 \text{ mm.}$$

The CLA (centre-line average) value can be taken roughly as,

$$R_a \cong \frac{h}{4} = 0.04315 \text{ mm}$$

$$= \mathbf{43.15\ \mu m}$$

PROBLEMS

1. Define machining process.
2. With the help of a sketch, explain the machining process.
3. Explain the various elements of a single point cutting tool, with the help of a neat sketch.
4. What is meant by 'hand' of a single point cutting tool ?
5. With the help of a sketch, discuss the principal surfaces and planes in metal cutting.
6. Name the two systems of designating the cutting tool.
7. Sketch a single point cutting tool under ASA system. Define various tool angles and discuss their importance.
8. Why negative rake angle is normally employed for cutting hard and strong materials ?
9. What is meant by 'tool designation' or 'tool signature' ?
10. What is Orthogonal Rake System ?
11. Show the ORS of tool angles with the help of a sketch.
12. Write the relations between ASA and ORS of tool angles.
13. Sketch and explain the two methods of metal machining.
14. Discuss the various types of chips produced during metal machining.
15. Why are discontinuous type chips preferred over continuous type ?
16. Explain, why build up edge on a cutting tool is undesirable ?
17. What is the use of a 'Chip breaker' ?
18. Name and discuss the principal elements of metal machining.
19. Define Tool life.
20. Discuss the variables affecting tool life.
21. What is meant by orthogonal cutting and oblique cutting ?
22. How does the rake angle affect the life of the cutting tool ?
23. What two pressure areas of the cutting tools are subjected to wear ?
24. With the help of a sketch, show crater wear and flank wear on a cutting tool.
25. Name the factors that contribute to flank wear.
26. Name the factors that contribute to crater wear.
27. Name the factors that contribute to the formation of discontinuous chips.
28. Name the factors that contribute to the formation of built-up edge.
29. Differentiate between positive and negative rake angles.
30. How is the nose radius of a cutting tool selected ?
31. Define cutting ratio.
32. What is the approximate thickness of shear zone in metal cutting ?
33. Discuss the two methods of metal cutting.
34. Discuss the various forces encountered in metal cutting.
35. Explain 'Merchant force circle'.
36. Why tool wear is important in metal cutting ?
37. Discuss various types of tool wears.
38. Enumerate the factors on which tool wear and tool life depend.
39. Discuss Taylor's relationship for cutting speed-tool life.
40. Derive an expression for optimum value of cutting speed.

41. In an orthogonal cutting operation, the following data have been observed :

Cutting speed	=	0.223 m/s
Uncut chip thickness	=	0.06 mm
Width of cut	=	3.83 mm
Chip thickness ratio	=	0.51
Rake angle	=	20°
Cutting force	=	363 N
Thrust force	=	127 N

Determine : Shear angle, friction angle, shear stress along the shear plane, chip velocity, shear strain in chip, shear strain rate and the power for the cutting operation.

42. The following equation for tool life has been obtained for H.S.S. tool :

$$VT^{0.13} \cdot f^{0.6} \cdot d^{0.3} = C$$

A 60 min. tool life was obtained using the following cutting conditions :

$$V = 40 \text{ m/min}, f = 0.25 \text{ mm}; d = 2.0 \text{ mm}$$

Calculate the effect on tool life if speed, feed and depth of cut are together increased by 25% and also if they are increased individually by 25%.

43. During machining of C–20 steel with an orthogonal tool having a rake of 10° at a feed of 0.2 mm/rev., the value of the shear angle has been observed to be 30° under a shear angle microscope. If the principal cutting edge angle is 90°, what is the value of cutting ratio or the chip reduction co-efficient ?

44. In a turning operation, it was observed that the tool life was 100 minutes and 50 minutes at cutting speeds of 25 m/min. and 100 m/min respectively. Find out the tool life at 200 m/min under the same cutting conditions.

45. The end of a tube is being turned on a lathe at 200 rev/min and at a feed rate of 25.4 mm/min.

The tube is 15 cm in diameter and 2.5 mm thick. The tube material obeys the equation,

$$\bar{\sigma} = K(\bar{\varepsilon})^n$$

where $n = 0.26$ and $K = 540$ N/mm^2.

V_c/V is $\frac{1}{3}$ and $\alpha = 5°$. The friction energy is 25 per cent of the total energy consumed. Determine the power in operation.

46. In an orthogonal cutting operation, the orthogonal rake is 12°, while the principal cutting edge angle is 75°. Calculate,

(*i*) back rake (*ii*) side rake

If the principal cutting edge angle is 90°, how much are

(*i*) back rake (*ii*) side rake?

47. During machining of C–20 steel, a double carbide cutting tool of 0–10–6–6–8–75–1 mm (ORS) shape has been used. Feed is 0.15 mm/rev., depth of cut of 1.5 mm at a cutting speed of 120 m/min., a chip thickness of 0.30 mm have been obtained. Calculate :

(*i*) the chip reduction co-efficient (*ii*) the shear angle.

48. Why do carbide tools employ negative rake angles more often than H.S.S. tools ?

49. When the use of positive rake angles and negative rake angles is recommended ?

50. Give the significance of providing nose radius on tool tip.

51. Describe the tool represented by 10, 10, 6, 6, 8, 8, 1 mm in ASA system.

52. In orthogonal cutting operation, the feed is 0.10 mm and the chip thickness is 0.25 mm. The cutting force is 1360 N and the feed thrust force is 770 N. The rake angle of the tool is + 10°. Find :

(*a*) The shear angle.

(*b*) The size of the force exerted by the tool on the chip.

(*c*) The coefficient of friction on the face of the tool.

(*d*) The sizes of the friction force and the normal force, on the tool face.

(*e*) The sizes of the shearing force and the normal force, on the shear plane.

53. An orthogonal cut 2.5 mm wide is made at a speed of 0.5 m/s and feed of 0.26 mm with a H.S.S. tool having a 20° rake angle. The chip thickness ratio is found to be 0.58, the cutting force is 1400 N and the feed thrust force is 360 N. Find :

(*a*) Chip thickness.

(*b*) Shear plane angle.

(*c*) Resultant force.

(*d*) Co-efficient of friction on the face of the tool.

(*e*) Friction force and normal force on the chip.

(*f*) Shearing force and normal force on the shear plane.

(*g*) Specific energy.

54. In orthogonal cutting, the feed is 0.127 mm and the depth of cut normal to the plane of the paper is 2.54 mm. The cutting speed is 4 m/s. The cutting force is found to be 1800 N and the feed thrust force 900 N. The rake angle of the tool is + 8°. Find :

(*a*) The power required for the cut in kW.

(*b*) The rate of metal removal, mm^3/s.

(*c*) The unit power in $W/mm^3/s$.

55. A workpiece is being cut a 1.25 m/s and the power is found to be 2.05 kW. The feed is 0.25 mm/rev. and the depth of cut is 5 mm. Estimate :

(*a*) Cutting force in N.

(*b*) Unit power consumption $W/mm^3/s$.

56. A tool making an orthogonal cut has a rake angle of – 10°. The feed is 0.10 mm, the width of cut 6.35 mm, the speed 2.7 m/s, and a dynamometer measures the cutting force to be 1800 N and the normal thrust force to be 1540 N. A high-speed photograph shows a shear plane angle of 20°. Estimate :

(*a*) Chip thickness.

(*b*) Co-efficient of friction.

(*c*) Shear and normal stresses on the shear plane.

(*d*) Shearing strain.

(*e*) Power to shear the metal.

57. What is understood by tool life ? What is the significance to an engineer who is interested in productivity ? What different criteria are used to identify that the tool has reached its limiting life ?

58. Establish the simple tool-life equation with magnitudes of constants from the following data (Taylor's tool-life equation) :

A tool life of 100 min is obtained from a cutting tool at a cutting speed of 25 m/min, and 10 min, at 33.3 m/min. What is the cutting speed for a 60 min, tool life ?

59. (*a*) State the general form of the Taylor's equation for tool life.

(*b*) Discuss the usefulness of Taylor's equation in selection of proper cutting speed for machining operations.

(*c*) Derive an expression for the most economic cutting speed.

(*d*) Is the most productive cutting speed the same as the most economic speed. Give reasons for your answer.

60. Following data were collected from an orthogonal machine test on steel :

Cutting speed	=	18 m/min.
Rake angle	=	20°
Clearance angle	=	10°
Width of cut	=	3.2 mm
Underformed chip thickness	=	0.10 mm
Deformed chip thickness	=	0.25 mm
Cutting force in the cutting velocity direction	=	800 N

Normal force in a direction normal to cutting velocity = 500 N. Draw the circle diagram of forces and evaluate : shear angle, shear strain, friction co-efficient against chip flow, friction force on the rake face.

61. Show that during orthogonal cutting with a zero degree of rake angle, the ratio of the shear strength, τ_s, of the work material to specific cutting energy, *e*, is given by

$$\tau_{s/e} = (1 - \mu r) \cdot \frac{r}{1 + r^2}$$

62. During orthogonal turning of M.S. tube of 5 mm wall thickness, the following data were recorded :

Speed	25 m/min.
Tool	H.S.S. 15° rake
Feed	0.15 mm/rev.
Chip thickness ratio	0.35
Co-efficient of friction	0.60

The friction force on the tool-chip interface was measured by means of a special device and was equal to 480 N. Determine the components of the cutting forces, shear angle and the specific energy of deformation.

63. In an orthogonal cutting operation, the cutting speed is 2.5 m/s, rake angle is 6° and the width of the cut is 10 mm. The underformed chip thickness is 0.2 mm. 13.36 gms of steel chips with a total length of 50 cm are obtained. The tool post dynamometer gives cutting and thrust forces 1134 N and 453.6 N. Determine :

(*a*) slip plane angle.

(*b*) friction energy at the tool-chip interface, as the percentage of total energy.

(*c*) shear energy as the percentage of total energy. (33°, 30.9%, 69.3%)

64. In orthogonal cutting of a low carbon steel, the specific cutting energy is 4080 N/mm². The uncut chip thickness is 0.2 mm and the chip width is 5 mm. The cutting speed is 1.1 m/s, and the rake angle of the tool is 10°. Assuming co-efficient of friction at the tool-chip interface as 0.7, determine :

(*a*) the cutting force

(*b*) the average shear stress in the shear plane

(*c*) the normal stress on the shear plane

(*d*) the average shear strain in cutting

(*e*) the average shear strain rate.

Use the relation of Lee and Shaffer to find the shear angle.

(4080 N, 861.8 N/mm², 861.8 N/mm², 3.906, 4.44 × 105 s⁻¹)

65. A 0.2% C–steel is machined with a tripple carbide cutting tool having 0–10–6–6–8–75–1 mm ORS shape. A feed of 0.2 mm/rev and depth of cut of 2 mm at the cutting speed of 150 m/min have been employed. A chip thickness of 0.36 mm has been obtained. Calculate the chip reduction co-efficient and shear angle.

66. Select the speed in rev/min for turning a round steel bar of diameter 320 mm with a H.S.S. T–1 tool, having a tool life of 60 minutes. A feed of 0.2 mm/rev. and depth of cut 1 mm have been chosen. The cutting speed equation is given in the following Taylorian form.

$$V = \frac{273}{T^{0.2} f^{0.2} d^{0.15}} \text{ m/min.}$$

f and d are in mm. (**Ans.** 165.2 rev./min)

67. A H.S.S. tool runs at 24 m/min. for one hour. Circumstances are such that it becomes desirable to run the tool for 1000 min. Estimate the suitable speed. Take n = 0.25 in the Taylor equation. (15.74 m/min)

68. A bar 76 mm diameter is reduced to 71 mm diameter by means of a cutting tool for which approach angle is 90°, and for which the cutting edge lies in the plane containing the work axis of rotation. The mean length of the cut chip is 73.9 mm and a feed of 0.2 mm/rev. is used. Find the cutting ratio. (0.32)

69. When turning 19 mm diameter bar on an automatic lathe employing sintered carbide tools, the value of n is 0.2 and the value of V_{60} is 104 m/min. At what speed should the spindle run to give a tool life of 6 hours? If a length of 50 mm per component is machined and the feed used is 0.16 mm/rev, what is the cutting time per piece and how many pieces can be produced between tool changes? (1220 rev/min., 16.5 s, 1300)

70. When cutting steel with a H.S.S. cutter, the tool life at a cutting speed of 30 mpm is 60 min. and at a cutting speed of 45 mpm, it is 30 min. What is the tool life equation of the cutter on the material? ($T = 21115/V^{1.724}$)

71. A HSS tool requires regrinding after 3 hours and 20 minutes when machining steel at a cutting speed of 40 m/min. Calculate the tool life if the speed is increased to 70 m/min. (2.7 min.)

72. C-steel is machined by a WC cutting tool. The tool has a life of 100 min. between consecutive grinds when operating at 80 m/min. and a life of 33 minutes at a speed of 100 m/min. Determine the values of 'n' and 'C' in Taylor equation. (n = 0.2, C = 201)

73. In Taylor equation, if n = 0.25 and C = 300 m/min., what should be V for a tool life of one hour ? (**Ans.** 107.8 m/min)

74. C-40 steel is turned with a H.S.S. tool of back rake angle of 8°. The uncut chip thickness is 0.3 mm, depth of cut (chip width) is 1.5 mm and the cutting speed is 0.6 m/s. One metre long chip weighs 5.7g. The width of the chip remains constant. Calculate the cutting ratio, taking density of material as 7.9 g/cm³. Also determine the chip compression ratio. (**Ans.** 0.625, 1.6)

75. For problem 74, calculate MRR, power requirement and the cutting force. (270 mm³/s, 891 W, 1485 N)

76. Using Taylor equation and taking n = 0.12, calculate the percentage change in cutting speed required to give an 80% reduction in tool life. (Ans. 21.4%)

77. The following data was obtained from the tool-life cutting test of H.S.S. tool material, used to cut die-steel :

Cutting Speed, m/min :	49.74	49.23	48.67	45.76	42.58
Tool life, min :	2.94	3.90	4.77	9.87	28.27

Determine the constants of the Taylor tool life equation $VT^n = C$ (**Ans.** n = 0.07, C = 54)

78. The power required to turn a medium C-steel is approximately 3.8 W/mm³/s. The power available at the machine spindle is 3.73 kW. Determine :

(*a*) Maximum MRR. (*b*) Cutting force.

(c) Depth of cut if cutting speed is 36 m/min. and the feed is 0.25 mm / rev.

(**Ans.** 982 mm^3/s, 6.216 kN, 6.216 kN, 6.55 mm)

79. A lathe running idle consumes 325 W. The power input increases to 2.58 kW when cutting an alloy steel at 24.5 m/min. If the spindle speed is 124 rev./min., depth of cut = 3.8 mm and feed is 0.2 mm/rev., determine :

(a) Cutting force. (b) Torque at the spindle.

(c) Specific power consumption. (**Ans.** 5.522 kN, 174 Nm, 7.27 $W/mm^3/s$)

80. In an orthogonal cutting operation on a workpiece of width 2.5 mm, the uncut chip thickness was 0.25 mm and the tool rake angle was zero degree. It was observed that the chip thickness was 1.25 mm. The cutting force was measured to be 900 N and the thrust force was found out to be 810 N.

(a) Find the shear strength of the workpiece material. (*GATE 1992*)

(b) If the co-eff. of friction was 0.5, what is the machining constant?

(**Ans.** 227 N/mm^2, 49.2°, Machining constant = $2\phi + \beta$)

81. While turning a mild steel bar on a lathe, the following data were obtained :

Machined length	:	150 mm
Speed	:	170 rev./min.
Feed rate	:	0.5 mm/rev.
Cutting force	:	2000 N
Dia. of workpiece	:	500 mm

Calculate : Power consumed in machining, machining time and the total amount of heat generated during cutting. (*Nagpur University*)

(**Ans.** 8.90 kW, 1.765 min., 942.51 kJ)

82. Draw Merchant's circle in order to analyse the mechanics of metal cutting and deduce various force angle relationships. Determine shear angle, friction angle, shear and normal stress on shear plane, shear strain by using the following data :

Uncut chip thickness	=	0.125 mm
Chip thickness	=	0.250 mm
Width of cut	=	6.5 mm
Cutting speed	=	100 m/min.
Rake angle	=	10°
Cutting force	=	70 N
Thrust force	=	25 N

(*Nagpur University*)

(**Ans.** 28.33°, 29.65°, 61.24 N/mm^2, 67.96 N/mm^2, 2.1866)

83. In an experimental set-up for the tool life of the cutting tool, following data have been noted :

No. of components machined between regrinds	Rev./min.	Feed in mm/min.
700	450	0.25
80	710	0.25

Find out the tool-life equation if the length of the work-piece is 100 mm and the mean diameter at the cutting edge is 25 mm. (*Nagpur University*)

(**Ans.** $VT^{0.14} = 87$)

84. A 50 mm diameter bar of steel was turned at 284 rev./min. and the tool failure occurred in 10 min. The speed was changed to 232 rev./min. and the tool failed in 60 min. of cutting time. Assuming a straight line relationship exists, what cutting speed should be used to obtain a 30 min. tool life ? (**Ans.** V_{30} = 39.39 m/min.)

85. For orthogonal cutting with a cutting tool having a rake angle of 10°, the uncut chip thickness is 0.15 mm and chip thickness is 0.4 mm. Determine (a) the cutting ratio, (b) the shear plane angle and (c) the shear strain. (**Ans.** 0.375, 21.55°. 2.734)

86. In an orthogonal cutting operation, the following data have been observed :

Uncut chip thickness,	t =	0.25 mm
Chip thickness,	t_c =	0.75 mm
Width of cut,	b =	2.5 mm
Rake angle,	α =	0°
Cutting force,	F_c =	950 N
Thrust force,	F_t =	475 N

Determine : shear angle, co-efficient of friction, the friction angle, ultimate shear stress of the material. (**Ans.** 18.4°, 0.5, 26.57°, 379.4 MPa)

87. In the machining of mild steel, the following data have been observed :

Uncut chip thickness,	t =	0.2 mm
Width of cut,	b =	2 mm
Cutting speed,	V =	200 m/min.
Rake angle,	α =	10°
Co-efficient of friction,	μ =	0.5
Ultimate shear stress of material,	t_s =	400 MPa.

Determine : Shear angle and the cutting and thrust components of the machining force.

(**Ans.** 36.7°, 420 N, 125 N)

Hint : Use eqn. 1.19, $\phi = 45° + \dfrac{\alpha}{2} - \dfrac{\beta}{2}$, to determine, ϕ.

88. Solve the above problem, by assuming the machining constant to be 70° in Merchant's second equation for anlges. Also, find out the results by using relation given by Lee and Shaffer relation. (**Ans.** Merchant Theory : F_c = 468.7 N ; F_t = 139.46 N
Lee and Shaffer theory : F_c = 463.92 N ; F_t = 154.6 N)

89. Determine the cutting force during orthogonal machining of mild steel with an uncut chip thickness of 0.25 mm and width of cut being 2.5 mm. Take a = 0°, μ = 0.5, and τ_s = 400 MPa.

(**Ans.** 1000.84 N)

90. Determine the cutting force and the thrust force and the cutting ratio when machining M.S. Take :

Uncut chip thickness	= 0.25 mm
Width of cut	= 2 mm
Rake angle	= 0°
μ	= 0.5
τ_s	= 400 MPa

Use Lee and Shaffer relation for shear angle. (**Ans.** 800.65 N ; 400.3 N, 0.333)

91. (*a*) Determine the various componenets of the machining force when machining a C.I. block on a shaper with depth of cut = 4 mm, feed = 0.25 mm/stroke, Normal rake angle of the tool = 10°, Principal cutting edge angle = 30°, μ = 0.6, τ_s = 340 MPa.

(*b*) Determine the average power consumption if the operation takes with 60 strokes/min. and the length of the job is 200 mm.

Solution : (*a*) Using Lee and Shaffer relation for shear angle,

$$\phi = \frac{\pi}{4} - (\beta - \alpha)$$

Now $\beta = \tan^{-1} \mu = \tan^{-1} 0.6 = 30.96° \cong 31°$

$\therefore$ $\phi = 45° - (31 - 10) = 24°$

Now, cutting force (it acts along velocity vector),

$$F_c = \frac{\tau_s . b . t}{\sec(\beta - \alpha) . \cos(\phi + \beta - \alpha) . \sin\phi}$$

Now, $b = \dfrac{d}{\cos C_s}$ and $t = f \cos C_s$

$\therefore$ $b . t = d . f = 4 \times 0.25 = 1 \text{ mm}^2$

$\therefore$ $F_c = \dfrac{340 \times 1}{\sec 21° \times \cos 45° \times \sin 24°} = 1103.8 \text{ N}$

Thrust force, $F_t = F_c \tan(\beta - \alpha)$, it acts normal to tool-job interface. (Ecn. 1.18a)

$= 1103.8 \times \tan 21° = 423.7 \text{ N}$

Now feed force component, $F_f = F_t \sin\lambda$ (normal to the velocity vector)

$= 423.7 \times \sin 60° = 366.92 \text{ N}$

Normal force component, $F_n = F_t . \cos\lambda = 311.85 \text{ N}$ (normal to the machined surface)

(*b*) Refer Fig. 3.39. Work done $= F_c \times$ length of stroke $= 1103.8 \times \dfrac{200}{1000} = 220.76 \text{ J}$

$\therefore$ Average Power $= 220.76 \times \dfrac{60}{60} =$ **220.76 watts**

2

CUTTING TOOL MATERIALS AND CUTTING FLUIDS

2.1. GENERAL

We have seen in the last chapter, that the cutting tool is subjected to : static and dynamic forces, high temperatures, wear and abrasion. To get a reasonable tool life, the tool material should meet the following requirements :

1. Hot hardness, so that the tool does not loose its hardness and strength at the high temperatures developed during machining.

2. Wear and abrasion resistance, so that the tool retains its shape and cutting efficiency for a reasonably long time before it is reconditioned or replaced.

3. Impact toughness, so that the fine cutting edge of the tool does not break or chip, when the tool is suddenly loaded.

In addition to the above basic requirements, the tool material should possess the following properties : increased thermal conductivity, lower co-efficient of thermal expansion, lower chemical and mechanical affinity for the work-material and it should be easy to form, grind and sharpen to the desired tool geometry.

The selection of a proper tool-material depends upon a number of factors such as : type of cutting operation, material of the work piece, machine tool to be used and surface finish required. Usually, a compromise has to be made in the selection of tool-material, since the requirements to be met by tool-material are often contradictory in nature. Over the years, a wide variety of cutting tool materials have been developed to meet the ever increasing demand of machining harder and harder materials. The various cutting tool-materials can be grouped as follows :

1. Plain Carbon Steels.
2. Medium Alloy Steels.
3. High Speed Steels (H.S.S.).
4. Non-ferrous Cast Slloys.
5. Cemented Carbides.
6. Ceramics or Oxides.
7. Cermets.
8. Diamond.
9. Cubic Boron Nitride (CBN).
10. UCON.
11. Sialon.
12. Coronite

Plain carbon steels, Medium alloy steels and High speed steels are known as "Tool Steels". Medium alloy steels and high speed steels contain one or more alloying elements to impart the desired properties to the cutting tools. The function of each alloying element is given below.

(*i*) **Carbon.** Carbon combines with iron to form carbide which makes it respond to hardening, thus increasing the hardness, strength and wear resistance. The percentage of carbon varies from 0.6 to 1.4%.

(*ii*) **Manganese.** It is added to steels as a deoxidizing and desulphurizing agent. It lowers the critical range of temperature. It increases the time required for transformation, so that, oil quenching, becomes practicable. Its content is about 0.5 to 2%.

(*iii*) **Chromium.** The addition of chromium results in the formation of various carbides of chromium which are very hard, yet the resulting steel is more ductile than a steel of the same hardness produced by a simple increase in carbon content. Chromium also refines the grain structure so that, these two combined effects result in both increased toughness and hardness. The addition of chromium increases the critical range of temperature and raises the strength at high temperatures. Alloy of chromium resists abrasion and wear. Its content ranges from 0.25 to 4.5%.

(*iv*) **Molybdenum.** Molybdenum is a strong carbide forming element and its action is very much like chromium but is more powerful. It increases strength, wear resistance, hardness penetration and hot hardness. It is always used in conjunction with other alloying elements. Its content ranges upto about 10%.

(*v*) **Cobalt.** Cobalt is commonly used in high speed steels to increase the hot hardness so that the cutting tools can be used at higher cutting speeds and temperatures and still they retain their hardness and a sharp cutting edge. Its content ranges from 5 to 12%.

(*vi*) **Vanadium.** It increases hot hardness and abrasion resistance. As vanadium has a very strong tendency to form carbides, hence, it is used only in small amounts (0.2 to 0.5% in alloy carbon tool steels and 1 to 5% in H.S.S).

(*vii*) **Tungsten.** It is widely used in tool steels because the tool maintains its hardness even at red heat. Tungsten produces a fine dense structure and adds both toughness and hardness. Its effect is similar to molybdenum except that it must be added in greater quantity(1.5 to 20%).

Note : Most of the tool-steels contain two or three alloying elements, as the combined action of several elements is more effective than that of one element even when its content in steel is considerable.

2.2 CUTTING TOOL MATERIALS

2.2.1. Plain Carbon Tool Steels. These are the oldest type of tool steels. The material is inexpensive, can be easily formed and ground. The properties of the material will depend upon the percentage of carbon content. Low carbon steels are tough and shock resistant, whereas, high carbon steels are abrasion resistant. They are basically water hardening materials (Type W tool steels), that is, they are hardened by heating followed by quenching in water to obtain hardness of H_{RC} 60 – 67. Their hardness decreases rapidly above 200°C and so they are useful for low speed operations — drilling, tapping, reaming, broaching etc. They are also used to manufacture ; woodworking tools, cold chisels, hammers, knives and punches. These materials are prone to deformation and cracking when hardened.

2.2.2. Alloy Tool Steels. The various alloying elements (discussed above) are added to plain carbon steel to impart the desired properties to the tool steel. These alloying elements slow down the transformation rates. Due to this, the materials can be hardened in oil or air, and the alloy steels become less susceptible to cracks while quenching. These steels have greater wear resistance and hot hardness than the plain carbon steels. These are widely used for drills, taps, reamers etc., but do not have sufficient hot hardness to be used in high speed turning or milling. Depending upon the method of quenching, these steels are of two types.

(*a*) **Type – O Tool Steels.** These steels are hardened by quenching in oil. Type O – 1 is most commonly used which has the following composition :

C : 0.90%, Mn = 1.00%, W = 0.5%, Cr = 0.5%

These steels find use in cold-working applications, such as punching and blanking, shearing, forming and drawing dies.

***(b)* Type – A Tool Steel.** These steels are hardened by cooling in air. In these steels, the content of alloying elements is higher as compared to type O steels. Usually, C is 1.0% and Cr is 5%. Chiefly used for cold working applications, for example thread rolling dies, coining dies and gauges.

The alloy tool steels can operate upto cutting temperature of about 300°C.

2.2.3 High Speed Steel (H.S.S.). This tool material is basically high carbon steel, to which the various alloying elements (Tungsten W, Molybdenum Mo, Chromium Cr, Vanadium V and Cobalt Co) have been added in larger amounts as compared to alloy tool steels, to improve hardness, toughness and wear resistance properties. These materials are deep hardening and can be quenched in oil, air, or salt. They are capable of retaining their hardness upto 600°C and so can be operated at much higher cutting speeds as compared to alloy tool steels, hence the name "high speed steel". This tool material was developed in 1905.

There are two basic types of high speed steels :

1. Tungsten type steels (T series) which has tungsten as the major alloying element (12 – 20%).

2. Molybdenum type steels (M Series) in which tungsten is partially or completely replaced by molybdenum.

The molybdenum type of H.S.S. is cheaper than the tungsten based material and is more readily sharpened. Also, it generally has a slightly greater toughness at the same level of hardness.

The popular 18-4-1 H.S.S. contains 18% tungsten, 4% chromium and 1% vanadium. Cobalt (5 to 8%) may also be added to increase red hardness. Carbon is about 0.75%. This material is designated as T-4. The ISI designation of this steel is (T 75 W 18 Co 5 Cr 4 V 1 Mo 70). The most commonly used types of H.S.S. steels are listed in Table 2.1, with their designations and compositions.

The most commonly used grades of high speed steels are : $M-1$, $M-2$, $M-7$, $M-10$, $T-1$ and $T-2$. $M-2$ tool steel has the applications of : turning carbon and alloy steels of hardness up to 375 *BHN*, turning nitriding steels of hardness upto 350 *BHN*, turning ultra high-strength steels of hardness upto 300 *BHN*, turning tool steels, cast steels of hardness upto 300 *BHN*, turning armour plate of hardness upto 325 *BHN*, and turning non-ferrous materials such as copper, brass, aluminum, magnesium and plastics. Broaches to cut ferrous materials of hardness upto 260 *BHN* are made of $M-2$ grade.

Milling cutters for cutting f errous materials of hardness upto 350 *BHN* are made of $M-2$ and $M-7$ grades. Gear hobbing, shaping and shaving tools are also made of $M-2$ and $M-7$ grades. Drills and reamers for steels of hardness upto 325 *BHN* are made of $M-1$, $M-2$, $M-7$ and $M-10$ grades. Taps are mostly made from $M-1$ grade. Form tools are usually made from $M-2$ or $T-2$ grade of H.S.S.

High speed steels with cobalt (for example grades : $M-33$, $M-36$, $T-4$, $T-5$ and $T-6$) have high hot hardness and wear resistance but lower toughness as compared to $M-2$ H.S.S. Tool bits for planing and heavy duty turning are made from $T-4$, $T-5$ and $T-6$ grades. Grades $M-33$ and $M-36$ are used for : drilling and milling hard alloy steels, stainless steel, titanium and heat resistant materials. $T-6$ grades are also called "Super H.S.S". They have a high W (18 to 22%) and high Co (10 to 12%) content. They are less tough than other types and need to be well supported on a very rigid machine. This type is the most expensive H.S.S.

H.S.S. with high percentage of vanadium (For example $M-3$, $M-4$, and $T-15$ grades)

Table 2.1 High Speed Steels

Designation	*Percentage of Constituents*						*Relative properties from 1 (low) to 10 (high)*			
	C	*W*	*Mo*	*Cr*	*V*	*Co*	*Wear Resistance*	*Toughness*	*Hardness*	*Cost*
M – 1	0.80	1.75	8.50	3.75	1.15	–	4	10	5	3
M – 2	0.85	6.00	5.00	4.00	2.00	–	5	10	5	3
M – 3	1.05	6.00	5.00	4.00	2.40	–	6	7	6	4
M – 4	1.30	6.00	4.50	4.00	4.00	–	9	5	6	4
M – 7	1.00	1.75	8.75	4.00	2.00	–	6	8	5	3
M – 10	0.90	–	8.00	4.00	2.00	–	5	8	5	3
M – 33	0.88	1.75	9.50	3.75	1.15	8.25	5	5	8	5
M – 36	0.85	6.00	5.00	4.00	2.00	8.25	5	5	8	5
M – 41	1.10	6.75	3.75	4.25	2.00	5.00	6	4	8	5
M – 42	1.10	1.50	9.50	3.75	1.15	8.25	6	4	9	5
M – 43	1.20	2.75	8.00	3.75	1.60	8.25	6	4	9	5
M – 44	1.15	5.25	6.50	4.25	2.00	12.00	6	3	10	6
M – 45	1.25	8.25	5.00	4.25	1.60	5.50	6	3	8	5
M – 46	1.25	2.00	8.25	4.00	3.20	8.25	8	3	9	5
T – 1	0.75	18.00	–	4.00	1.00	–	4	8	5	5
T – 2	0.85	18.00	–	4.00	2.00	–	5	6	5	5
T – 4	0.75	18.00	–	4.00	1.00	5.00	5	4	7	6
T – 5	0.80	18.00	–	4.25	2.00	8.00	–	4	8	6
T – 6	0.80	20.00	–	4.50	1.75	12.00	5	2	9	8
T – 15	1.50	12.00	–	4.50	5.00	5.00	10	3	9	6

possess increased red hardness and wear resistance but their toughness and grind ability is reduced. Due to increased red hardness and wear resistance, these steels have greater tool life as compared to conventional H.S.S. They are used for machining conventional alloys, super alloys and refractory materials. Single point lathe tools, flat and form cutters, broaches and drills are made of these steels.

H.S.S. with cobalt (grades $M-41$, $M-42$, $M-43$, $M-44$, $M-45$ and $M-46$) have high hot hardness and wear resistance. Their grind ability is better as compared to H.S.S. with vanadium. These steels are used for machining : heat treated steels, titanium alloys and aerospace materials of high hardness (cobalt and nickel base alloys). Drills, milling cutters, form tools, broaches, hobs, shavers, taps and tool bits are made of these steels.

In the conventional H.S.S., there is every possibility of large dispersion and segregation of carbides. Due to this, there will be local variations in the chemical composition and structure of the material, resulting in difficulty in producing the tool and also poor performance of the tool. These drawbacks are overcome by using powder metallurgy (*P/M*) for the manufacture of tool steel. The powder of the steel alloy is made by atomizing the molten metal. The powder is then compacted under pressure in dies, to produce billets. These billets are then transformed to the desired shape and size by conventional hot forging. In this tool steel, there is better and more uniform distribution of carbides and the alloying elements. H.S.S. produced by *P/M* prossesses higher wear resistance greater toughness, greater impact strength, better hot workability, improved grindability, higher material removal rates and good dimensional stability.

2.2.4 Non Ferrous Cast Alloys (Stellite). This material which was introduced in 1915 is an alloy of Cobalt, Chromium and Tungsten with composition : Cobalt, 38 to 53 per cent, Chromium, 30 to 33 per cent, Tungsten, 10 to 20 per cent and Carbon, 1 to 3 per cent. 1 per cent Carbon content gives a relatively soft and strong tool and 3 per cent Carbon content gives a hard and more wear resistant grade. Cast alloy tools are cast and ground to any desired shape. Cast alloys bridge the gap between H.S.S. and Carbides (the next tool material). They have properties intermediate between H.S.S. and cemented Carbides. This material maintains great hardness at high temperatures and has good wear resistance but is not as tough as H.S.S. and is sensitive to shock loading. The tool material is available in the form of tool bits for use in holders and in the form of tips brazed to a medium carbon steel shank. The material is recommended for deep roughing operations at relatively high speed and feed rates and it can machine more difficult materials such as high tensile steels, stainless steels and heat resistant steels, and C.I. This material is used at surface speeds above those of H.S.S. and below those of Carbides, and can withstand a cutting temperature in the range of 900°C.

The introduction of Cast alloys as cutting tool materials somewhat overlapped that of the Tungsten carbide and since these carbides were, in general, superior cutting tools, the cast non-ferrous alloys never caught on to the extent anticipated.

2.2.5 Sintered or Cemented Carbides. The first sintered carbide cutting tool (tungsten carbide) was marketed in 1926. Since then lot of research has been done and many types of sintered carbide materials have been produced to improve their performance. This has been achieved by improving the methods of their manufacture and their composition. For example, carbides of titanium, tantalum, niobium and columbium etc. can be added to straight tungsten carbide to extend the range of their application. Sintered carbides are produced by *P/M* technique and have the following properties. Very high red hardness (of about 1000°C), very high wear resistance, high modulus of elasticity, low thermal expansion and high thermal conductivity.

There are three general groups of cemented carbides in use :

1. Straight tungsten carbide with cobalt as a binder.

2. Tungsten carbide with cobalt as a binder and having large percentages of carbides of Ti, Ta and Nb, which along with *WC* form a solid solution of WC – TiC – TaC – NbC.

3. Titanium carbide with nickel or molybdenum as the binding material.

Manufacture. The cemented carbides are manufacture by *P/M* technique involving the following steps :

1. Firstly, we should get the ingredients. Tungsten oxide is reduced in hydrogen to get tungsten metal powder. Similarly, to obtain titanium and tantalum, titanium oxide, and tantanlum oxide are reduced in hydrogen atmosphere, respectively.

2. Tungsten powder is then mixed with lamp black and the mixture is heated at about 1600°C to form carbide of tungsten, and so on.

3. The lump of the carbide so produced is crushed to a powder. Then, powdered cobalt metal which acts as a binder to hold together the particles of carbide powder, is thoroughly mixed with the latter.

4. To increase the mouldability of *WC – Co* mixture, it is mixed with a lubricant such as paraffin, ethylene glycol or camphor. Sometimes, only water is used.

5. After drying and refining, the mixture is compacted in a press to get the desired blocks.

6. The strength of the green compacts so obtained is quite low. To give the necessary strength to the compact, it is heated at 1300°C to 1865°C (sintered process) depending upon the composition and the cobalt content. As a rule, the sintering temperature is 90°C to 100°C below the melting point of pure cobalt. The sintering process is done for about 1.5 to 2 hours and it should be done preferably in an inert atmosphere to avoid oxidation or decarburization.

During sintering, there is about 15 to 25% shrinkage and this factor must be taken into account while designing the product shape.

7. The product is finally ground or lapped depending on the requirements.

Properties and Uses. The properties of the carbide tool material depend upon :

(*i*) Chemical composition

(*ii*) Method of manufacture

(*iii*) Micro-structure of the tool material.

The finer the grain size, the higher is the hardness. Out of these, the major effect is of the chemical composition (the method of manufacture and the grain size can be properly taken case of).

Straight tungsten carbide contains cobalt from 3 to 20%. With increase in the percentage of cobalt, the hardness, the brittleness and compressive strength of the material decreases. However, with increase in cobalt content, the transverse rupture strength of the tool material increases. This material has high abrasive wear resistance and high strength in respect of a given hardness. The main drawback of straight tungsten carbide is its affinity for steel. Due to this, the steel chips tend to weld to the tool surface resulting in crater wear. Hence, this tool material is not suitable for cutting steel, but gives superior performance with non-ferrous, non-metallic material and cast iron. This material is called as 'C' grade cemented carbide material.

To reduce the tendency of the metal chips to weld to the tool and to decrease the diffusion of the tool material to the chips, TiC is added to WC–Co system. However, with increase in the percentages of TiC, compressive, transverse and impact strengths and also the thermal conductivity and the modulus of elasticity of tool material decrease. These drawbacks of TiC are overcome by adding TaC, which increases the transverse strength and the hot hardness of the tool material. The effect of adding NbC is similar to that of TaC. These mixed carbides are used for machining steel. This cemented carbide material is known by 'S' grade. It contains : about 16% TiC, O to 10% TaC. Cabalt content varies from 3 to 16%.

With TiC tool material nickel and molytdenum are used as binding materials. Molybdenum is added to check the very high grain growth with TiC-Ni system. This material has low wettability. This improves its crater wear resistance and hence it is used when cutting temperatures are high, because of high cutting speeds or hard workpiece material.

According to ISO (International Standards Organization), the various grades of carbide tool material have been grouped in three series :

1. Carbide tools used for cutting cast iron and non-ferrous metals are designated from : K_{01} to K_{40}.
2. Grades of carbide tools used for machining steel are designated as : P_{01} to P_{60}.
3. Grades of carbide tools used for general purpose applications are designated as : M_{10} to M_{30}.

In all the above three series, the harder and brittle materials have the lowest number and the less hard and most tough materials have the higher numbers.For example in K-series, K_{01} is the hardest and most brittle and K_{40} is the least hard and the most tough material. The K-grades of carbides are essentially straight tungsten carbides with cobalt as the binder, and are used for machining cast iron, non-ferrous metals, plastics and similar materials. P-grade carbides are combined carbide tool materials (carbides of W, Ti, Ta and Nb) with cobalt as the binder. These materials are similar in composition to the K-grade, but have different properties due to different manufacturing methods. These are used to cut heat resistant steels and stainless steels.

All carbides, when finished, are extremely brittle and weak in their resistance to impact and shock loading. Due to this, vibrations are very harmful for carbide tools. The machine tools should be rigid, faster and more powerful. Light feeds, low speeds and chatter are harmful. Due to the high cost of carbide tool materials and other factors, cemented carbides are used in the form of inserts or tips which are brazed or clamped to a steel shank, Fig. 2.1.

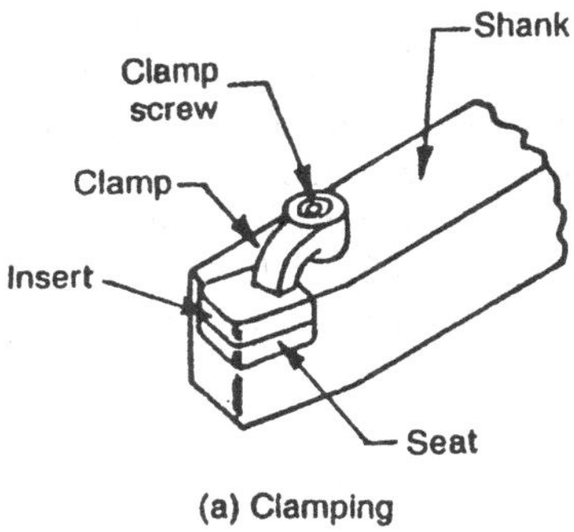

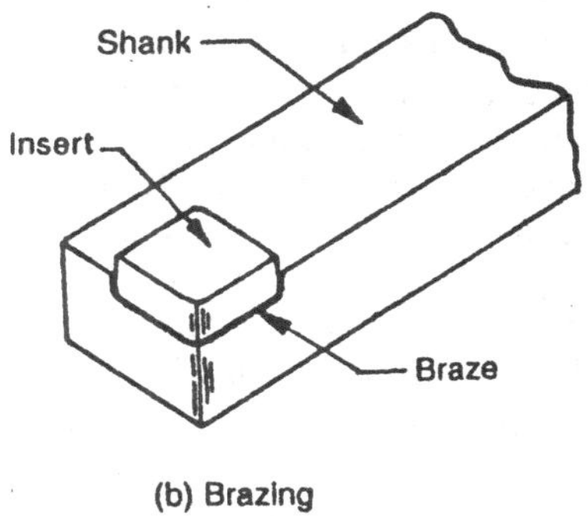

Fig. 2.1. Methods of Attaching Inserts to Tool Shanks.

In the case of a brazed tip, when it is worn out, it is resharpened with the help of special grinding wheels on a Tool and Cutter grinder. The main drawbacks of a brazed tip are : For resharpening purposes, the tool will have to be removed from the machine involving a resetting operation. Also, because of the difference in co-efficients of expansion of tip material and tool shank material, the brazing has to be done very carefully.

Mechanically clamped tips are known as "Indexable tips or inserts", because these have more than one cutting edge which are used one by one by indexing the tip. These tips are also known as "Throw-away" or "disposable", because once all the edges of the tip have been used, the tip or insert is removed from the tool shank seat and thrown away or disposed-off. In this way, a rectangular tool bit (tip or insert) can be used upto eight times before disposal and requires no resharpening cost. The inserts are available in a variety of shapes, such as square, triangle, diamond and round, as shown in Fig. 2.2.

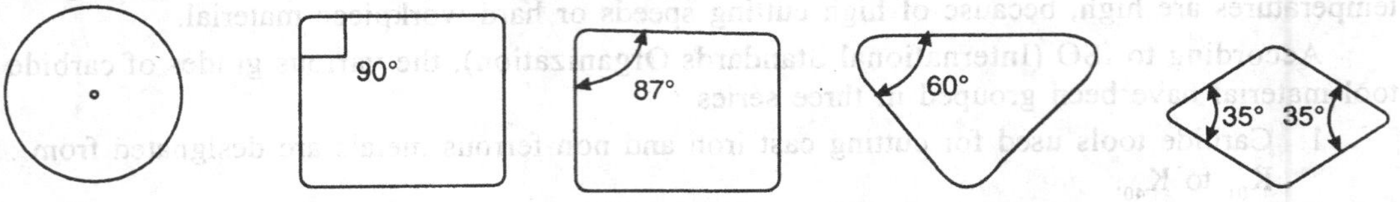

Fig. 2.2. Different Shapes of Inserts.

The various parameters depend upon the shape of the insert. For example,

(*a*) Higher the cutting edge angle, higher is the cutting edge strength. That is, cutting edge strength increases from right to Left in Fig. 2.2.

(*b*) Power reguirement decreases as the cutting edge angle decreases. (Left to Right).

(*c*) Tendency to vibration increases with increase in cutting edge angle. (Right to Left).

(*d*) Versatility and accessibility increases as thecutting edge angle decreases. (Left to Right).

There are certain rules of thumb to be followed when selecting an insert for a particular Job :-

(*i*) For strength and economy, the largest possible cutting edge angle should be selected.

(*ii*) When there is a tendency to vibration during operation, strength versus versatility and assessibility through the use of smaller cutting edge angles should always be considered.

To improve the cutting edge strength, the insert edges are usually honed and chamfered or produced with a negative land. The radius of honing may be about 0.025 mm. When using ceramics, re-chamfering is recommended to minimize the risk of burrs when the insert exits the cut. It also has a positive effect on the insert while entering.

Laminated and Coated Carbides. We have seen above that the resistance of straight tungsten carbide to crater wear can be increased by the addition of TiC. The same result can be obtained by laminated and coated carbides. In laminated carbides, laminates consisting of a hard thin surface layer of TiC and in the form of throw-away tips, are bonded by epoxy resin to the rake face of a tip body of WC. Coated carbides have a thin coating of TiC on all faces of the tip. The coating thickness is of the order of a few microns (0.0025 to 0.005 mm). These tools resist the diffusion wear on the crater and give a tough shock resistant tool. Laminated and coated carbides are shown in Fig. 2.3.

Coating is done by Chemical Vapour Deposition (CVD).

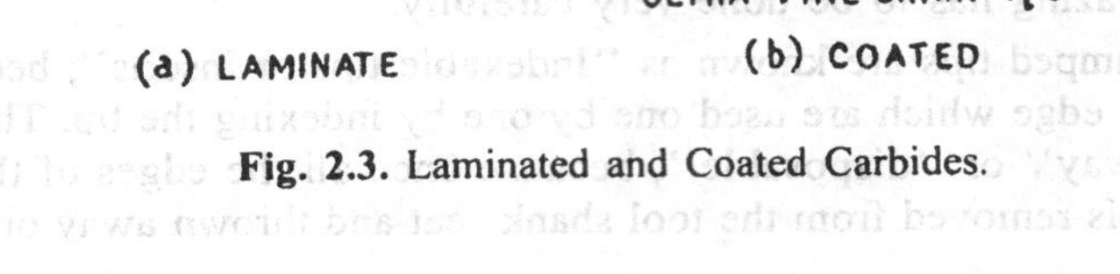

Fig. 2.3. Laminated and Coated Carbides.

Table: 2.2

Classification of carbide tips according to their range of application (IS: 2428–1964)

Designation	Increasing direction of the characteristic of		Range of application	
Indentifi-cation colour	Carbide tip	Cutting	Material to be machined	Machining conditions
P01	↕ Resistance to wear ↕ Toughness	↕ Cutting speed ↕ feed	Steel, steel casting	Precision turning and fine boring Cutting speed: high Feed: Low
P10			Steel, steel casting	Turning, threading and milling Cutting speed: high, Feed: low or medium
P20			Steel, steel casting, malleable cast iron forming long chips	Turning, milling. Cutting speed and feed: medium planing: with low feed rate
P30			Steel, steel casting, malleable cast iron forming long chips	Turning, planing, milling. Cutting speed: medium to low. Feed: medium to high even if operating conditions are unfavourable
P40			Steel, steel casting with sand inclusions or shrinkage cavities	Turning, planing, milling. Cutting speed: low, Feed: high. Rake angle: high, for machining under unfavourable conditions and work on automatic machines
P50			Steel, steel casting, manganese steel, grey cast iron, alloyed cast iron	Turning, planing, shaping. Cutting speed: low. Feed: high. Rake angle: high, for machining under unfavourable conditions and work on automatic machines
M10	↕ Resistance to wear ↕ Toughness	↕ Cutting speed ↕ feed	Steel, steel casting, manganese steel, grey cast iron, alloyed cast iron	Turning. Cutting speed: medium to high. Feed: low to medium
M20			Steel, steel casting, austenitic steel, manganese steel, grey cast iron, sphrodised cast iron and malleable cast iron	Turning, milling. Cutting speed: medium. Feed: medium
M30			Steel, steel casting, austenitic steel, grey cast iron, heat resisting alloys	Turning, milling, planing. Cutting speed: medium Feed: medium or high
M40			Free cutting steel, low tensile strength steel, brass and ligh alloy	Turning, profile turning, parting off especially in automatic machines
K01	↕ Resistance to wear ↕ Toughness	↕ Cutting speed ↕ feed	Very hard grey cast iron, chilled castings of hardness up to 60 HRC. Alluminium alloys with high silicon content, hardened steel, plastics of abrasive type, hard board and cermics	Turning, precision turning and boring, milling, scraping
K10			Grey cast iron of hardness more than 220 HB, malleable cast iron forming short chips, tempered steel, aluminium alloys containing silicon, copper alloys, plastics, glass, hard rubber, hard cardboard, porcelain, stone	Turning, precision turning and boring, milling, scraping
K20			Grey cast iron of hardness up to 220 HB, non-ferrous metals, such as copper, brass, aluminium, laminated wood of abrasive type	Turning, milling, planing, reaming, broaching
K30			Soft grey cast iron, low tensile strength steel laminated wood	Turning, planing, shaping, milling, rakeangle large, even under unfavourable conditions
K40			Soft or hard natural wood, nonferrous metals	Turning, milling, planing, shaping. Rakeangle: large, even under unfavourable maching conditions

Oxide Coating of Carbide tools. The diffusion of atoms between the tool and chip material (which is the major cause of carbide tools) can be retarded by coating the tool surface of the carbide tools with oxides of aluminium and zirconium. This considerably increases the tool life. Coated Carbides are used for machining super alloys.

These tools operate at cutting speeds which are about 5 times of that for H.S.S. Cemented carbide tool materials also include : nitrides, borides and silicides which are compounds of nitrogen, boron and silicon with such metals as Tungsten, Titanium, Tantalum, Niobium and Molybdenum.

Table 2.2 gives classification of carbide tips

2.2.6 Ceramics and Oxides. Ceramics or sintered oxides were developed as cutting tool materials during 1950 – 1960. These materials are basically aluminium oxide (Al_2O_3) containing additions like MgO, NiO, Cr_2O_3, TiO and TiC etc. to improve the grain structure, cutting properties and sintering.

These materials are produced in the same manner as sintered carbides, that is, by P/M technique. Th aluminium powder is ball milled to a suitable grain size. Water is added and the ingredients are thoroughly mixed to make a stiff paste. This is then pressed into pallets of required form in punch and die. The pallets are dried and then sintered at temperature of 1500°C to 1700°C. The grain particles then get bonded to one another by the process of diffusion. The pallets are then cut and ground to the required shape, size and surface finish, with the help of resinoid bonded grinding wheels. Ceramics are always used in the form of indexable inserts of standard shapes and sizes. At present there is no satisfactory method of brazing the tool tips to steel shank. So, the ceramic tool tips are clamped mechanically to the steel tool shank.

Ceramic tool materials have a very high abrasion resistance, are harder than cemented carbides and H.S.S. and have less tendency to weld to metals during machining. However, they lack impact toughness, so, vibration and chatter are fatal to these tools. Due to this, the tool mountings should be rigid and the machine tools should be rigid. These materials are especially effective at very high cutting speeds (2 to 5 m/s) and for uninterrupted cutting operations. These materials are particularly successful for machining cast iron, and high tensile materials, at cutting speeds which are 2 to 3 times of that of comented carbides. These tools can retain cutting edge hardness upto about 1400°C and exhibit uniform strength upto 1200°C.

Another ceramic tool material is Silicon nitride (Code named S-8), which is particularly used for cast iron machining applications. These ceramic tools have a tool life that was effective over 1500 C.I. pieces, where coated tungsten carbide tools lasted only 250 pieces before dulling. Ceramics are less expensive than carbides and the trend now is to replace the latter with the former whenever possible and reasonable. Ceramics are mainly used for finishing and super-finishing. Poor results are obtained if Al_2O_3 tool material is used to machine Al or Ti alloys, because strong bonds tend to form between the chip and the tool.

2.2.7 Cermets : The cermets are combinations of ceramics and metals, bonded together in the same manner in which P/M parts are produced. They combine some of the high refractoriness of ceramics and toughness and thermal shock resistance of metals. For cutting tool materials, the usual combination is Al_2O_3 plus metal additions (W, Mo, Boron, Ti etc.) in an amount upto 10%. These additions reduce the brittleness to some extent, but they reduce the wear resistance of the material as well.

2.2.8 Diamonds. Diamond is the hardest of all the cutting tool materials. Diamond has the following properties : extreme hardness, low thermal expansion, high heat conductivity, and a very low co-efficient of friction. This is used when good surface finish and dimensional accuracy are desired. The work-materials on which diamonds are successfully employed are the non-ferrous one, such as copper, brass, zinc, aluminium and magnesium alloys. On ferrous materials, diamonds are not suitable because of the diffusion of carbon atoms from diamond to the work-piece material.

Diamond tools have the following applications : single point turning and boring tools, milling cutters, reamers, grinding wheels, honing tools, lapping powder and for grinding wheel dressing. Due to their brittle nature, the diamond tools have poor resistance to shock and so, should be loaded lightly. For fine turning, the following values are recommended : cutting speed 200 to 500 m/min, feed 0.01 to 0.05 mm/rev. and depth of cut 0.10 to 0.15 mm.

Diamonds are available either as naturally or as man made (synthetic). The natural diamonds are, however, of low grade. For metal cutting applications, polycrystalline diamond instead of single crystal diamond, has been recently introduced. This tool material known as ''Compacts'' has increased strength and shock resistance. This material is made of diamond powder by sintering into moulded shapes by high pressure and temperature technology.

Polycrystalline diamond has been successfully used for machining tough, abrasive non-ferrous materials, plastics, ceramics and glass. The material can be moulded into standard shaped inserts and used as conventional indexable inserts. Indexable inserts can also be made by brazing compacts of polycrystalline diamond to each cover of a carbide insert. This material can also be used by bonding its thin layer (about 0.5 mm thick) to a WC substrate. These blanks are brazed on to a steel shank in position, ground and then used. WC being tougher than diamond will increase the shock resistance of the tool. such tools are called as ''Compax'' tools. Diamond tools can with stand heat upto 2000°C with highest tool life (50 to 100 times more than that of WC).

2.2.9 Cubic Boron Nitride (CBN). Next to diamond, CBN is the hardest material currently available. This material, which consists of atoms of nitrogen and boron, was produced in the early 1970's by high pressure, high temperature processing. As a cutting tool material, CBN is used in the polycrystalline form. CBN has high hardness and high thermal conductivity. It has much higher tensile strength (1000 N/mm^2) as compared to diamond (300N/mm^2). CBN being chemically inert, is used as a substitute for diamond for machining steel. Other applications are : as a grinding wheel on H.S.S. tools, for machining high temperature alloys, Titanium, Nimonic, Stainless steel, Stellites and Chilled C.I.

In dealing with iron-based alloys and hardened steels, the life of a CBN tool is 4 to 5 times higher than that of a diamond tool.

CBN, as a cutting tool material, can be used in different ways : In ''Compax'' tools, a 0.5 mm thick layer is bonded to a cemented carbide substrate at high temperature and pressure. These tools combine the high hardness and wear resistance of CBN with the high shock resistance and toughness of WC. CBN can also be made in the form of indexable inserts and blanks of standard shape and size. The blanks can be brazed on to steel shanks, form ground and then used. This material is known by the trade name ''Borazon'', in U.S.A. and 'Elbor' in Russia.

2.2.10 UCON. This is also new cutting material developed by Union Carbide, U.S.A. Its constituents are : Columbium 50%, Titanium 30%, and Tungsten 20%. This tool material is manufactured according to the following steps :

1. The powders of columbium, Titanium and Tungsten are thoroughly mixed and blended.
2. The mixture is compacted in a punch and a die.
3. The compact is then melted in an electric arc furnace and the alloy is cast into ingots.
4. The ingot is rolled into sheets which are then cut into strips.
5. The strips are then cut into blank of desired shape and size.
6. The blanks are then can ground and honed subsequently by tumbling to produce a radius of 0.05 to 0.075 mm at cutting points.
7. Lastly, the blanks are nitrides in a nitrogen atmosphere at a very high temperature.

UCON has the following properties : High hardness, High toughness, excellent shock resistance and excellent resistance to diffusion and adhesion wear. This is a basically steel cutting material and is not preferred for cutting cast iron, stainless steel and super alloys containing Ni, Co and Ti as base materials. Cutting operations recommended for UCON are : roughing, semi-roughing and finishing, turning, facing and boring operations. It permits 60% increase in cutting speed when compared with WC.

2.2.11 Sialon (SiAlON). The research on this tool material has been going on for the last about 14 years. The material is produced by milling together Si_3N_4, Aluminum oxide, Al_2O_3 and yttria. The powder is dried, pressed to shape and sintered at a temperature of about 1800°C. This material has been found to be considerably tougher than ceramics, and thus can be successfully used for machining with interrupted cuts. Cutting speed can be 2 to 3 times, those with carbides. At present, the field of application of this tool material (in the form of tips) is for machining aerospace alloys, Ni-based gas turbine blades etc. at cutting speed in the range of 3.3 to 5 m/s.

2.2.12. Coronite. It is a new cutting tool material whose properties lie in between those of H.S.S. and cemented carbides.

It combines the toughness of H.S.S. with hardness and wear resistance of cemented carbides. This improves tool life, reliability and surface finish. Cutting tools made from this material are mainly endmills used for machining grooves, pockets and for profiling in majority of the workpiece materials.

The material consists of fine grains of TiN evenly dispersed in a material of heat treatable steel. The hard grains of TiN form about 35 to 60% of material's volume. The properties of the material are attributed to : very small size of hard grains of TiN (about 0.1 micron) as compared to 1 to 10 microns in H.S.S. and Cemented carbides and the proportion of hard grains in the material (which is higher than in H.S.S. but less than in cemented carbides). The material is produced by particle metal technology.

Majority of the tools are not produced from solid coronite but by compound and coating technology as follows :-

1. A core of H.S.S or spring steel.

2. A layer of about 15% of diameter of core is created over the core by extrusion process at about 540°C. The bar thus produced is the raw material for coronite cutting tools.

3. A thin coating (about 2 microns) of TiCN or TiN is created on the material by PVD method.

2.3. CUTTING FLUIDS

In any metal cutting operation lot of heat is generated due to : plastic deformation of metal, friction at the rake face of the tool between the tool and the chip and also the friction between the workpiece and the flank of the tool. This increases the temperature both of the workpiece and the tool point, resulting in decrease in hardness and hence life of the tool. The machined surface will also be rough and the possibility of built up edge increases. So, the use of a cutting fluid during a machining operation is very essential. Its application at the workpiece-tool interface produces the following effects :

1. Friction at the workpiece -tool interface is reduced, due to lubricating action.
2. Heat is reduced due to cooling action at the interface.
3. Chips are washed off.

So, the cutting fluid performs the following functions :

1. Reduces heat generation.
2. Provides lubricating action.
3. Carries away the heat generated and so provides cooling action, thus, reducing workpiece temperature and distortion.
4. Provides flushing action in washing off the ships.

5. Reduces friction and wear, which improve tool life and surface finish.
6. Reduces force and energy consumption.
7. Protects the newly machined surface from environmental corrosion.

Due to reduction in friction because of lubricating action : shear angle increases, chips produced are thin, cutting force is decreased, less heat is produced and there is low built up edge. Due to reduction in heat produced and cooling effect of the cutting fluid, the tool and the workpiece remain cooler. This results in : maintenance of tool hardness, less tool wear and longer tool life, less distortion and easy handling of the job. Washing off the chips helps in better surface finish and use of higher feed rate.

2.3.1. Lubrication and Cooling Action of Cutting Fluids. During metal cutting, the area of contact between tool and job is very large and also the ratio of real to apparent area of contact is very nearly equal to unity. Again, the contact pressure at the tool-workpiece interface is very high. Due to the above two factors, the type of lubrication in metal cutting can never be full fluid film lubrication. It can only be boundary lubrication. Because of this, the chemical properties of a cutting fluid are more important than its physical properties. Additives like chlorine, phosphorous, sulphur and fatty acids in the cutting fluid react with workpiece material and form a layer of solid lubricant. This lubricant has a low shear strength, and can withstand high temperature. Due to this lubrication action, the amount of heat generated is considerably reduced, the formation of weld between the tool and the chip is prevented and the chips shear easily in sliding. However, at cutting speeds of more than 60m/min, the lubricating effect of chemically active cutting fluids is only marginal and for such high speed cutting operations, the cooling action of the cutting fluid is more predominant than its lubricating action.

2.3.2. Requirements of A Cutting Fluid. A cutting fluid should have the following properties :

1. It should wet the surfaces of cutting tool and workpiece for better cooling and lubricating effects.
2. Good lubricating property.
3. High heat absorbing capacity.
4. High flash point.
5. It should not damages or react with the materials of machine tool parts.
6. It should not stain or leave residues on the workpiece surface.
7. It should not emit toxic vapours.
8. It should be stable, that is, it should not get oxidised or decomposed when left in air.

2.3.3 Types of Cutting Fluids. There are basically two main types of cutting fluids :

1. Those which are mixed with water, such as, soluble oils and soaps. These are emulsions of oil and water or soap and water.
2. Those which are not mixed with water, called cutting oils, which can be pure oils or a mixture of two or more oils.

Soluble Oils. Water increases the cooling effect and oil provides the best lubricating properties. By mixing various proportions of water with soluble oils or soaps, cutting fluids with a wide range of cooling and lubricating properties, can be obtained. The ratio of oil to water depends upon the application of the cutting fluid and ranges from about 1 : 5 to 1 : 50. The usual ratios for the various machining operations can be :

Turning : 1 : 25

Milling : 1 : 10

Drilling and reaming : 1 : 25

Grinding : 1 : 50

Cutting Oils. These are fixed oils and mineral oils. Fixed oils consist of animal, fish and vegetable oils. Chiefly used fixed oils are : lard oil, sperm or whale oil, and olive, cotton seed and linseed oil. Turpentine oil distilled from vegetable oils is also used. Fixed oils have greater oiliness than mineral oils but are not so stable as mineral oils and tend to become gummy and decompose when heated. Mineral oils come from crude petroleum oils, for example, paraffin.

To combine stability of the mineral oils with good lubricating properties of fixed oils, they are often mixed. Sometimes chlorine and sulphur is added to give the property of ''Wetting'' the metal with a highly adhesive oil film. This imparts the antiwelding properties which help to prevent the formation of built up edge on the cutting tool. Such cutting oils are known as ''sulphonated oils''. To prevent the growth of bacteria and fungi which makes the cutting fluid to become injurious to the human skin, a phenol disinfectant is added.

During machining of heat resistant and stainless steels, nimonic alloys etc. the pressure and temperature at the cutting edge can be very heavy. The normal cutting oil may not be able to support this heavy presure at the cutting edge, resulting in welding of metal chips of the tool face. To prevent this, extreme pressure (EP) additives are added to the cutting oils. The most commonly used EP additive is sulphur and next is chlorine. These additives form solid films of iron chloride and iron sulphide between the tool face and the chips. These films are easily sheared and have high melting points. Thus, they help in preventing the chips from welding to the nose of the tool where the pressure is maximum.

So, depending upon the working conditions, four types of cutting oils can be had :

(*i*) Straight mineral oils. These are suitable for light duty and high speed work.

(*ii*) Mixture of mineral oils and fixed oils. These are suitable for light and medium duty.

(*iii*) Mineral oils with EP additives suitable for heavy duty.

(*iv*) Mixture of mineral oils and fixed oils with EP additives suitable for the heaviest duty.

2.3.4. Application of Cutting Fluids. The cutting fluid may be applied to the cutting tool in the following ways :

1. By hand, using a brush.
2. By means of a drip tank attached to the machine body.
3. By means of a pump.

For the effective use of cutting fluid and for heavy and continuous cutting, the fluid should penetrate into the cutting zone. For this, the third method (using a pump) of supplying the cutting fluid is the most common. This is known as 'Flood application'. Here, a continuous stream of cutting fluid is directed at the cutting zone with the help of a nozzle or jet, (Fig. 2.4). This is also known as ''Hi-jet Method''. The used cutting fluid drops into a tank at the bottom.

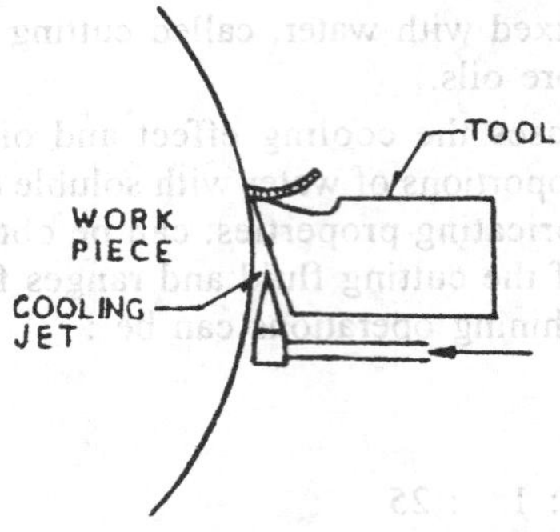

Fig. 2.4. Flood Application of Cutting Fluid.

Before it, is recirculated by the pump, it passes through many filters to remove chips and dirt. To avoid excessive splashing and vapourization of the fluid, it is supplied to the cutting zone from the top also, (Fig. 2.5).

For some applications, the cutting fluid is supplied through the tool itself and directed along the flank face of the tool.

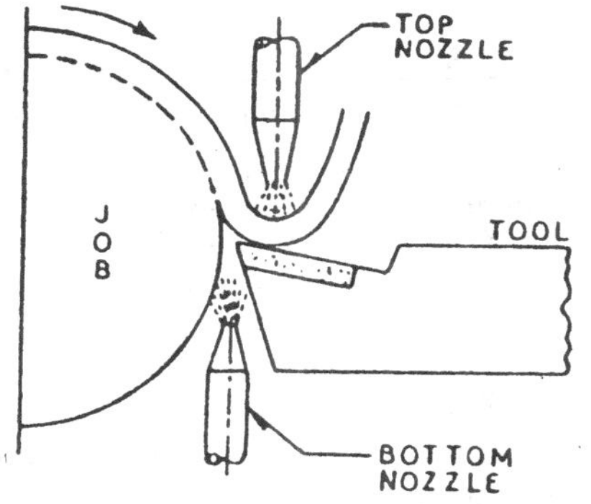

Fig. 2.5. Use of Two Nozzels.

Under suitable circumstances, the 'high-jet' method gives a significant increase in tool life. However, the apparatus for applying the cutting fluid is expensive and also the high-presure just may be of some danger to the operator. Thus, it may be an economical procedure of coolant application, it is not universally adopted.

Mist method of coolant application. The application of the cutting fluid as a fine atomized mist (combination of the carrier air and very fine drops of fluid) has received considerable attention in the last 25 years. The size of the fluid drops is of the order of 10 to 25 μ m. The mist is sprayed onto the cutting zone at high velocities of about 300 mps and more and under pressures of the order of 0.28 to 0.42 N/mm^2. The beneficial cooling effects of mist cooling result from the following facts :

1. Due to the high velocity of fluid application the dispersal of heat by convection currents, is intensified. This maintains the desirable high temperature gradient near the tool surface.
2. The surface area of the coolant is much greater in the form of mist than in flood applications. This increases the cooling capacity of the coolant.
3. Due to the expansion of the mist in the issuing nozzle, its temperature falls down considerably.

Mist cooling can be applied to almost all cutting operations, but it is generally more useful with high-hardness work materials.

It has been proved by research workers that mist cooling is less effective than flood cooling at significantly higher and lower temperatures (480° and 150°C). At low temperatures, the cooling is by heat conduction into the coolant and in this case, the stream of coolant is more effective than the mist. At very high temperatures, the mist produces a substantial blanket of vapours around the cutting zone. Its heat-transfer co-efficient becomes poor and its cooling efficiency descreases. Thus, mist cooling is more effective than flood cooling within a limited temperature range where the flood cooling loses its effectiveness due to the formation of a vapour layer under it.

The process of producing mist of the cutting fluid is based on the venturi principle, (Fig. 2.6). The high pressure air flowing by a siphon tube, draws the fluid into the nozzle. The shape and the density of the mist is then controlled by the nozzle. The basic components of a mist producing system are : air pump with an air storage, the cutting fluid container, piping and the spray nozzle.

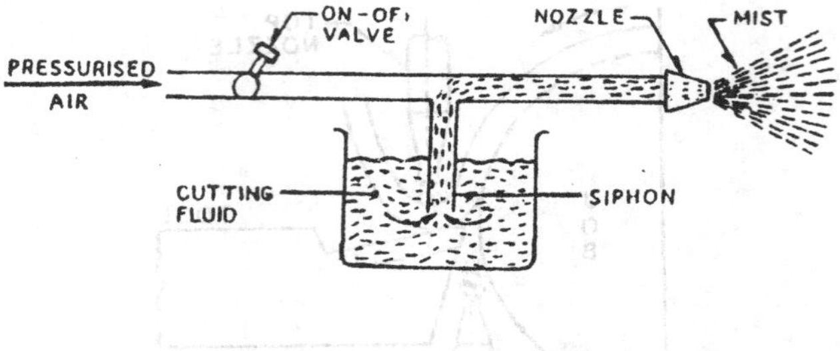

Fig. 2.6. Production of Mist.

2.3.5. Selection of a Cutting Fluid. The type of cutting fluid to be employed depends upon the work material and the characteristics of the machining process. For some machining process, a cutting fluid which is predominantly a lubricant is desirable, while with other machining processes, a cutting fluid which is predominantly a coolant should be used. For machining free cutting steels and "yellow" metals, soluble oils are used as cutting fluids. For expensive cutting tools, such as, form tools, gear cutters, broaches and milling cutters, a carefully selected cutting oil which does not effect the material of the cutting tool is selected. For machining difficult materials and for severe cutting conditions, a cutting oil is preferred to a soluble oil. Cast iron is usually machined dry, because, the usual cutting fluid forms a sort of slurry or mud with the powdery chips and this can enter the grooves and recesses in the machine tools. The graphite flakes within gray cast iron provide excellent lubrication. Brass, bronze and aluminum may be machined dry or wet with water soluble oils. Steel should always be machined with a lubricant. Magnesium and zincalloy die castings are generally machined dry when light cuts are taken or when cutting is done at low speeds. For heavy stock removal and for cutting at high speeds, cutting oils are used for magnesium and for zinc, kerosene lard oil mixture (50 : 50) is most efficient.

By far, the most common of all cutting fluids is the use of soluble oils which are emulsions of mineral oil with an emulsifier such as soap in water. These are combination of fatty oils, fatty acids, wetting agents, softening agents, emulsifiers, sulphur, chlorine, rust and foam inhibitors, germicides and water.

Water as a cutting fluid is the best cooling medium and is most effective fluid for high speed cutting, but it has little lubricating value, it does not spread well over a surface to wet it because of high surface tension, it causes rust and corrosion. To improve it as a cutting fluid, small proportions of inorganic additives are mixed with it. These additives include amines and nitrites to prevent rust, nitrates for nitrite stabilization, phosphates and borates to act as water softeners, wetting agents, and in some cases phosphorous, chlorine, or sulphur to prevent extreme pressure lubrication.

The general working rules for cutting-fluid selection are :

1. Select oil with extreme pressure properties for low cutting speeds (less than 30 to 60 mpm). At low cutting speeds, lubrication is the important factor in order to reduce the tendency to produce B.U.E.

2. At higher cutting speeds, temperatures increases substantially and cooling is the important consideration. So, select coolants (oil-in-water emulsions or water with rust inhibitor) for medium to high speed cutting.

Recently, pure compounds like carbon tetrachloride (CCl_4), chloroform ($CHCl_3$), trichloroethane and certain other chlorinated hydrocarbons have been found to be efficient lubricants when cutting many metals at low speeds (few cms. per min). However, their main drawback is their toxicity and their use involving high temperatures is out of question, since, when heated they give off poisonous gases.

Table 2.3. gives types of cutting fluids used with different work materials and for different machining operations.

Table 2.3. Types of Cutting Fluids

Material being machined	*Machining operation*			
	Turning	*Drilling*	*Tapping*	*Milling*
Cast Iron	Dry	Dry	Dry or 25% Lard oil + 75% Mineral oil	Dry
Alloy Steels	25% sulphur base oil + 75% mineral oil	Soluble oil	30% Lard oil + 70% mineral oil	10% Lard oil + 90% mineral oil
Low carbon and Tool steels	25% Lard oil + 75% Mineral oil	Soluble oil	25–40% Lard oil with mineral oil	Soluble oil
Malleable Iron	Soluble oil	Soluble oil	Soluble oil	Soluble oil
Bronze	Soluble oil	soluble oil	30% Lard oil with mineral oil	Soluble oil
Copper	Soluble oil	Soluble oil	Soluble oil	Soluble oil

Effect of Coolants on Cutting Variables, i.e., cutting speed, feed and depth of cut : Tool life is a direct function of cutting temperature (temperature at the work - tool interface). Excessive temperature is the most serious limitation to tool life, because cutting tool materials markedly soften at sufficiently high temperature, thus adversely affecting the tool life. The cutting temperature increases with increase in three process variables. Increased speed and feed and depth of cut magnify the thermal problems of the tool and the surface finish of the job. However, the work-tool interface temperature can be sufficiently reduced and hence the tool life can be enhanced by the effective use of a cutting fluid. Reduced average temperature at the tool-work interface permits a sufficiently high cutting speed and feed. This inhibits the formation of built-up edge, increases the metal removal rate and shortens the cycle time. Also, higher cutting speed leads to a higher shear-plane angle and areduced co-efficient of friction, both advantageous with respect to optimum cutting conditions.

PROBLEMS

1. Enumerate the essential requirements of a tool-material.
2. Name the various cutting tool materials.
3. Discuss the role of carbon, manganese, chromium, molybdenum, cobalt, vanadium and tungsten in tool steels.
4. Give the field of application of type 'O' tool steels and type 'A' tool steels.
5. What are the significant characteristics of high-speed steels ?

6. Discuss the T- series and M- series high-speed steels.
7. Enumerate the advantages of high-speed steel produced by powder metallurgy.
8. How are carbide tools made ? Describe the process.
9. Which sintered carbides are employed for machining steels and for machining non-ferrous metals ?
10. Name the various binding materials uscd in the manufacture of sintered carbides.
11. What do you understand by "laminated" and "coated carbides" ?
12. Give the advantages of coating the face of tungsten carbide tools with a thin coating of titanium carbide.
13. Enumerate the advantages of ceramic cutting tools over tungsten carbide tools.
14. Give the constituents of stellite tool material. How this is manufactured ?
15. How 'UCON' is produced ?
16. List the various tool materials used in industry. State the advantages and disadvantages of each material.
17. State the optimum operating temperature of each of the tool materials used in industry.
18. What are the main functions of a cutting fluid ?
19. List the essential characteristics of a cutting fluid.
20. Discuss the various types of cutting fluids.
21. Discuss the various methods of applying the cutting fluid at the cutting zone.
22. Give the constituents of ceramics. How these materials are produced ?
23. What are cermets ? Give their composition and properties.
24. Give the advantages of C.B.N. as a cutting tool material.
25. What are EP additives ? Why are these added to a cutting fluid ?
26. Write a note on "selection of a cutting fluid'.
27. Why were cutting tool inserts developed ?
28. What are "Indexable Inserts" and "Throw away inserts" ?
29. Discuss the two methods of attaching inserts to tool shanks.
30. Why indescable inserts are better than brazed tool tips ?
31. What is the composition of Sialon ?
32. Explain the application and limitation of ceramic tools ?
33. Why are tools coated ?
34. What is 18-4-1 type of cutting tool material ?
35. What are "Compax tools"?
36. Write on "Oxide Coating" of carbide tools.
37. What is "Compact" diamond tool material ?
38. Why C.I. is usually machined dry ?
39. Why gray C.I. does not need any lubrication during machining ?
40. List the additives added to water to improve its properties as a cutting fluid.
41. What is "Coronite" cutting tool material ?
42. Write about the use of H.S.S. in the form of inserts and also about coated H.S.S. tools.

    Note : Till recently, H.S.S. tools have been used as solid tools. However, of late, there has been an increasing tendency to use H.S.S. in the form of inserts just like inserts of WC, Al_2O_3, Sialon etc. This has insulted due to the high cost of H.S.S. tool material. These inserts are attached to the tool shank (Carbon or low alloy steel) by clamping or by brazing (See Fig. 2.1), or welded to it. For example, H.S.S. cutting ends of drills --- welded to the steel shank by "Friction Welding". Similarly, in band saws, narrow H.S.S. strips are welded to the steel bands by Electron Beam Welding (EBW).

Coated H.S.S. tools : Just like coated carbides, coated H.S.S. tools have recently been introduced in be market. The cutting tool (in the form of insert) is given a thin coating (2 to 6 μm thick) of a refractory carbide or nitride, for example TiC, TiN, Hafnium nitride and Alumina (Al_2O_3).

Cost of coated tool = 2 to 5 times the cost of uncoated tool.

However, life of coated tool = 5 to 10 times the life of uncoated tool.

Coated tools perform much better while cutting general materials as well as hard to machine alloy such as *Cr – Mo* steels.

44. Write about whisker reinforced tool materials

Whisker reinforced tool materials :

The performance of cutting tool materials has been further enhanced by the development of whisker reinforced cutting tool materials (composite Materials). Whiskers have been used as reinforcing fibres in composite cutting tool materials. For example, SiC whiskers are used as reinforcement in Al_2O_3 matrix. The resultant material is wear resistant and very tough. The material is used in the form of inserts. This cutting tool material is superior to coated carbides, beause it is resistant to ship

Cutting speed = 2 to 3 times that for plain carbide tools. Such cutting tool materials are suitable for intermittent cutting and machining of Ni-based alloys.

3

MACHINE TOOLS

3.0. GENERAL

A machine tools is a power driven machine for making articles of a given shape, size and accuracy (according to the blueprints) by removing metal from workpieces in the form of chips. Machine tools are factory equipment for producing machines, instruments and tools of all kinds. So, it can be said that the machine tool is the mother of all machines. Hence, the size of a country's stock of machine tools and their technical quality and condition largely characterizes its industrial potential.

Most machine tools perform the following four functions:

1. Hold the job.
2. Hold the cutting tools.
3. Move one or both of these (rotary motion or reciprocating motion).
4. Provide a feeding motion for one of these.

Classification of Machine tools. Machine tools can be classified in various ways. From the point of view of their field of application, machine tools are classified as:

1. General purpose machine tools : General purpose or universal machine tools are used for performing a great variety of machining operations on a wide range of workpieces. These are employed chiefly in piece and small-lot production and for repair work. Machine tools used for a particularly wide range of work are known as multipurpose machine tools. Especially versatile machine tools are also called "Omniversal". General purpose machine tools include : plain turning lathes, turret lathes, milling machines, drilling machines, grinding machine etc.

2. Single purpose machine tools : These machine tools are designed to perform a single definite machining operation, e.g., broaching, thread cutting, gear shaping and hobbing machines, machines for machining pistons, crank shafts, camshafts and for turning the cam contours on camshafts etc.

3. Limited purpose machine tools : These machine tools are capable of a narrow range of operations on a wide variety of workpieces, *e.g.*, automatic cutting off machines.

4. Production machine tools : These are mainly used in batch and mass production and feature high power and rigidity. These machine tools include : multi-tool lathes, single-and multi-spindle automatics, and semi- automatic lathes, plunge-cut cylindrical grinders, centreless, planer-type milling machines, thread rolling machines for tap production, numerically controlled machine tools etc.

5. Specialized machine tools : These are used for machining articles similar in shape but different in size. This group includes unit built machine tools. These machine tools allow the machining of several surfaces in different planes. Their advantage is that they are readily changed over from one job to another. This is done by mounting additional unit heads, positioning them at an angle to a horizontal or vertical plane or other-wise. These machines are used mainly in large-lot production.

6. Special machine tools : These machine tools are designed and manufactured individually and are intended for performing a certain definite operation in machining a certain definite workpiece. These machine tools include : machines for sharpening round threading dies, for grinding relief surfaces at the chamfer of round threading dies, for marking round threading dies and shank-type tools, for threading by die taps, for grinding flutes on taps and reamers, tap chamfers, flutes on twist drills etc. These machines find applications in large-lot and mass production.

The general purpose machine tools have the following characteristics:

(*i*) Usually less initial investment in equipment.

(*ii*) Greater machine flexibility.

(*iii*) Fewer machines may be required.

(*iv*) Less maintenance cost.

(*v*) Less set up and debugging time.

(*vi*) Less danger of obsolescence.

The special purpose machines have the following characteristics:

(*i*) Uniform product flow.

(*ii*) Reduced in-process inventory.

(*iii*) Reduced manpower requirements.

(*iv*) Reduced factory floor space.

(*v*) Higher output.

(*vi*) Higher product quality.

(*vii*) Reduced inspection cost.

(*viii*) Reduced operator skill requirements.

According to accuracy the machine tools are divided into five classes :

(*a*) **Normal accuracy :** Machine tools of normal or standard accuracy include the majority of the general purpose machine tools.

(*b*) **Higher accuracy :** Machine tools of higher accuracy are manufactured to the same drawings as the normal accuracy models, except that higher requirements are made to the accuracy with which the critical parts are manufactured, as well as to the quality of assembly and adjustments.

(*c*) **Precision** machine tools have certain parts that have been specially designed with the aim of maintaining the high accuracy standards. In addition, narrow tolerances are stipulated for the machining of all the parts, as well as for their assembly and adjustment as a whole.

(*d*) **High precision** machine tools are manufactured according to even more rigorous accuracy requirements than ''Precision'' class machine tools.

(*e*) **Super-high precision** or Master machine tools are intended for machining the parts which determine the accuracy of machine tools belonging to ''precision'' and ''high precision'' class.

To ensure the required accuracy of their operation, the machine tools of the last three classes are to be installed in special constant-temperature rooms, with automatically controlled (Constant) temperature and humidity.

According to weight, machine tools can be classified as :- Light weight (upto 1 tonne), medium-weight (upto 10 tonnes), and heavy-weight (over 10 tonnes). The last group can be further divided into subgroups as: Large size (10 to 30 tonnes), Heavy (30 to 100 tonnes) and Extra-heavy (over 100 tonnes).

According to the type of processing operations they perform or the tools they employ, all machines can be divided into nine groups, as under:–

1. Lathes : Engine and facing lathes, cutting-off lathes, multiple-tool lathes, Turret lathes, Automatics and Semi-automatics (Single spindle, multiple-spindle). Vertical turning and boring mills, and specialized machine tools.

Vertical turning and boring mills are used to machine blanks of large diameter and relatively small height. Most of the larger machines (some can accommodate work 25 m in diameter) are called vertical boring mills and the smaller models are usually known as vertical turret lathes.

2. Drilling and Boring machines : Upright drill presses, Semi-automatic single-spindle drilling machines, Semi-automatic multiple-spindle drilling machines, Jig borers, Radial drills, Boring machines, Precision boring machines, and horizontal drilling machines.

3. Planers, Shapers, Slotters and Broaching machines : Open side Planers, Double-housing Planers, Shapers, Slotters, Horizontal broaching machines, Vertical broaching machines.

4. Milling machines : Horizontal knee-type milling machines, Vertical knee-type milling machines, Tracer controlled milling and engraving machines, Continuous milling machines, Vertical-spindle compound-table milling machine, Fixed-bed and planer-type milling machines, and Ram-head milling machines.

5. Grinding and micro-finishing machines : Cylindrical grinders, Internal grinders, Snagging grinders, Specialized grinders, Tool and cutter grinders, Surface grinders, and Micro-finishing machines.

6. Gear and thread cutting machines : Shapers and Planers for Spur and Helical gears, Bevel gear generators, Hobbers for Spur and helical gears and splined shafts, Worm Wheel and Worm cutting machines, Gear tooth Chamfering machines, Thread-milling machines, gear finishing machines, Gear and thread grinders.

7. Combination machine tools : General purpose machines, Semi-automatic machines and Automatic machines.

8. Cutting-off machines : Cutting off lathes, Abrasive cutting machines, Circular-saw friction cutting machines, Straightening and cutting off machines, Bond saw cutting-off machines, Circular Cold sawing machines, and Power hacksawing machines.

9. Miscellaneous machine tools : Coupling and pipe thread cutting machines, Saw-cutting machines, Centreless bar turning and straightening machine, Tool Testing machines, Dividing machines, and Balancing machines.

Motions in Machine Tools : To obtain a finished workpiece on a machine tool, certain co-ordinated motions must be imparted to the work and the cutting tool. These motions are of two types : Primary or working motions and auxiliary motions. Primary motions consist of principal or cutting motions and feed motions. They serve the purpose of removing metal from the workpiece. The speed of the principal motion depends on the optimum cutting speed, while the speed of the feed motion depends on the required degree of surface finish.

Auxiliary motions help in the completion of the machining process and include such motions as : handling and clamping the work in the machine, advance and withdrawl of the cutting tool, engagement and disengagement of working motions, and changing their speeds etc. Working motions are power driven. However, certain small machines have hand feeds. Auxiliary motions may be either hand or power operated. On automatic machine tools, practically all auxiliary motions are automated.

Principal motions are of three types : rotary, reciprocating or a combination of these. This motion can be imparted either to the workpiece or to the tool.

Rotary motion to work : Lathes

Rotary motion to tool : Milling, grinding and drilling machines.

Both simultaneously : In drilling small diameter holes.

Reciprocating motion of tool : Shapers, slotters, Broaching machines etc.

Reciprocating motion of work : Planers.

Feed motions may be : Continuous, intermittent or compound.

Continuous feed: lathes, milling, drilling machines etc. Intermittent feed : Shapers, planers etc.

Compound feed : Gear cutter for cutting helical geacs.

Combined feed : Cylindrical grinders.

Methods of Machining a Shape:

1. Forming. Here the shape of the tool is the finished shape of the workpiece and the tool is called "Form tool" To finish the job, all that is necessary, in addition to the relative movement required to produce the chip (primary motion) is to feed (plunge) the tool in depth, *e.g.*, turning as shown in Fig. 3.1(*a*). Other examples are : drilling, plunge cut method of cylindrical grinding and plunge cut method of thread grinding.

2. Generating. Here, the required shape of the workpiece is obtained by combining several

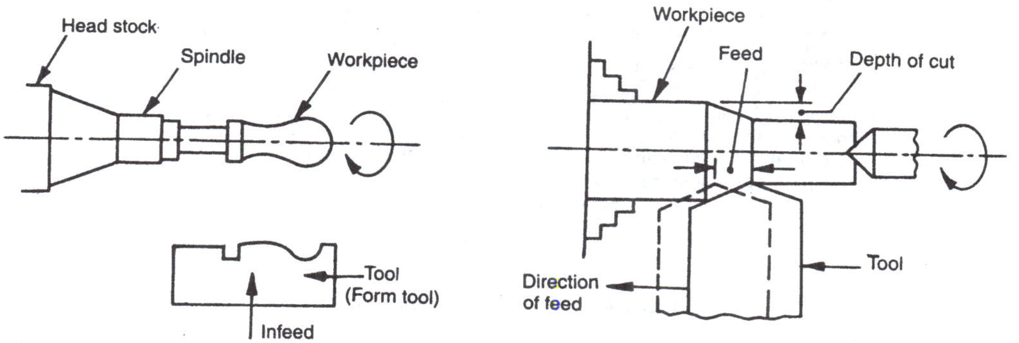

(*a*) Forming (*b*) Generating

Fig 3.1. Producing a Shape

motions that not only accomplish the chip forming process (primary action) but also move the point of engagement (of the job and the tool) along the surface (called as feed motion), *e.g.*, in cylindrical turning on a centre lathe, Fig. 3.1(*b*), the tool is set to cut a certain depth of cut and then traverses along the job to produce the chips. Other examples are : shaping and planning a flat, surface grinding, peripheral milling and so on.

3. Combined Forming and Generating: Thread cutting with a single point tool and hobbing etc.

When discussing the various machine tools, the various motions of the tools, the various motions of the tool/workpiece will be represented as given below:

(*i*) **Reciprocating Motion** :

Right : → ; Left : ←

Combined Right-Left : ↔

Up : ↑ ; Down : ↓

Combined Up-Down : ↑↓

In : ⊕ ; Out : ⊖

(*ii*) **Rotary Motion : About,**

Horizontal Axis :  ; Vertical Axis : 

3.1. LATHE MACHINES

A lathe is one of the oldest and perhaps most important machine tools ever developed. The job to be machined is rotated (turned) and the cutting tool is moved relative to the job. That is why, the lathes are also called as "Turning machines". If the tool moves parallel to the axis of rotation of the workpiece, cylindrical surface is produced, while, if it moves perpendicular to this axis, it produces a flat surface.

A lathe was basically developed to machine cylindrical surfaces. But many other operations can also be performed on lathes, for example, facing, parting, necking, knurling, taper turning, thread cutting and forming etc., (Fig. 3.2). We also can perform operations of other machine tools on a lathe, for example, drilling, reaming, milling and grinding operations etc. No wonder, a lathe is called "the father of the entire machine tool family". About one half of machine tools operating in engineering plants are of the lathe group.

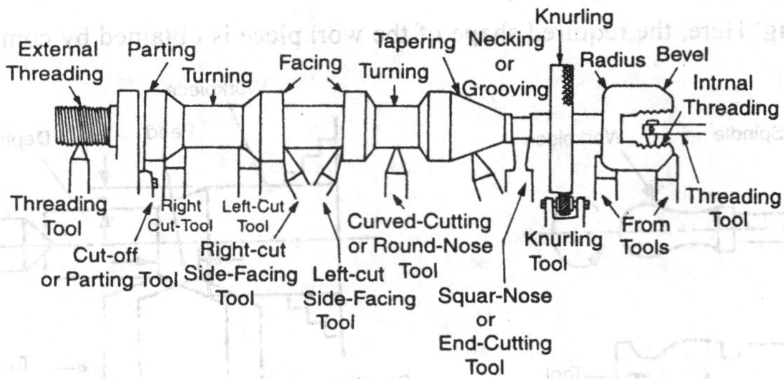

Fig. 3.2. Common Lathe Operations

On the basis of their purpose, design, number of tools accommodated, degree of mechanisation and other factors, lathe-type machine tools may be classified as :

1. **Limited or low-production Machines.** The lathes included in this category are : engine lathe (centre lathe), bench lathe, tool room lathe and speed lathe.

2. **Medium-production Machines.** Turret lathes and duplicating (or tracer controlled) lathes.

3. **High-production Machines.** Semi automatic and automatic lathes.

The construction and principle of lathe-type machine tools will be illustrated by the example of the most common representative of this class-the engine lathe.

3.1.1. Engine Lathe. It is so called because the first of this type of lathe was driven by a steam engine. It is also called "Centre lathe", because, it has two centres between which the job can be held and rotated. A very high percentage of all lathe work is turned between centres.

The main parts of a centre lathe are : Bed, Head stock, Tail stock, Carriage and the Electric drive, (Fig. 3.3).

1. **Bed.** The bed is the base or foundation of the lathe. It is a massive (heavy) and rigid casting made in one piece to resist deflection and vibrations. It holds or supports all other parts, that is, head stock, tailstock and carriage etc. The top of the bed is planed to form "guides" or "ways". Ways are accurate rails which support carriage and the tailstock. More expensive lathes have a combination of V ways and flat ways. Less expensive lathes have flat ways. Directly under the front way on the bed is a rack. A pinion gear meshes with the rack for moving the carriage when the handwheel is turned. The bed is usually fastened to steel legs so that the lathe can be bolted to the shop floor.

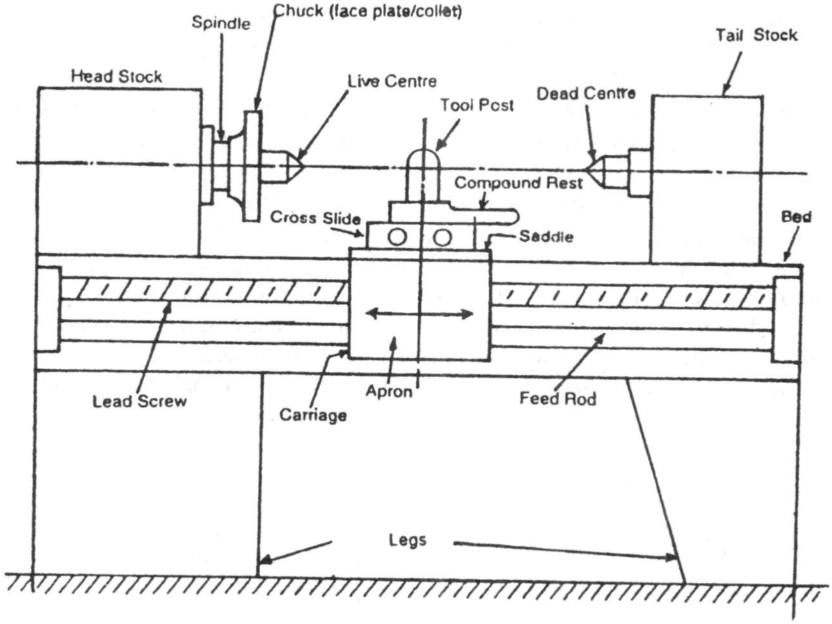

Fig. 3.3. Block Diagram of an Engine Lathe.

2. Head stock. The headstock assembly is permanently fastened to the left hand end of the lathe. It serves to support the first operative unit of the lathe, that is, the spindle. The spindle revolves in bearings, one at each end of the headstock. The spindle is rotated by a combination of gears and cone pulleys or by gears alone. Present day lathes have individual motor drives and most of them have geared headstocks. The steel spindle is hollow to take long bar stock. The spindle has a definite taper at the front end for holding centres and other tools having a tapered shank. The hole through the spindle makes it possible to use a knockout bar to remove such tools. Work-holding attachments such as driving plate, face plate or various types of chucks may be mounted on the threaded spindle nose. Some type of work may also be held in a collet which is inserted into the hollow headstock spindle. A taper sleeve fits into the taper spindle hole. The headstock or live centre fits into the sleeve. This centre is called live centre because it turns with the work. The centre is a tapered metal part with a pointed end. This supports the end of the workpiece as it turns between the centres. All centre points have a 60-degree included angle.

3. Tail stock. Tail stock is on the other end of the bed from the head stock. Its chief function is to hold the dead centre so that long workpieces ($L/D > 4$) can be supported between centres. It can be moved along the bed and clamped to the bed at the various desired locations to suit the length of the workpiece.

Tailstock consists of two main parts. The lower part rests directly on the bed ways, and the upper part rests on the lower part. Adjusting screws hold the two parts together. The upper casting can be moved toward or away from the operator to offset the tailstock for taper turning and to re-align the tailstock centre for straight turning. The body of the tailstock has a bore for the hollow cylindrical sliding member, known as a "quill". This quill is sometimes called as "tailstock spindle" even though it can not rotate. The quill moves in and out of the tailstock bore when the tailstock handwheel is turned. Once set, the quill may be clamped to remain in a desired position. The quill has a taper hole into which the dead centre is fitted. Drills, reamers, taps and other end cutting tools are held and fed to the work piece by the quill, the shanks of the tools being held in the tapered hole of the quill.

4. **Carriage.** In between the headstock and the tailstock is the carriage. It is movable on the bed ways and its purpose is to hold the cutting tool and to impart to it either longitudinal or cross feed. It has five major parts :

(*a*) **Saddle.** The base of the carriage is the saddle which slides along the ways of the lathe bed.

(*b*) **Cross slide.** The cross-slide is mounted on the saddle. It provides cutting tool motion which is perpendicular to the centre line of the lathe itself. The cross-feed movement may be controlled by manual (with cross slide handle) or by power feed.

(*c*) **Compound Rest.** It is mounted on top of the cross slide. The compound rest has a graduated base and can be swivelled around a vertical axis. In this way, its slide can be set at any angle with the axis of the workpiece. It can be clamped to remain at any angular setting. The range of compound rest is only limited and is used for obtaining angular cuts and short tapers, as well as convenient positioning of the tool to the work. Both the cross slide and the compound rest screws are equipped with micrometer collars. These are used in making accurate adjustments when turning workpieces to close measurements, and when cutting screw threads. There is no power feed for the compound rest.

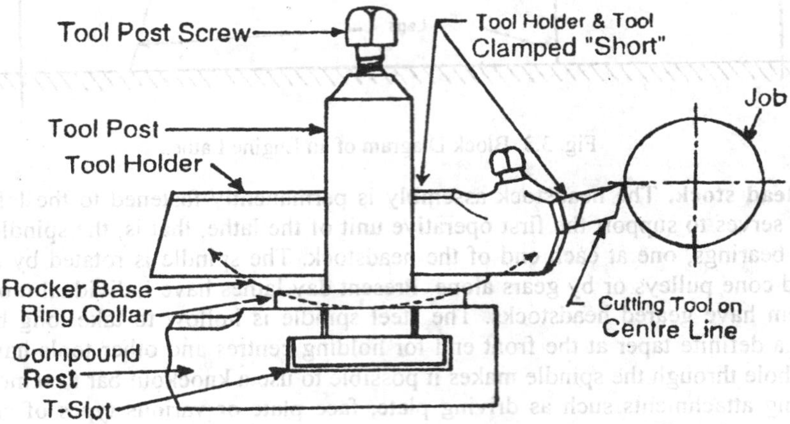

Fig. 3.4. Tool Post

(*d*) **Tool Post.** The tool post is mounted on the compound rest and slides in a T-slot. Cutting tool/tool holder is firmly held in it. The tool can be swivelled as well as tilted by means of a rocker and a concave ring collar, (Fig. 3.4).

(*e*) **Apron.** The apron is secured underneath the saddle and hangs over the front of the bed. It contains the gears, clutches, and levers for operating the carriage by hand and power feeds. The apron hand wheel can be turned to move the carriage longitudinally by hand. This hand wheel is attached to a pinion that meshes with the rack under the front of the bed. The apron also contains friction clutches for automatic feeds and a splitnut or halfnut. The split nut can be closed over the lead screw threads and is used only when cutting screw threads.

3.1.2. Drive. The centre lathe has primarily two motions : the primary cutting motion and the feed motion. These motions are accomplished by means of corresponding drives, which are systems of mechanisms for transmitting power from its source (the electric motor) to the operative units of the lathe, that is, to the spindle for primary cutting motion (main drive) and to the carriage for feed motion (feed drive).

1. **Main Drive.** The function of main drive is to drive the spindle and to change its speed so as to obtain the most expedient cutting speeds. Power for driving the spindle is provided by an electric motor.

Two general types of main drives are used on lathes : Stepped drive and Step -less drive.

In the first case, the maximum, minimum and a series of intermediate spindle speeds in definite steps are available. In the step-less system, or the infinitely variable spindle speed system, any spindle speed from the minimum to the maximum can be obtained. This feature allows the most suitable cutting speeds for each workpiece diameter to be set up, thereby maintaining the specified surface finish without decreasing the rate of production.

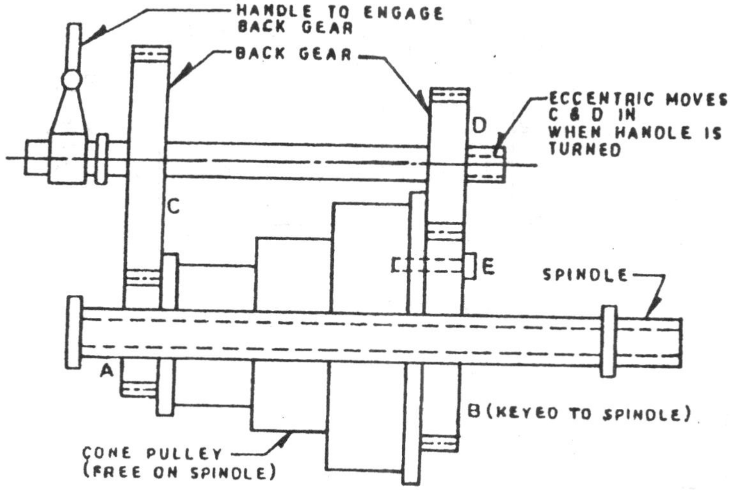

Fig. 3.5. Step—Cone Pulley with Back Gear Arrangement.

Stepped spindle speed variation can be achieved in two ways :

(*a*) **Cone Pulley System.** In the smaller lathes, we have the cone or step pulley arrangement. The power flows from the motor in the base to the cone pulley attached to the spindle, by means of a belt. The spindle speed is changed by moving the belt to different positions on the step pulley. To obtain slower speeds and more power, the back gears are used.

To understand how the back gears operate, (see Fig. 3.5). The gear B called the bull gear is fastened to the spindle. The spindle is a loose fit over a sleeve. The step pulley is firmly secured to the sleeve. The small end of the pulley has a pinion A attached to it. With the above arrangement, the bull gear can never turn free of the spindle and the pinion A always turns when the pulley turns. The step pulley and the pinion A are connected with the bull gear by a sliding pin E called the bullgear lockpin. At the back of the headstock are two gears C and D, called the back gears, mounted on the same shaft (an eccentric one). They are spaced to line up or mesh with the bull gear B and the pinion A. To make the back gears mesh with bull gear B and the pinion A, the back gear handle is pulled forward. When the back gears are not in mesh and the bull gear lock-pin is in place, we will get a direct drive. The speed of the spindle will then be the same as that of the cone. If a reduction in speed is desired, the bull gear lock-pin is disengaged and the back gear lever is pulled forward. Now the power from the step pulley is delivered to the bull gear through the pinion A and the back gears, which turns the spindle. So, there will be a direct as well as a back-gear speed for every step on the cone pulley.

The cone pulley drive is very simple, and cheap in design, but it occupies lot of space and the number of speed steps are limited. The flat belt drive is not positive and shifting of belt from one step to the other takes time affecting the rate of production.

(*b*) **All Geared Drive.** In this arrangement, we have a completely geared headstock with a combination of shifting levers. The power from the motor is delivered to the spindle through a belt drive and speed gear box. A geared headstock has got the following advantages :

(*i*) It is more efficient and compact than the cone pulley drive.

(*ii*) Possibility of transmitting high power.

(*iii*) The available power on the spindle remains almost constant for the different speeds.

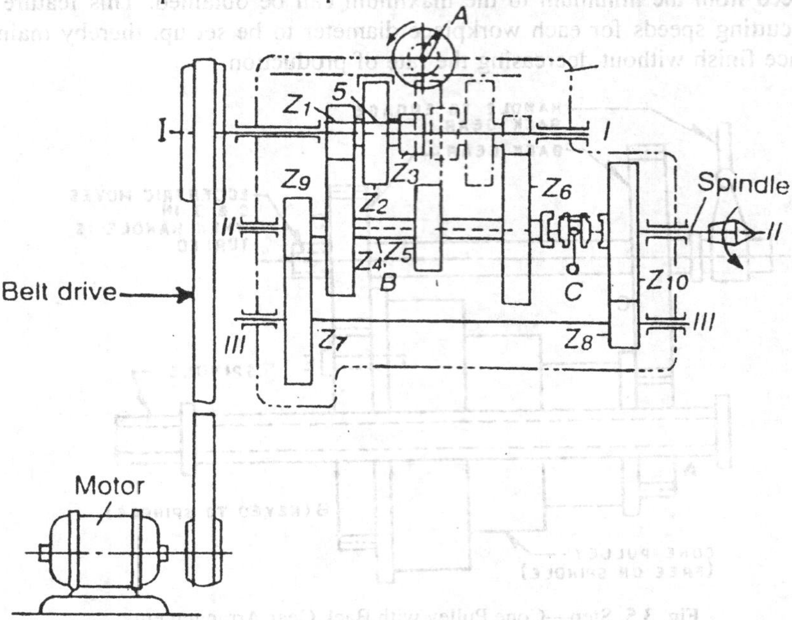

Fig. 3.6. Geared Main Drive

Fig. 3.6 shows a simple geared headstock. Power is transmitted from the motor through a belt drive to shaft I - I. This shaft carries three cluster gears Z_1, Z_2 and Z_3. The cluster gear can be shifted with help of lever A, on key 5 along the shaft I - I, so that its three rims mesh in turn with the gears Z_4, Z_5 and Z_6 which are rigidly mounted on sleeve B. This sleeve is freely mounted on the lathe spindle II - II. The gears Z_9 and Z_{10}, mounted directly on the spindle are constantly in mesh with gears Z_7 and Z_8 of the countershaft III - III. When the clutch C is shifted into engagement to the left, countershaft III - III is disengaged. The spindle will obtain one of the three speeds - n_1, n_2 and n_3 – depending upon the position of the cluster gear.

If n_1 = Speed of the shaft I - I, then

$$n_1 = n_0 \times \frac{Z_1}{Z_4},\ n_2 = n_0 \times \frac{Z_2}{Z_5},\ \text{and}\ n_3 = n_0 \times \frac{Z_3}{Z_6}.$$

When the clutch C is shifted to the right, the counter shaft is also included in the gear train. Its gearing ratio will be,

$$\frac{Z_{10}}{Z_8} \times \frac{Z_7}{Z_9}$$

With this, the spindle obtains three more speeds. Therefore, the countershaft enables additional speeds to be obtained, in the present case 6 spindle speeds can be obtained. Geared headstocks are commonly designed for 3, 4, 6, 8, 12, 16 and 24 spindle speeds. In properly designed, geared head stock, the values of the spindle speeds in the available range vary in a geometrical progression. The constant ratio has one of the following values : 1.06, 1.12, 1.26, 1.41, 1.58 and 2. The less the constant ratio, the less the difference will be between successive speeds and the more exactly the required speed can be approximated. In geometrical progression, the various spindle speeds will be given as :

$$n_1, \varphi n_1, \varphi^2 n_1, \varphi^3 n_1, \ldots \varphi^{Z-1} . n_1$$

where $n_1 = n_{min}$. ; $\phi^{Z-1} . n_1 = n_{max}$

and φ = step ratio and Z = number of spindle speed steps

$$\therefore \qquad \phi^{Z-1} = \frac{n_{max}}{n_{min}}$$

$$\therefore \qquad \phi = \left(\frac{n_{max}}{n_{min}}\right)^{\frac{1}{Z-1}}$$

The "Step-less drive" is discussed in volume II (A Textbook of Production Engineering).

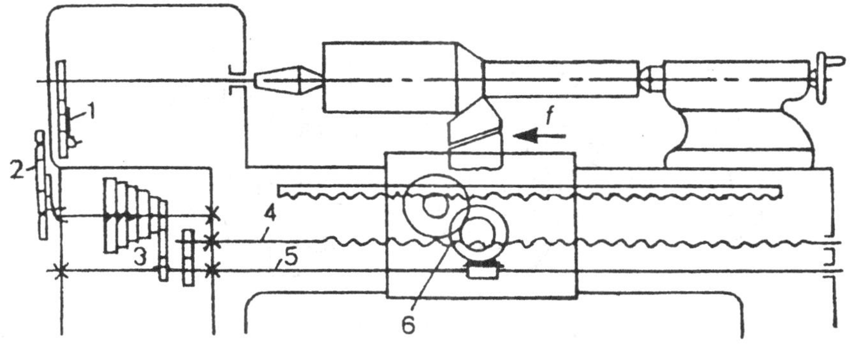

Fig. 3.7. Feed Drive of Engine Lathe

2. **Feed Drive.** The feed drive serves to transmit power from the spindle to the second operative unit of the lathe, that is, the carriage. It, thereby, converts the rotary motion of the spindle into linear motion of the carriage. It also enables the specific rate of feed as well as its direction to be selected. The feed drive not only provides the different feeds required in machining but also provides a wide range of pitches for thread cutting operations.

The feed drive of an engine lathe consists of : the reversing mechanism (1), change gear quadrant (2), quick-change gear box (3), leadscrew (4), and feed rod (5), and apron (6), (Fig. 3.7).

(*i*) **Reversing mechanism.** Its function is to reverse the direction or movement of the lead screw/feed rod. The most frequently used reversing mechanism for engine lathe comprises : four spur gears with consecutive engagement of the bracket holding the reverse gears, (Fig. 3.8).

The gear Z_1 is mounted on the spindle I and the gear Z_4 - on shaft II which is usually a shaft of the change gear quadrant. Gears Z_2 and Z_3 are freely mounted on studs III and IV of the bracket. P is the feed reverse lever. The bracket is mounted on shaft II and the feed reverse lever can be moved to three positions 1, 0, or 2. Gears Z_2, Z_3 and Z_4 are

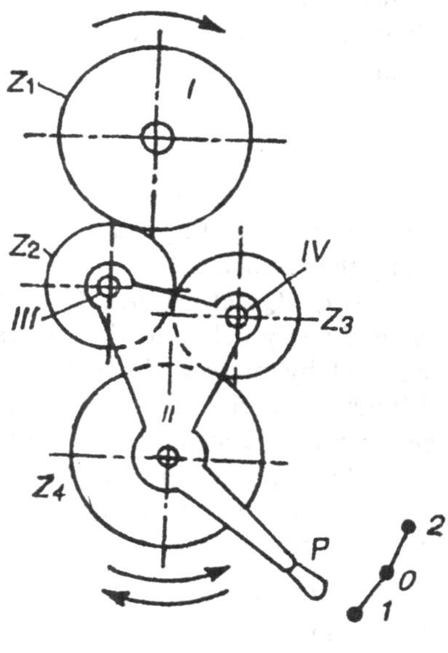

Fig. 3.8. Feed Reversing Mechanism

constantly in mesh with each other. If the lever is shifted to position 1, gears Z_1, Z_2, Z_3 and Z_4 are put into consecutive mesh and with the automatic feed engaged, the shaft II begins to rotate in a direction opposite to that of the spindle. With lever in position 2, only gears Z_1, Z_3 and Z_4 will be in mesh and shaft II will rotate in the same direction as the spindle. With lever in the centre position O, both gears Z_2 and Z_3 are pulled out of mesh with gear Z_1 and no motion is transmitted from the spindle to shaft II. The revers- ing bracket is thus disengaged.

(*ii*) **The change gear quadrant.** The change gear quadrant of an engine lathe consists of a set of change gears and the device known as the quadrant proper. It serves to set up the feed drive to different speeds of the lead screw and feed rod. The quadrant also transmits motion to the quick-change gear box or to leadscrew.

A frequently used change-gear quadrant for an engine lathe is shown in Fig. 3.9. It is a

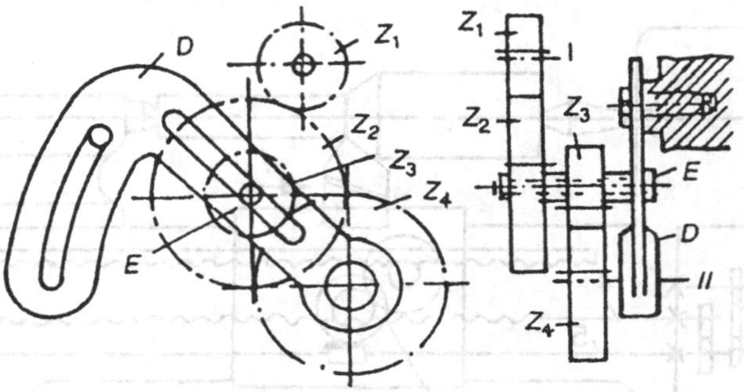

Fig. 3.9. Change Gear Quadrant.

two-pair arrangement. There can also be single-pair or three-pair-arrangement. The two pair arrangement has four gears Z_1, Z_2, Z_3 and Z_4. Gear Z_1 is mounted on shaft I of the reversing mechanism and the gear Z_4 - on shaft II of the quick- change gear box. The intermediate gears Z_2 and Z_3 are keyed on a sleeve which, in turn, is mounted freely on stud E. The stud E can be adjusted and clamped as required along a straight slot in quadrant *D*. The latter is mounted on shaft II and can turn freely in reference to this shaft. The gearing ratio of the change gear quadrant is

$$\frac{Z_1}{Z_2} \times \frac{Z_3}{Z_4}$$

For new gearing ratios with new sets of change gears, the centre distances are changed as required by adjusting stud *E* along the straight slot and swivelling the quadrant about shaft II along the circular slot.

(*iii*) **Quick-change Gear Box.** The quick-change gear box is located on the front of the lathe, directly below the headstock assembly. It allows a variety of feeds to be easily the rapidly selected by merely shifting the corresponding levers. This gear box contains a number of different-size gears, which provides a means to change (1) the rate of feed and (2) the ratio between revolutions of the spindle and the movement of the carriage for thread cutting.

An index chart or plate is attached to the front of the gear box. It tells the position of the levers to use to obtain the desired feed or threads per cm. The most widely used design of quick-change gear box is the tumbler gear, (Fig. 3.10).

It comprises a cone of gears 1 to 8 mounted on shaft S_2. The tumbler gear can slide on shaft S_1. It can mesh with any gear on shaft S_2 through an intermediate gear which is located on a swinging and sliding lever so that it can engage gears 1 to 8 of different diameters, on shaft S_2. The lever can be fixed in any desired ratio position with the help of a stop pin. The drive is usually

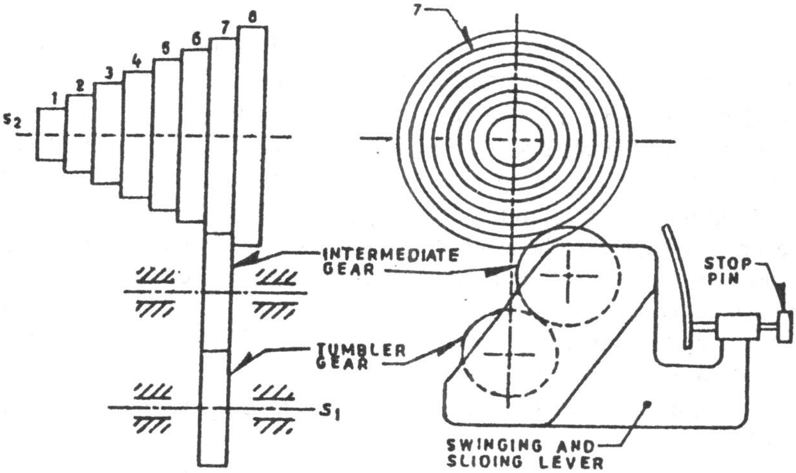

Fig. 3.10. Norton type Tumbler-gear quick-change Gear box.

from the driving shaft S_1 to the driven shaft S_2.

(*iv*) **Apron.** The function of the apron has already been explained. Power comes to the apron for feeding the carriage (from lathe spindle) through feed reversing mechanism, change-gear quadrant or quick-change gear box and the lead screw/feed rod. The apron encloses clutches, systems of spur and worm gearing, Fig. 3.11. Its feed mechanisms convert rotary motion of the lead screw or feed rod into linear motion (feed) of the carriage.

The lead screw 1 is only used to move the carriage for cutting threads. For other lathe operations, feed rod 2 is used so as to save the lead-screw. Worm 3 with its key slides along the

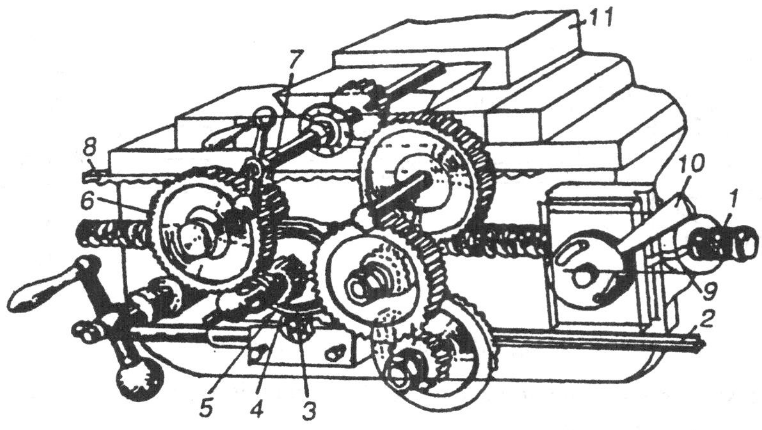

Fig. 3.11. Lathe Apron Mechanism.

feed rod which has a key way along its whole length. When the feed rod rotates, the worm turns. The worm meshes with a worm wheel 4 and the motion is transmitted to spur gear 5 mounted on the same shaft. When the automatic feed knob is turned to the right, the clutch engages another gear (gear 5 meshes with gear 6) which transmits rotation to pinion 7 since the gear 6 and pinion 7 are mounted on the same shaft. Pinion 7 runs along the stationary rack 8, fastened to the bed, thus feeding the carriage 11, rigidly secured to the apron, along the bed. Thus, the rotary motion of the feed rod is converted into linear motion of the carriage.

Other gearing arrangements in the apron convert the rotary motion into cross feed of the cutting tool by rotation of the cross-feed screw in the cross slide. For this, the feed-change lever on the apron is moved to the "down" position and the automatic feed knob is turned to the right.

For thread cutting, the feed rod is disengaged from the apron gearing by placing the feed-change lever on the apron in the centre or neutral position. The lead screw feeds the carriage lengthwise through two half nuts 9 mounted on the rear wall of the apron. Upon engaging lever 10, the half nuts are closed and they engage the rotating lead screw. Then the half nuts and, with them, the carriage are fed along the bed. Disengagement of lever 10 spreads the half nuts to release the lead screw and the carriage stops.

3.1.3. Work-holding devices. The common work holding devices used for a centre lathe are discussed below :

1. Centres. A very high percentage of all lathe work is turned between centres. Long work pieces (shafts and axles) with L / D ratio > 4, are turned length wise between centres. The work-piece, in whose ends centre holes have been previously drilled, and on which a driving dog has been clamped on one end, is mounted between the headstock (live) and tailstock (dead) centres. The workpiece is rotated by driving the lathe dog. The lathe dog is driven with the help of a drive plate. The drive plate is mounted on the threaded nose of the spindle. It has one open slot and three closed slots. A pin is inserted in the open slot which engages with the tail of the lathe dog, (Fig. 3.12 *a*). When the spindle rotates, the workpiece will rotate through this pin and the dog. If the dog has a bent tail, it will fit into one of the slots, (Fig. 3.12 *b*).

The two types of the lathe dog are shown in Fig. 3.13.

Since the tailstock centre does not rotate, it acts as a bearing. It, therefore, must be lubri-

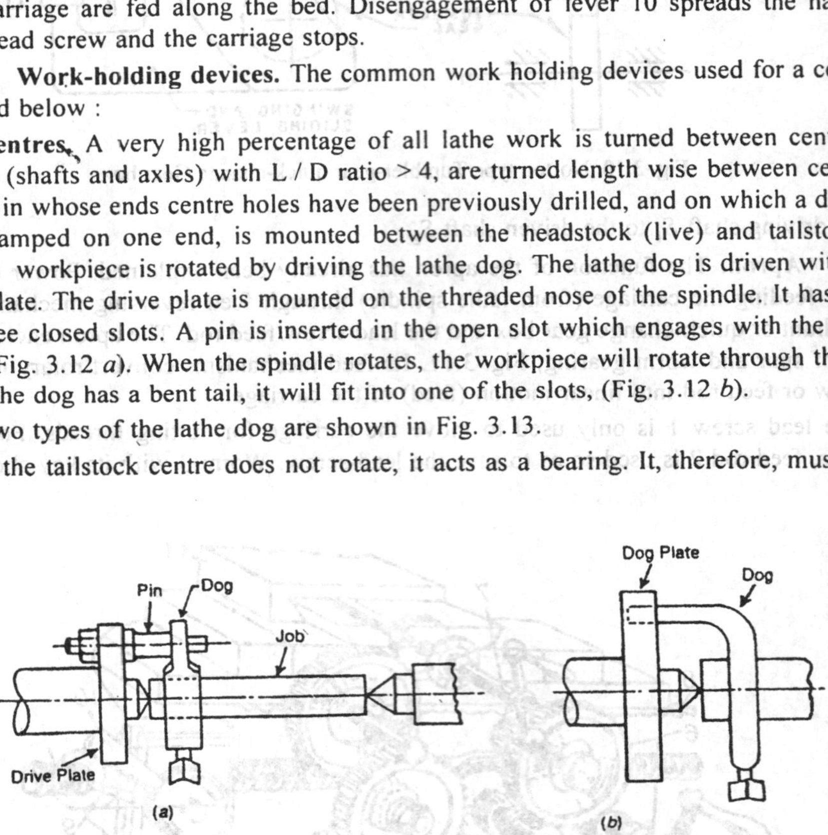

Fig. 3.12. Turning between centres.

cated and excessive pressure should be avoided. To avoid this process of lubrication in production work, a revolving centre is used.

Three designs of lathe centres are illustrated in Fig. 3.14. An ordinary centre lathe is shown in Fig. 3.14 *a*. A ball- ended centre (Fig. 3.14 *b*) is used when the tail stock is set over to turn a taper. A half centre (Fig. 3.14 *c*) enables external straight turning to be combined with facing the end of the workpiece. The wear resistance of the centres can be substantially increased by tipping the point with cemented carbide or metallizing it with a hard-facing alloy.

2. Chucks. Work piece with a length L < 4D may be clamped in a chuck without the need of additional support of the free end. Three-and four - jaw chucks, screwed on the spindle nose,

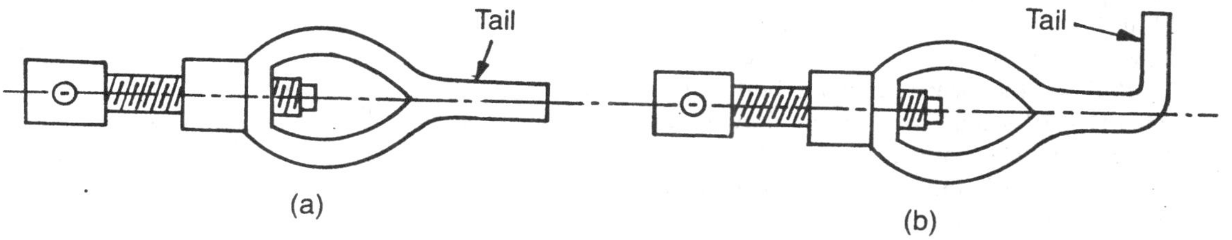

Fig. 3.13. Lathe Dogs.

are employed for this purpose. The chuck may be adapted to the screwed nose spindle by using a back plate, whilst on the flanged spindle, it is bolted direct.

Among the chucks used in a lathe, the most widespread are three-jaw universal self-centring chucks. These are used for holding symmetrical work as a rule. Here, the three jaws move together towards or away from the centre of the chuck.

In the 4 -jaw chuck, each of the four jaws is moved by its own screw independent of the other jaws along radial slots of the chuck body. These chucks are used for holding complex and non-symmetrical work pieces.

Pneumatically and hydraulically operated chucks are used to speed up and facilitate handling operations of certain parts in mass and large-lot production.

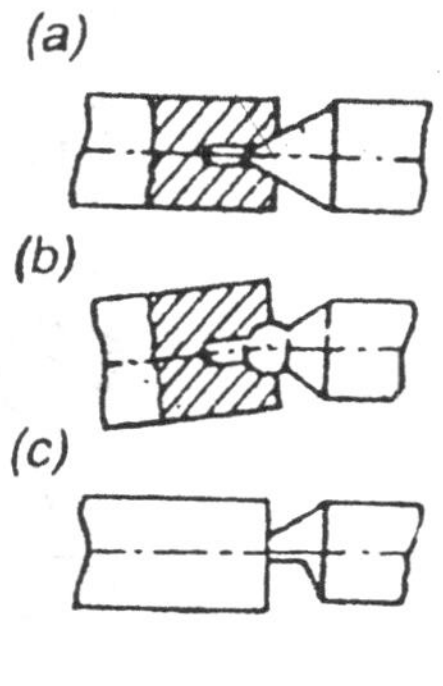

Fig. 3.14. Types of Lathe Centres.

3. Collets. They are the most accurate of the chuck family. They are like three jaw chucks and are used primarily for bar stock or other sections upto about 63 mm, for example round, square, hexagonal etc.

4. Face Plate. A face plate is larger than a drive (dog) plate and is screwed on the spindle nose. It usually has four T-slots and a number of plain radial slots, (Fig. 3.15). It is highly efficient in machining asymmetrical work or work of complex and irregular shape which is inconvenient or even impossible to clamp in jaw-type chucks. The workpiece is clamped to the faceplate with bolts and straps. Sometimes, it is more convenient to mount an angle plate on the face plate and to clamp the work on the angle plate.

5. Mandrels. A mandrel is used to locate and held a workpiece with a central hole, such as, gear blanks, pulleys etc. A mandrel is a solid hardened bar, with centres and flats on each end. The mandrel is usually tapered so that the work can be forced on it with a press fit and then removed after working. A mandrel is held between centres and rotated with a lathe dog clamped on its flat, Fig. 3.16. The taper is about 0.005 mm per cm. length.

6. Steady Rests. When very long workpieces (L/D > 10 or 12) or long slender workpieces of low rigidity are machined between centres, steady rests are used to additionally support the workpiece and prevent it from bending due to the pressure of cut. There are two types of steady rests used on a lathe :

(*a*) **Fixed Steady Rest.** This type of steady is fixed on both bed ways of the lathe, between the headstock and tailstock. The

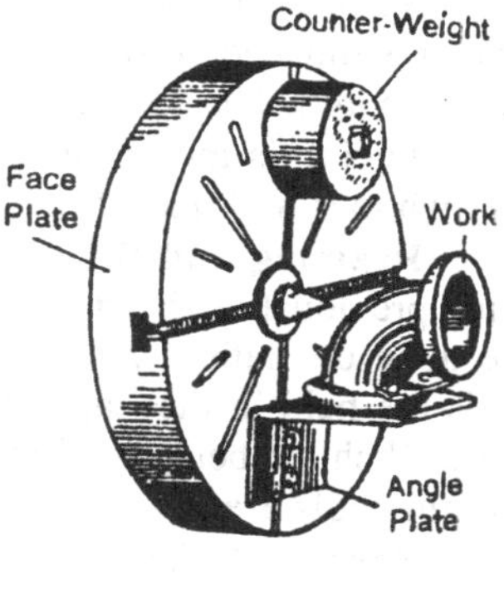

Fig. 3.15. A Face Plate.

workpiece is supported by three adjustable jaws, (Fig. 3.17 *a*). Steady rests for high-velocity turning have ball or roller bearings contacting the rotating workpiece, the bearings being built into the jaws. This type of steady rest has the drawback that since the carriage can not pass it, the job will be turned in two stages by being reversed end for end after half its length has been machined.

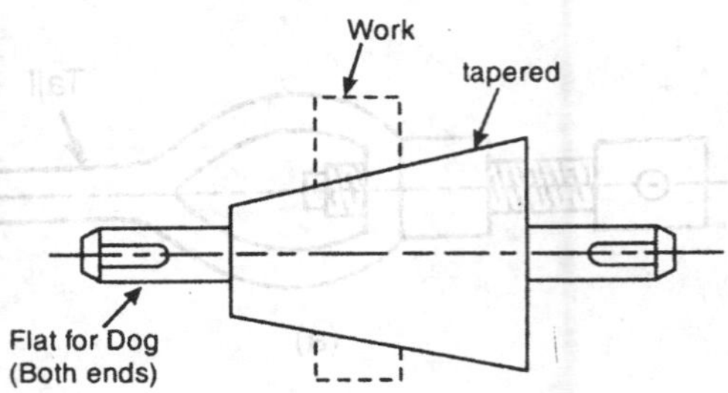

Fig. 3.16. A Mandrel.

(*b*) **Follower or travelling Steady Rest.** This steady is mounted on the saddle and moves together with the tool. It has two jaws which support the work opposite the tool, (Fig. 3.17 *b*) This design does not have the drawback of fixed steady rest.

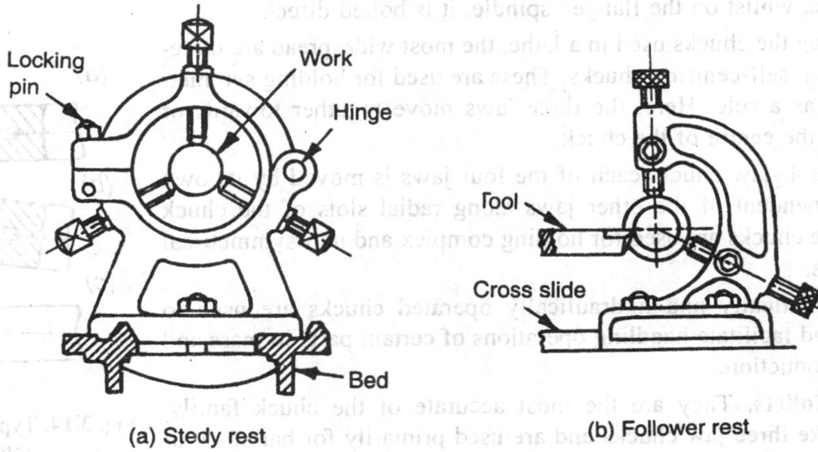

Fig. 3.17. Steady Rests.

7. Milling Vise. Milling vise is mounted in place of the compound rest. The work is held in the vise and an end mill or stub-arbor mounted cutter is inserted in the lathe spindle. Work movement is limited.

8. Special Fixtures. When a job can not be held in one of the above discussed work holding devices, a fixture is designed, made and used for that job.

3.1.4. Lathe Operations. A large variety of operations can be performed on an engine lathe, as shown in Fig. 3.2.

Turning is the operation to remove material from the outside diameter of a workpiece to obtain a finished surface. The finished surface may be of continuous diameter, stepped, tapered or contoured. The feed of the tool for turning operation is along the axis of the lathe.

Facing is the operation of machining the end of a workpiece to make the end square with its own axis and that of the lathe. The tool moves perpendicular to the axis of the lathe.

Reaming and Drilling. Drilling is the operation of making a hole in a work piece where none previously existed. Reaming is the operation of finishing the drilled hole. These operations are done on lathe by holding the drills and reamers in the tailstock quill. The shanks of these tools are held in the tapered hole. For smaller tools, drill chucks can be used to hold the tools, the drill chuck being held in the tapered hole of the quill by its shank. The job is held in a chuck and the tools are fed to the revolving workpiece by the quill by rotating the tailstock handle.

Boring. Boring is the operation of enlarging the drilled hole. The workpiece is held in a chuck in the lathe spindle and the boring bar is mounted in the tool post. Boring is done by moving the carriage towards the headstock.

Knurling. It is the operation of plastically displacing metal into a particular pattern for the purpose of creating a hand grip or roughened surface on a workpiece. The knurling tool is held in the tool post and is pressed against the surface of the workpiece by cross feed.

Milling. For the milling operation, small milling cutters are held in the headstock and revolved while the work is clamped in a vise mounted over the top of the compound rest, instead of the tool post. The operation is used only for small work.

Grinding. Cylindrical and internal grinding can be done on a lathe, with a tool-post grinder. This is a holder containing a spindle for mounting a grinding wheel and a motor for driving it. A small wheel is used for internal grinding and a larger wheel is used for external grinding. The tool post grinder is mounted on the compound rest in place of tool post. For external grinding, the workpiece is held in a chuck or between centres. For internal grinding, the workpiece is held in a chuck or mounted to the faceplate. Tool post grinder is used for simple jobs like grinding mandrels or reamers, truing chuck jaws, sharpening lathe centres and milling cutters etc.

Taper turning. Tapered surfaces can be turned by employing one of the following methods:

1. Compound rest Method. The compound rest is swivelled to the required angle, (Fig. 3.18 *a*). The angle is determined by the formula :

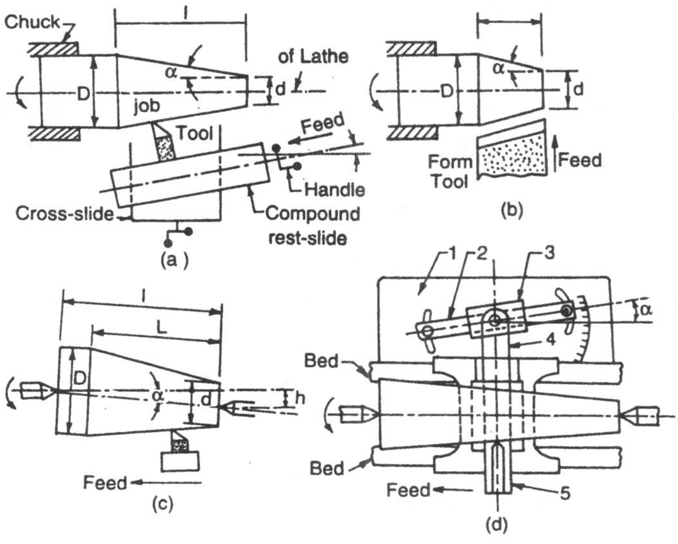

Fig. 3.18. Taper Turning Methods.

$$\tan \alpha = \frac{D - d}{2l}$$

where α = Half taperangle.

D = largerdiameter of the taper

d = smallerdiameter

l = lengthof the taper.

After swivelling the compound rest to this angle about the vertical axis, it is clamped in position. The taper is turned by hand wheel by rotating the handle. The method can be employed for turning short internal and external tapers with a large angle of taper. The workpiece is commonly held in a chuck.

2. With a Form tool. Short external tapers with various angles of taper can be turned with a form tool, Fig. 3.18 *b*, using cross feed. The width of the form tool slightly exceeds that of the taper being turned. The work is held in a chuck or clamped on a faceplate.

3. Setting over the Tail-stock. Long workpieces with a small angle of taper (not exceeding 8°) are usually turned by setting over the tailstock centre, (Fig. 3.18 *c*). Tailstock setting over or offset can be determined from the formula,

$$h = \frac{L(D-d)}{2l} \text{ mm}$$

where, L = full length of the workpiece

l = length to be taper turned.

4. Taper turning attachment. Long tapered work is frequently turned with a taper turning attachment, Fig. 3.18 *d*. Bracket 1 is attached to the lathe bed. It carries guide bar 2 which can be turned to the required angle and clamped in place. Guide block or slide 3, linked to the lathe cross slide 5 by tie member 4, is free to slide along the guide bar. The cross slide is disengaged from the cross feed screw. Upon travel of the saddle along the bed ways, the guide block slides along the guide bar, and through the tie member, forces the cross slide with the cutting tool to travel parallel to the guide bar at the given angle and the work is turned to the specified taper. The angle of taper can be found out by the relation :

$$\tan\alpha = \frac{D-d}{2l}$$

In some taper turning attachments, the bracket carries divisions in mm, instead of in degrees. In such cases, the guide bar has to be swivelled through mm divisions, which can be found out as :

$$S = \frac{D-d}{2l} \times lg$$

where lg = Half of the total length of the guide bar or plate.

Conicity or taper, T, is defined as the ratio of the difference in large and small diameters to the length of taper.

$$\therefore \quad T = \frac{D-d}{l}$$

In the case of compound rest method,

$$T = 2\tan\alpha$$

In the "Setting over tail stock" method,

$h = \dfrac{L \times T}{2}$, when the length of taper is less than length of workpiece

and $h = \dfrac{T}{2} \times l$, when the length of taper is equal to length of work.

In the case of "Taper turning attachment" method, again

$$T = 2\tan\alpha.$$

Thread cutting on a Centre lathe. The first step in cutting threads on a lathe is to get an accurately shaped and mounted tool. The form and setting of the tool is checked with the help of a thread template or centre gauge, (Fig. 3.19). The job is either mounted between centres or held in a chuck (for external threads) and held in a chuck for internal threads. When mounting the tool in the tool post, it must be ensured that the top of the tool is horizontal and is in line with the axis of rotation of the job, Fig. 3.20. After this, the second step is to establish a specific relationship between the longitudinal movement of the tool parallel to the axis of rotation, and the rotation, of

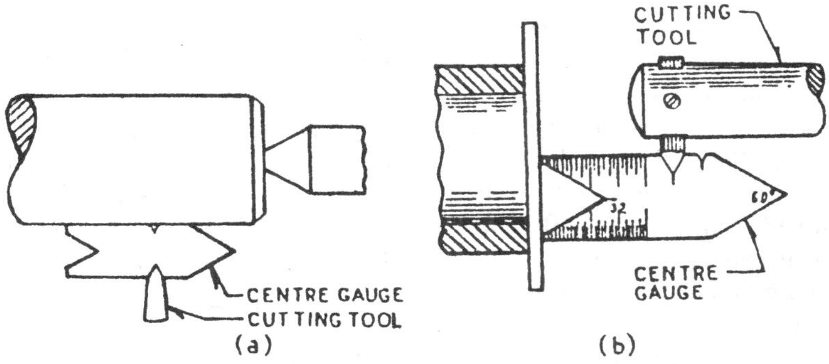

Fig. 3.19. Centre Gauge.

the job. This will determine the pitch or lead of the thread. This is achieved with the help of lead screw and a split nut. The two halves of the split nut are fastened to the carriage. When the nut is closed on to the lead screw, it acts as complete nut, and the carriage starts moving as the lead screw rotates. The lead screw is geared to the spindle and the proper speed ratio between the two is set by means of a gear-change box. Therefore, as the lead screw rotates, the carriage will move a predetermined distance (depending upon the pitch or lead of the thread) per revolution of the job. The third requirement is that the split nut must be engaged at an exact predetermined time, for taking successive cuts, so that the tool enters the helical groove of the cut previously produced, otherwise the tool may remove some of the desired thread. This is achieved with the help of a 'thread dial', which is mounted on the carriage and is driven by the lead screw through a worm gear. The face of the thread dial is graduated into an even number of full and half divisions, (Fig. 3.21). Whenever the lead screw rotates and the split nut is not engaged, the thread dial rotates. The split nut must be engaged when a particular line on the dial face coincides with the zero line For cutting even number of threads, the split nut should be engaged when any line on the dial coincides with zero line, and, for cutting odd-number threads, when any numbered line coincides with zero line.

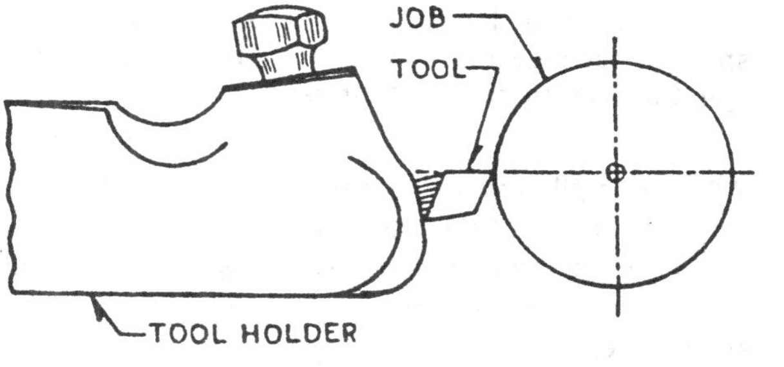

Fig. 3.20. Setting of Cutting Tool.

To start cutting a thread, the tool is fed inward until it first scratches the surface of the job. The graduated dial on the cross-slide is noted or set to zero. The split nut is then engaged and the tool moves over the desired job length. At the end of tool travel, it is quickly withdrawn by means of cross slide. The split nut is disengaged and the carriage is returned to the starting portion, for the next cut. These successive cuts are continued until the thread reaches its desired depth (checked on the dial of cross-slide). The depth of first cut is usually 0.25 to 0.40 mm. This is gradually decreased for the successive cuts until for the final finishing cut, it is usually 0.027 to 0.075 mm. The tool can be fed inward either radially or at an angle of 29° by swiveling the compound rest, (Fig. 3.22). The drawback of the first method is that the absence of side and back rake will not produce proper cutting except on brass and cast iron. In the second method, the cutting mainly

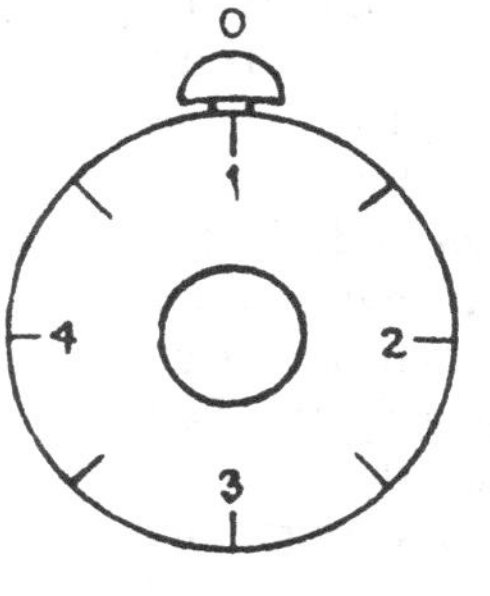

Fig. 3.21. Thread Dial.

brass and cast iron. In the second method, the cutting mainly takes place on one face of the tool and some side rake can be provided. Also, the chip will curl more easily. For cutting square, acme and worm threads, the first method is used. For cutting L.H. threads, the tool is moved from left to right and for cutting right hand threads, it is moved from right to left. Thread cutting on a lathe is a slow process, but it is the only process of producing square threads, as other methods develop interference on the helix.

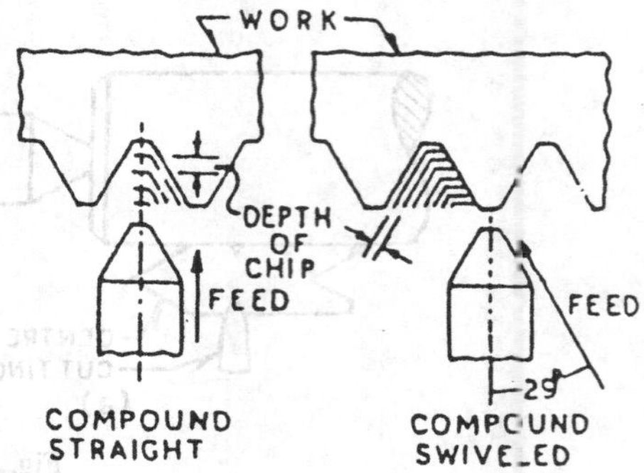

Fig. 3.22. Feeding the Tool into the Job.

Cutting Screw threads on a Lathe without a quick - change gear box. When an engine lathe does not have a quick-change gear box or when the thread pitch is non-standard and not covered by the given gear box or when precise threads are required, the screw threads can be cut by selecting the change gears and setting up these on the change-gear quadrant.

As a rule, the gearing ratios of the pair of gears linking the spindle to the reversing mechanism and that of the mechanism itself are equal to unity. Now in one revolution of the spindle, carriage must travel the pitch of the screw thread to be cut. Thus, the equation of the spindle-to-carriage gear train is :

$$P = L \cdot i_{cg}$$

where P = pitch of the screw thread to be cut

L = pitch of the lead screw

i_{cg} = gearing ratio of the change-gear quadrant.

For quadrants with two pairs of gears the set up formula for change gear selection in cutting screw threads will be

$$i_{cg} = \frac{Z_1}{Z_2} \times \frac{Z_3}{Z_4} = \frac{P}{L}$$

where Z_1, Z_2, Z_3 and Z_4 are number of teeth of the change gears, (Fig. 3.23). Gears with teeth Z_1 and Z_3 are driver gears and those with number of teeth Z_2 and Z_4 are driven gears.

Principle of thread cutting in a Lathe (Fig. 3.23). In thread cutting practice a thread pitch is said to be even if the lead screw pitch is an exact multiple of this pitch and odd if it is not. An even screw thread can be cut by disengaging the half nuts at the end of each pass, retracting the tool and returning the carriage back to the beginning of the thread manually and without stopping or reversing the lathe. When the half nuts are engaged for the next pass and the tool is advanced again it will exactly follow the previously cut thread groove. In cutting an odd thread, the half nuts are not disengaged after

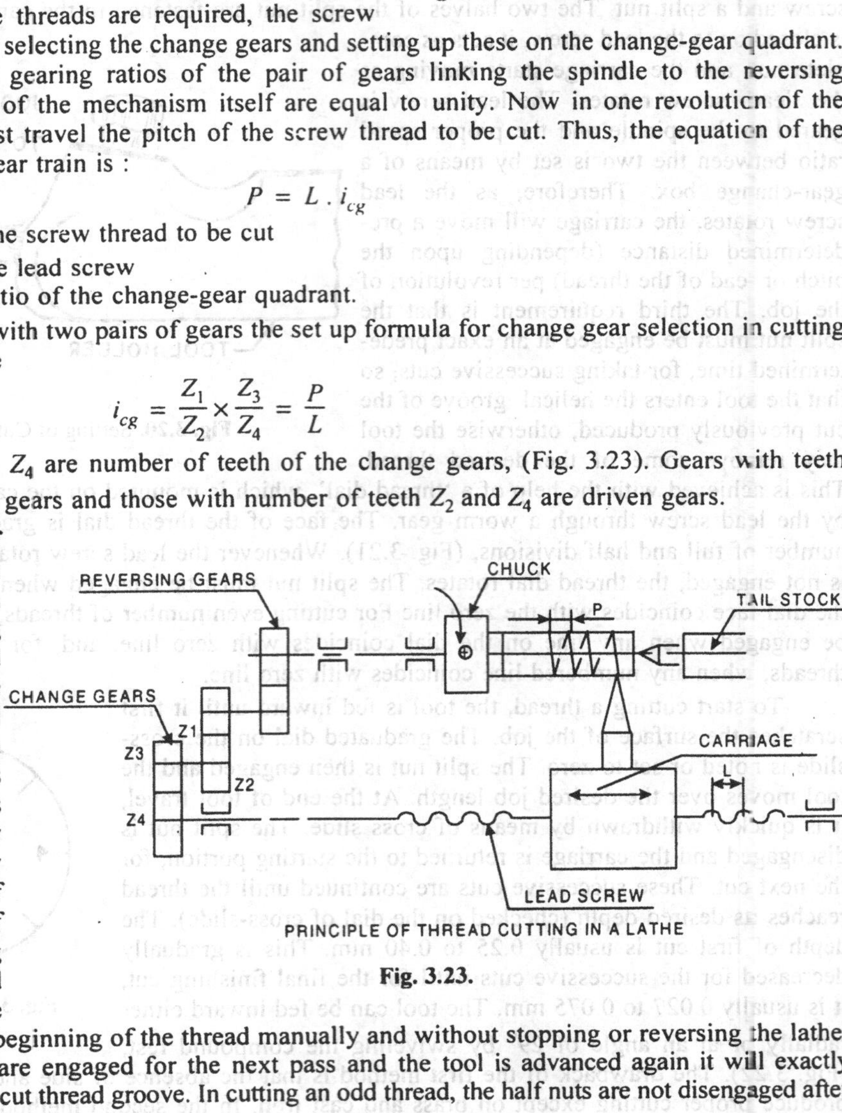

Fig. 3.23.

each pass. The lathe is stopped at the end of the pass, the tool is withdrawn and the carriage is returned to the beginning of the thread by reversing the lathe, otherwise the tool will not fallow the previously cut thread groove in the subsequent pass.

In thread cutting, the job is mounted between centres or clamped in a chuck, depending upon its length. The form of the thread cutting tool must coincide with the profile of the thread being cut.

Cutting external threads with Button Dies and a Die Holder. For smaller diameter threads, a fast method of cutting threads is to use adjustable round split (button dies) in a die holder. Select a button die of the correct size. Select a lathe die holder to match . Insert the die in the die holder. Place the die holder in the tailstock spindle. The workpiece is held in a chuck or a collect and rotated at a very slow speed. Turn in the tailstock hand wheel as the die cut the threads. Use plenty of cutting fluid. Be careful to stop the machine as soon as the correct length of the thread is machined. The external threads can also be cut by screwing the button die (held in a die stock) on the workpiece held and rotated between centres.

Cutting internal threads. Internal threads are cut in the same general way as the external threads. However, the cutting is more difficult because one can't see the cutting operation. Also, it is not possible to hold the tool so rigidly.

A forged boring tool can be used with the end ground the same way as for external thread cutting. The cutting edge may need a little more end relief to keep the bottom of the tool from rubbing on the work. A standard tool bit ground as a threading tool can be used with a boring bar if the hole is large enough. A centre gauge is used for grinding the tool and setting it.

Setting up the Lathe. (*i*) The size of the hole to be threaded must be no smaller than the minor diameter of the thread. This would give a full depth of thread.

(*ii*) If the thread is cut only part way through the hole, it is good idea to cut a recess at the end of the thread. This recess should equal the depth of the thread. This will permit you to release the split nut without turning in the cross feed.

(*iii*) Set the compound rest to 30 degrees to the left (clockwise) of the cross-feed. This is just the opposite of the setting for external threading.

(*iv*) Fasten the threading tool in a holder or boring bar.

(*v*) Hold the centre gauge against the face surface of the workpiece as shown in Fig. 3.19. This can also be done by holding the centre gauge against the side of the workpiece. Adjust the tool on centre and at right angles to the workpiece. To set the tool at centre height, use a surface gauge. Set the scriber point to the point of dead centre. Then set the tool bit to the surface- gauge scriber point.

(*vi*) If the thread is cut only partially through the hole, move the carriage by hand to the point where the thread will end. Mark a line on the lathe bed with a piece of chalk or a pencil. During thread cutting, this reference line will indicate that the tool bit has reached the end of the thread length so that the operator can open splitnut lever and stop carriage.

Cutting the threads. While cutting the internal threads, the following procedure is adopted:

1. Set the micrometer collar of the cross-feed a zero.
2. Turn the compound-rest feed out until the point of the tool just touches the workpiece. Set the micrometer collar at zero.
3. Then the same general procedure is followed as explained for cutting external threads. The following two points must be remembered :

(*a*) The cross-feed must be turned in rather than out when opening the split nut at the end of the threading.

(*b*) If a recess was already machined at the end of the threaded area, the split nut can be opened without turning in the cross slide to stop the carriage. However, the cross-feed must be turned in before moving the carriage back for the next cut.

Cutting internal threads on a Lathe with a Tap. The best and simplest way to cut internal threads is with a tap. The hand tap is held in the tail stock after removing the dead centre.

The tap is fed into the workpiece by moving the tailstock spindle with the help of hand wheel. The workpiece is held and rotated by means of a collet or chuck fastened to the headstock spindle. The hand tap can also be held in a tapstock which is supported on the dead centre. The tap engages the workpiece and is fed in the same manner as above.

Change gear ratio. There are several change gear sets available. A typical set contains the following change gears with number of teeth : 20, 25, 30, 35, 40, 45, 50, 55, 60, 65, 70, 75, 80, 85, 90, 95, 100, 110, 120, 125 and 127. The change gear ratio, i_{cg}, must be transformed by multiplying numerator and denominator by a suitable number, to obtain gears available in the change gear set.

The change gear ratio may result either in a "Simple gearing" or "Compound gearing".

3.1.5. Size of a Centre Lathe. The size of a centre lathe is given by the "Swing" and the length of the bed. The swing is twice the distance from the live centre point of the spindle to the top of the bed, that is,

Swing = 2 × Height of centres from the top of the bed.

In other words, it gives the largest diameter that can be turned over the ways of a lathe, but not over the carriage cross-slide. The length of the bed includes the part the headstock rests on. It indirectly gives the maximum length of the job that can be accomodated between lathe centres.

These machines are available with about 22.5 cm to 125 cm swing and with bed lengths of about 1 m to 5 m. Other factors to be considered while selecting an appropriate lathe are : Hole diameter in the headstock spindle, Power ratings, change gear equipment and type of attachments and accessories.

Designating the size of a centre lathe by 'Swing' is the American system. In the continental system, it is given by the height of centres from the top of the bed.

Accessories. Accessories of a centre lathe are common work and tool holding devices and supports, for example, chucks, collets, centres, drivers, dogs, face plate, dog plate, rests, fixtures and mandrels etc.

Attachments. These are special devices used to perform specific operations, for example, thread chasing dials, taper turning attachments, stops and devices that enable milling, grinding, boring, gear cutting and cutter relieving to be done on the lathe.

3.1.6. Other Low - Production Type Lathes

1. Bench Lathe. Small lathes with beds upto about 1.8 m long and swing upto about 30 cm are commonly set on benches. In most cases, they are as complete as larger lathes, but are smaller and lighter.

2. Toolroom Lathe. This lathe looks like a conventional engine lathe, but is built more accurately, has more speeds and feeds and will be equipped with all the accessories and attachments. It costs more and is mainly used for the manufacture of small tools, dies, gauges, fixture and precision parts of all kinds. It may either be a pedestal or a bench type.

3. Speed Lathe. A speed lathe is a lathe in its simplest form. It consists simply of a headstock, tailstock and a tool post, mounted on a light frame bed. Headstock may have a step-cone pulley arrangement or may be equipped with a variable speed motor. The lathe has no gear box, leadscrew, or carriage. The speeds of these lathes are much higher as compared to conventional lathes. Such lathes are used for making centre holes in workpieces prior to holding them between centres, for metal spinning, polishing and wood working.

3.1.7. Medium Production Lathes. The operation of an engine lathe (by hand) requires considerable skill. The skill of an operator is poorly utilized in repetitive production. Therefore, various efforts at automation have long been made. For medium production, we have the following types of lathes :

1. Turret Lathes. A turret lathe is a medium production, semi- automatic lathe to make parts in greater quantities, to close tolerances and faster. It is an adaptation of engine lathe where the tail stock is replaced by a turret slide (cylindrical or hexagonal). At least six tools can be held on the six faces of the turret. Tool post of the engine lathe is replaced by a square cross-slide

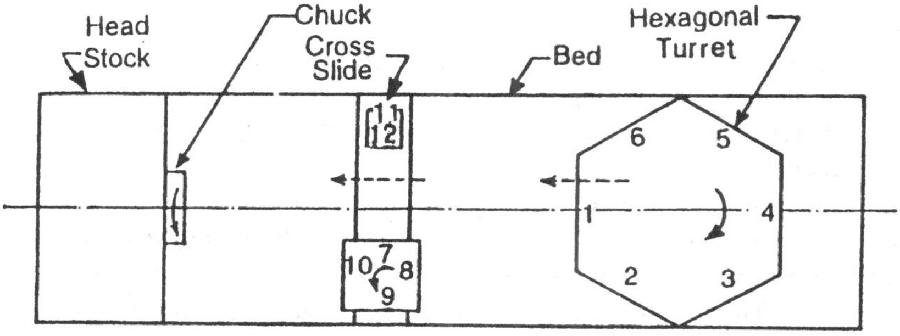

Fig. 3.24. Block Diagram of a Turret Lathe.

(front cross slide) which can hold four tools. Two more tools can be mounted on the rear cross-slide, (Fig. 3.24). The movement of the tools on the turret slide is axial while that of the tools on the cross-slides is perpendicular to the lathe axis. The feed movement of the tools is terminated when a preset stop is reached. The work is held in collet or chuck. The turret lathe can be "set up" to perform many different operations: drilling, boring, reaming, threading, turning and cutting off etc. These operations can be done in any order. Once the machine is "Set up", it requires relatively little skill to operate.

There are two types of Turret Lathes :

(*a*) **Ram Type Turret Lathe.** Here, the turret is carried on a ram (intermediate slide) which moves longitudinally on a saddle positioned and clamped on the ways of the bed, at any desired position. This lathe which is also called as "Capstan lathe" is used for small and medium sized work (mainly bar stock).

(*b*) **Saddle Type Turret Lathe.** Here the turret is carried directly on a saddle which moves on the bed ways. This design is used for heavier work (castings, forgings etc.). This design has no rear cross-slide.

The size of a turret lathe is given by two numbers. The first number gives the "swing" and the second number gives the diameter of the bar that can be passed through the hole in the spindle.

Another difference is the two features in the headstock : rapid shifting between at least two spindle speeds with a brake to stopthe spindle very rapidly and automatic feeding out and clamping the bar stock which passes through the spindle. Yet another constructional difference is that in place of the tool- past, there is square turret and there can be a rear tool slide also (in Ram type turret lathe).

The hexagonal turret can be indexed about its vertical axis, to bring each of its faces into working position. The turret, alongwith its saddle, can be moved longitudinally, either by a manual feed or a power feed. At the end of its working travel towards the spindle, when the turret is moved back with the help of the star wheel, it automatically indexes (or is manually indexed, in the case of heavy machines) at end of the movement, thus bringing each of the six faces into working position.

The square turret on the cross-slide can be indexed manually about a vertical axis, to bring each of the four tools into working position. This square turret can be moved longitudinally through power or manual operation of carriage and tranversally, either by power or manually by means of the cross slide. The tools held in the rear slide can be traversed only crosswire. As shown in Fig. 3.24, at least 12 tools can be held in the tool holders. By using special tool holders on the hexagonal turret, this number can be increased.

A turret lathe can perform all lathe operations on either piece balanks or bar stock. The tools on the front square turret can do the same operations as in an engine lathe, except for chasing thread. In a turret lathe, threads are usually cut from the hexagonal turret, using round dies or a die head for external threads and taps for internal threads. The tools in the rear cross slide are used to cut gooves, to face ends or shoulders and to cut off the workpiece, if it is made of bar stock.

It is clear from above that turret lathes differ from centre lathes in that they can perform operations with a whole set of cutting tools installed ina definite sequence on the turret and on the cross-slide. Therefore, turret lathes have higher production capacity than centre lathes and are used in lot production.

The application of turret lathes is justified economically for jobs that require consective or simultaneous turning and boring of external and internal surfaces, drilling, enlarging holes, tapping etc.; that is, when several tools can be in operation simultaneously, and also in machining workpiece in batches. The use of multiple-tool holders and simultaneous machining of workpiece with tools, both of the turret and the cross slide, help to reduce machining time, while the presetting of the machine for working with several tools helps to reduce non-cutting time. However, because of the lack of sufficient rigidity of the machine-fixture-tool workpiece, cutting speeds, feeds and the depths of cut are much lower in turret lathe. Setting up of the turret lathes is quite time-consuming, so these set ups are economical only when large batches are to be machined. This has led to the wide application of group set ups for turret lathes can be used with very slight readjustments to machine a whole ''family'' of selected similar parts.

2. Copying Lathe. A copying lathe (a tracer or a duplicating lathe) is one with an attachment that enables the machine to perform various operations from flat or round templates/masters (or models of the workpiece). The cutting tool is made to follow a path that duplicates the path of a stylus or tracer finger moving along the template. These machines produce parts by semi-automatic means. Except for tape controlled NC lathes, the operator is required to load and unload the workpiece and to start the automatic operation for each cutting cycle. The movement of the cutting tool is co-ordinated with that of the tracer finger either by mechanical, hydraulic, pneumatic or electrical device.

A number of lathes can be equipped with such attachments enabling repetitive work to be performed rapidly by copying the profile of a flat or round template, while some machines are made especially for the purpose.

These machines will machine irregular contours, including steps, tapers, right angle or tapered shoulders, recesses, grinding recesses (necks), radii formed surfaces, bore contours or many similar operations. Components like spindles, shafts, pistons etc. can be machines on these lathes. Such lathes are useful for volume production.

For large work, the template is mounted above and to the rear of the lathe centres, out of the way of the work and the tool. On some machines, intended for reproducing small templates, the template is mounted on the front.

Below, we discuss one design of a copying lathe, based on Pneumatic-Hydraulic Control, Fig. 3.25.

A flat template is positioned on a rail at the rear of the machine and is in contact with the pneumatic tracer attached to the overhead arm. The fixed end of the arm is carried by the hydraulically-operated tool slide set at 45° to the main cross-slide.

Air from an approximately constant-pressure source is fed through a restricting orifice to a flexible metal bellows and also to a pilot nozzle on the tracer. The movable end of the bellows is attached to the stem of a hydraulic value through which oil, supplied under pressure from a motor driven pump, can be fed to either side of a piston connected to the angular tool slide. Longitudinal power feed to the saddle operates in the normal way.

When the tracer is contacting a straight portion of the template, air escapes slowly from the nozzle, the pressure in the bellows being designed to centralise the hydraulic valve under these conditions. Oil can neither enter nor escape from the operating cylinder and the tool slide is accordingly locked in place, the oil from the pump being by-passed to the reservoir. If the longitudinal feed is engaged, the tool will turn a constant diameter on the workpiece so long as

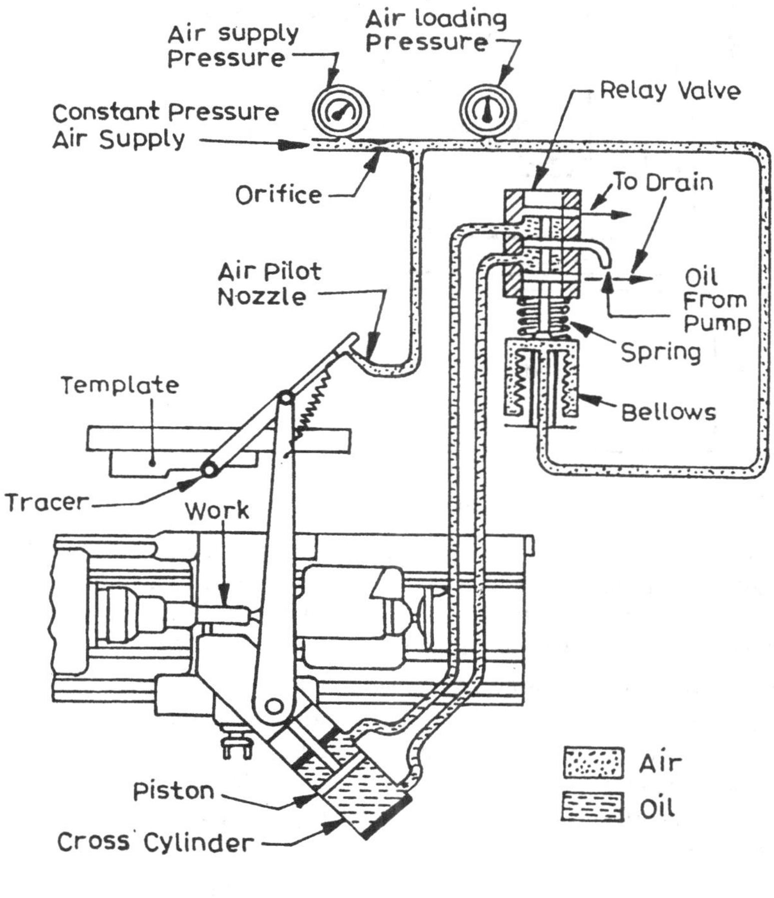

Fig. 3.25. Copying lathe Principle.

the tracer follows a straight path.

When the tracer contacts a shoulder, it comes to rest and the continued movement of the saddle causes the air valve to open. The fall in air pressure transmitted via the bellows, opens the hydraulic valve, allowing the oil to enter the cylinder behind the piston and escape from the opposite end. The tool slide, therefore, moves backwards and the combination of this movement and the saddle travel produces a square shoulder on the workpiece. The rearward movement of the slide carries the tracer with it, and as soon as the latter clears the template shoulder and again reaches in the bellows which centralises the hydraulic valve and relocks the slide in its new position.

The response of the tool to changes in air pressure is practically instantaneous.

3. **Vertical Turret Lathes.** These turret lathes, with vertical orien- tation, (Fig. 3.26), are

used for machining large and/or heavy parts, such as motor bodies, pinions, ring gears and

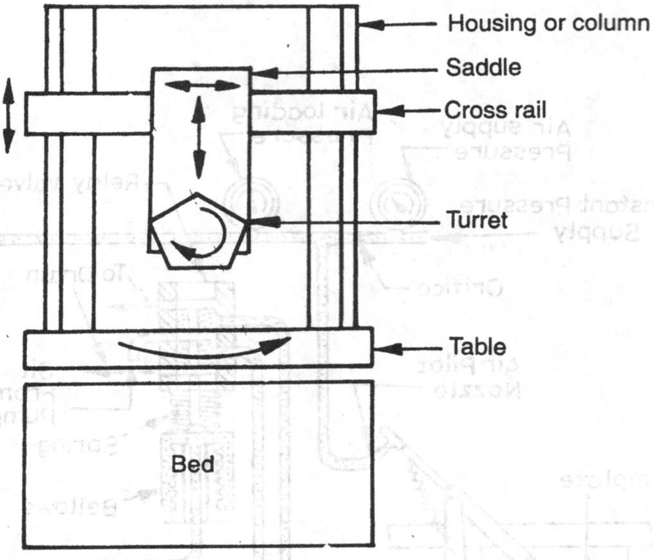

Fig. 3.26. Vertical Turret Lathe.

couplings etc. Such components can be mounted more conveniently on the rotary table of the vertical turret lathe than to be held in chucks or face plates in the case of horizontal turret lathes. Also, the turret of a vertical turret lathe has movement in the two axes as compared to a horizontal turret lathe, where the turret moves only in the longitudinal direction.

Vertical Turret Lathe, Turning and Boring Mill

Vertical turret lathes and vertical turning and boring mills are employed to machine work-pieces that are difficult to set up and machine in centre lathes and horizontal turret lathes, that is, heavy piece of large diameter and relatively small length. They are used for turning and boring cylindrical and tapered surfaces, facing, cutting annular gooves, drilling, countersinking, counter-boring and reaming, etc.

In design arrangement, smaller vertical boring and turning machines have a single upright column or housing, and are called as vertical turret lathes. They can machine workpieces from 300 to 2000 mm in diameter. Larger machines, which can machine workpiece from 1.5 to 2 metres and above, have double-upright column or housing and are called as ''Vertical turning and boring mills''. These machines have 18 or more table speeds. Special (unique vertical turning and boring mills) can machine huge workpiece of a diameter upto 20 metres. These machine tools can perform all lathe operations with the exception of thread chasing, and have as many as 4 or 5 tools in simultaneous operation.

A vertical turret lathes is shown in Fig. 3.26. The workpiece is placed on the revolving table, which performs the rotating principal movement. The main tool head is mounted on the cross rail, along which it travels horizontally and with which it travels vertically. The five sides turret or tool holder is mounted on a ram, which travels vertically in the cross-rail tool head. The turret can be easily and quickly indexed from hole to hole and clamped with a lever. Tools are clamped in the tool holes of the turret by the same types of holders as those used in horizontal turret lathes. A side tool head is mounted on the upright and carries a square turret or tool holder.

The side tool head accomodates one or two single-point tools. They can travel either horizontally or vertically. All the cross-rail tool heads and the side tool head have power feeds in the forward and reverse directions, as well as rapid traverse motions for tool approach and return motions.

Double-upright vertical turning and boring mills have some special features. Medium and heavy-weight models are characterized by portal-type frame formed by bed with revolving table, uprights secured to the bed, and top brace linking uprights at the top. Unique models are provided with two revolving tables (annular), that is, inner and external, which are driven separately or together. The inner table is rotated at a higher speed than the external one, which makes it possible to machine workpieces varying in diameter at optimum cutting speeds. The portal frame of these models is adjustable and is set up, depending on the diameter of the job to be machined. These machines are provided with two rail heads, placed on the cross rail and a movable side head, mounted on the guide ways of the right-hand upright. Some heavy models have a third rail head, travelling along the guideways of an additional overhanging cross rail fastened to the main one.

The main trends in vertical boring and turning machine development are : higher productive output through faster speeds and greater power, reduction of non-cutting, time through simplification and automation of handling, widening of capability and improvements in the design of machine units.

3.1.8. High Production Lathes. The high production lathes come in the category of "Automatic lathes". An automatic lathe is a kind of engine or turret lathe that operates automatically. It requires very little attention after the set up has been made and the machine loaded.

3.1.9. Special Purpose Lathes. The lathes included in this category are :

1. Gap Lathe. This is a special lathe used for machining extra large diameter pieces. To accommodate large swing of the workpiece, the bed ways are stopped short of the headstock, to provide a sufficient gap. To restrict the gap, the workpiece must be relatively short in thickness. The work is mounted primarily for the facing or some internal operations.

The size of the gap lathe is given by the maximum diameter and length of the job that the gap will accomodate.

2. Special-purpose Engine Lathes. These engine lathes are designed for machining special types of workpieces. Examples are : Wheel turning lathes for turning the tread on railroad-car and locomotive and so on.

3. Instrument Lathes. These lathes come in the category of all purpose, low-production lathes. These lathes are of smaller size than "Bench lathes" and are used by instrument makers.

4. Facing Lathes. These lathes are used to machine workpieces of large diameter, but short in length in single piece production and in repair shops. These lathes are similar to engine lathes, but the bed is shorter in length and the faceplate in much larger in diameter [upto 4 m]. The workpiece is clamped on the faceplate by means of jaws or clamps and hold-down bolts. When required, the workpiece is supported by tail-stock centre. However, many facing lathes do not have tail-stocks. The feed mechanism is powered by a separate motor.

Facing lathes are now rarely used, because of : inadequate accuracy, labour-consuming set-up of workpiece and low productivity. They have been replaced by more advanced boring and turning machines.

5. Flow Turning Lathes. A flow turning lathe is used for roll flowing, as a method of cold-flowing metal. A heavy pressure is applied spirally with two hardened rollers against a metal block. The lathe and the flow turn process may be used with flat plates and forgings and castings and for extruded and other shapes.

6. Heavy Duty Lathe. If the swing of centre lathes ranges from 250 to 1200 mm, these models are known as larger-sized lathes. As noted above, the smaller lathes are called Bench lathes and Instrument Lathes. A lathe, that has a swing of 500 mm or more and is used for roughing and finishing cuts, is often referred to as a Heavy-duty lathe.

SOLVED PROBLEMS

Example 1. *Find the angle at which the compound rest should be set up to turn taper on the workpiece having a length of 200 mm, larger diameter 45 mm and the smaller diameter 30 mm.*

Solution. Now half the taper angle is given as,

$$\tan\alpha = \frac{D-d}{2l}$$

where $D = 45$ mm, $d = 30$ mm and $l = 200$ mm

$$\therefore \quad \tan\alpha = \frac{45-30}{2\times 200} = \frac{3}{80} = 0.0375$$

$$\therefore \quad \alpha = \tan^{-1} 0.0375 = 2^\circ, 9'.$$

$\therefore$ Angle at which the compound rest should be set up is,

$$\alpha = \mathbf{2^\circ, 9'}$$

Example 2. *A taper pin of length 80 mm has a taper length of 48 mm. The larger diameter of taper is 83 mm and the smaller diameter is 73 mm. Determine (i) taper in mm/metre and in degrees (ii) the angle to which the compound rest should be set up (iii) the tail stock setting over.*

Solution. $L = 80$ mm, $l = 48$ mm

$D = 83$ mm, $d = 73$ mm

(*i*) Taper, $T = \dfrac{D-d}{l} = \dfrac{83-73}{48}$

$$= \frac{10}{48}$$

i.e. for a length of 48 mm, the taper is 10 mm

$\therefore$ Taper in mm/metre $= \dfrac{10}{48} \times 1000$

$= 208.33$ mm/metre length

Now $\tan\alpha = \dfrac{D-d}{2l} = \dfrac{10}{96} = 0.1042$

$$\therefore \quad \alpha = 5^\circ, 57'$$

(*ii*) The angle to which the compound rest should be set up is,

$$\alpha = 5^\circ, 57'.$$

(*iii*) Tail stock set over, $\quad h = \dfrac{D-d}{2l} \times L$

$$= \frac{10}{96} \times 80 = \mathbf{8.33\ mm}$$

Example 3. *Find the intermediate speeds of a head stock spindle for a lathe given the following data :*

Minimum speed = *45 rev/min*
Maximum speed = *2000 rev/min*
Number of speeds = *18*

Solution. Now $n_{min} = 45$ rev/min

$$n_{max} = 2000 \text{ rev/min}$$

$$\mathbf{Z} = 18$$

$$\therefore \text{ Step ratio,} \quad \phi = \left(\frac{n_{max}}{n_{min}}\right)^{\frac{1}{Z-1}}$$

$$= \left(\frac{2000}{45}\right)^{\frac{1}{17}} = \left(\frac{400}{9}\right)^{0.0588}$$

$$= (44.44)^{0.0588} = \mathbf{1.25}$$

∴ Speeds are **45, $\varphi \times$ 45, $\varphi^2 \times$ 45 .... 2000**

Example 4. *Calculate the change gears to cut a single start thread of 3 mm pitch on a centre lathe, having a lead screw of 6 mm pitch.*

Solution. $P = 3$ mm, $L = 6$ mm

∴ Change gear ratio, $i_{cg} = \frac{P}{L} = \frac{3}{6} = \frac{1}{2}$

Now from the available change gear set, this gear ratio can be obtained by selecting gears $Z_1 = 20$ and $Z_2 = 40$

or $Z_1 = 25$ and $Z_2 = 50$

∴ It will be a simple gearing.

Example 5. *Calculate the change gears to cut a single start thread of 0.5 mm pitch on a centre lathe having a lead screw of 12 mm pitch.*

Solution. $P = 0.5$ mm, $L = 12$ mm

∴ Change gear ratio, $i_{cg} = \frac{P}{L} = \frac{0.5}{12} = \frac{1}{24}$

Since no convenient change gears can be found in the change gear set, this ratio must be split into two factors.

So,

$$i_{cg} = \frac{1}{24} = \frac{1}{4} \times \frac{1}{6}$$

$$= \frac{25}{100} \times \frac{20}{120} = \frac{Z_1}{Z_2} \times \frac{Z_3}{Z_4}$$

∴ Driver gears Z_1 and Z_3 = 25 and 20 teeth

Driven gears Z_2 and Z_4 = 100 and 120 teeth

∴ It will be a compound gearing.

Example 6. *Calculate the change gears to cut a single start thread, four threads per inch, on a centre lathe with lead screw of 3 mm pitch.*

Solution. $P = \frac{1}{4}$ inch $= \frac{25.4}{4}$ mm, $L = 3$ mm

$$\therefore \quad i_{cg} = \frac{P}{L} = \frac{25.4}{4} \times \frac{1}{3} = \frac{12.7}{6} = \frac{127}{60} = \frac{Z_1}{Z_2}$$

$Z_1 = 127$ teeth

$Z_2 = 60$ teeth (A simple gear ratio)

Note. For a single start thread,

$$i_{cg} = \frac{P}{L}$$

For a multi-start thread,

$$i_{cg} = \frac{P \times \text{No. of start}}{L}$$

PROBLEMS

1. List some of the operations that can be done on the Engine Lathe.
2. How will you classify the Engine Lathes ?
3. Name the four main parts of a Lathe.
4. Sketch a Tool Post.
5. What are the four major parts of a Carriage ?
6. Describe how the back gears operate.
7. Sketch and describe the geared main drive of the Engine Lathe.
8. Sketch and explain the feed drive of the engine lathe.
9. Sketch and explain "feed reversing mechanism".
10. Sketch and explain "change gear quadrant".
11. What is the purpose of the quick-change gear box ? Sketch and explain a tumbler type gear box.
12. What is the difference between the 'live' and 'dead centre' ?
13. How do you obtain power cross-feed ?
14. How can the movement of the carriage be reversed ?
15. When is the split nut or half nut used ?
16. Sketch and describe the three types of Centres used in the Engine Lathe.
17. Sketch the two types of Lathe Dogs.
18. Sketch and explain the use of a Face Plate.
19. What is the difference between a 'face plate' and a 'drive plate' ?
20. Sketch and explain the working of a Mandrel.
21. What are the advantages of using a Collet Chuck ?
22. What is the use of Steady Rests ? What is the difference between a steady rest and a follower rest ?
23. Sketch and explain the various methods of turning a Taper.
24. Name some of the advantages of using a 'taper turning attachment'.
25. What is a centre gauge used for in threading ?
26. Why is the compound rest set at an angle of 29° for most threading ?

27. What is the use of a thread dial ?
28. When cutting odd number of threads per cm, when is the split nut or half nut closed ?
29. How is the size of the Engine Lathe given ?
30. What are : Bench lathe, Toolroom lathe, Speed lathe ?
31. Sketch a block diagram of a Turret Lathe. Name its major parts.
32. Sketch and describe a Vertical Turret Lathe.
33. What are : Tracer lathes and Gap lathes ?
34. What are : Special purpose engine lathes, Instrument lathes, Facing Lathes, Flow turning lathes and Heavy duty Lathes ?

3.2. SHAPERS, PLANERS AND SLOTTERS

Introduction. The main function of shapers, planers and slotters is the machining of flat surfaces by means of straight line reciprocating single point cutting tools similar to those used in lathe operations. The flat surfaces produced may be horizontal, vertical or inclined at an angle. These machine tools can also be arranged for machining contoured surfaces, slots, grooves and other recesses.

The processes of planing, shaping and slotting are similar to one another, but in the first case the main cutting motion is transmitted to the workpiece on the worktable which travels past the cutting tool (except in pit and edge planers where the work is stationary and the cutting tools reciprocate), while in the second and third cases, the workpiece is stationary and the cutting motion is transmitted to the tool. The cutting tool / workpiece travels horizontally in shaping and planing and vertically in slotting. In all cases, the feed motion is accomplished in a plane perpendicular to the cutting motion (of the workpiece or tool). Feed may be either parallel or perpendicular to the cutting edge, or in a circle.

Shaping is limited to small and moderate size workpieces, whereas planers are designed for machining large workpieces. They can take heavy cuts. Cutting is intermittent in all the three processes, since cutting is done in one direction only. The return stroke is idle stroke, and is faster than the cutting stroke.

3.2.1. Shapers. The shaping process can be defined as a process for removing metal from surfaces in horizontal, vertical and angular planes by the use of a single point cutting tool held in a ram that reciprocates the tool in a linear direction across the workpiece held on the table of the

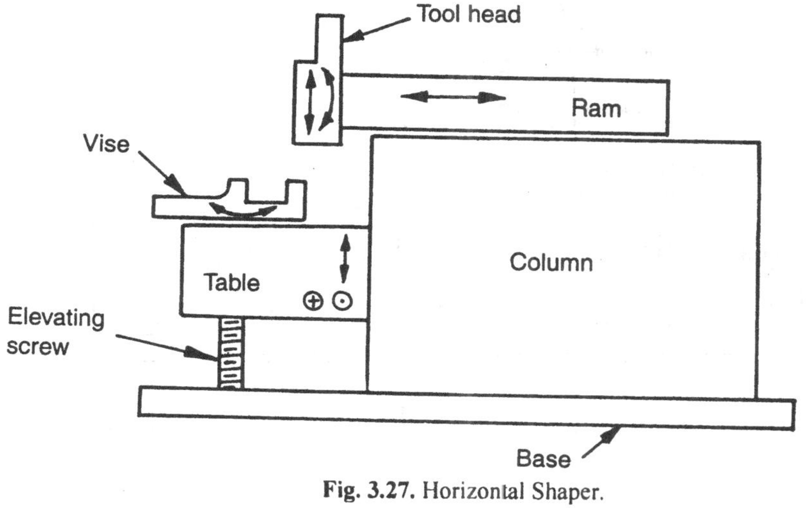

Fig. 3.27. Horizontal Shaper.

machine. The work is fed at right angles to the direction of the ram in small increments, at the end of the return stroke.

Types of Shapers. There are mainly two types of shapers : Horizontal shapers and vertical shapers depending upon whether the ram and the cutting tool reciprocates in the horizontal plane or in the vertical plane.

Construction details of a Horizontal Shaper. Fig. 3.27 shows the block diagram of the horizontal shaper.

1. Base. The base is the necessary bed or support required for all machine tools. All other parts are mounted on and above the base. The shaper is installed by rigidly bolting the bed to the shop floor. The bed takes up the total dead weight of the machine as well as the dynamic load during machining operations. Its material is usually gray cast iron.

2. Column. The column of the shaper is a hollow casting and is mounted on the base. It houses the drive mechanisms for the ram and the table. At the top of the column, the bearing surfaces known as "guide ways" are provided on which the ram reciprocates.

3. Table. The worktable of a shaper is fastened to the front of the column. The table moves across the column on crossrails to give the feed motion to the job. The cross rails are a part of the slide which can be moved up and down on the column ways (with the help of elevating screw) to accommodate various sizes of the workpiece.

Shaper tables are provided with T-slots so that most jobs are directly held on the table by means of bolts, blocks and clamp plates. Jobs can also be held in vises which are secured to the table. Sometimes special work holding devices (fixtures) may be needed to hold the jobs. Heavy duty shapers are provided with a table support bolted to the outer end of the table and supported by resting upon the bed. This makes the table rigid and prevents work deflection under heavy cuts.

4. Ram. The ram carries the tool head at its front end and travels in "guideways" to give straight line reciprocating motion to the tool. The ram is either mechanically driven or hydraulically operated. A single point tool is fastened in the tool post. As the ram moves forward, it completes the cutting stroke. As it returns and completes its idle stroke, the table feeds cross -

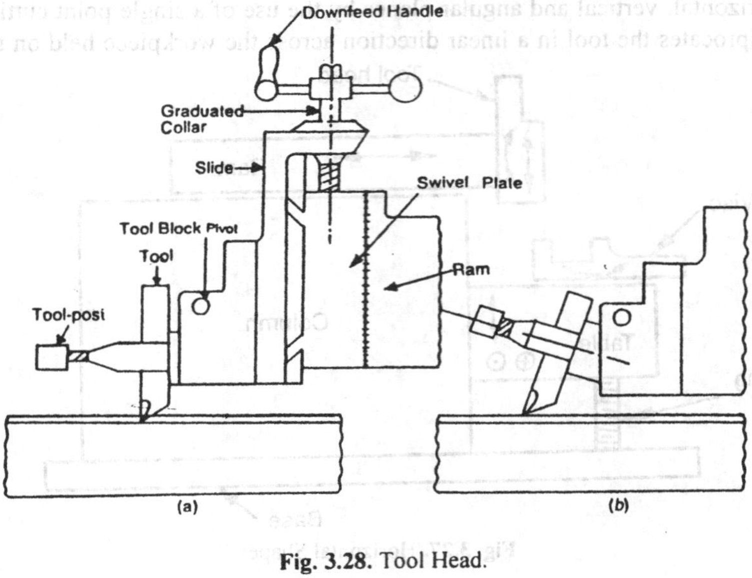

Fig. 3.28. Tool Head.

wise a preset incremental distance equal to the feed desired. The ram travels upto two times faster on the return stroke than on the forward stroke.

5. Tool Head. The tool head, (Fig. 3.28), holds the cutting tool and is fastened to the front of the ram. The tool is held in a tool holder/tool post similar to the lathe tool post. The tool post and the tool block fit snugly in the clapper box and is hinged at the upper edge. During the forward cutting stroke, the tool block is solidly supported against the clapper box due to the cutting force. On the return stroke, the tool block is free to swing forward from its pivot, (Fig. 3.28 (*b*)). This lifts the tool block out of the clapper box a sufficient amount to prevent the tool from rubbing the work surface. The tool head has a tool slide and feed screw rotated by a bellcrank handle for raising and lowering the tool to adjust the depth of cut. The down feed of the tool is usually manual, but power feed can also be provided. The slide is mounted on a swivel plate, which enables the tool head to be rotated or swiveled enabling it to take angular cuts to machine inclined surfaces. Then the tool slide will be parallel with the desired surface. When machining inclined surfaces, the clapper box must be swiveled to enable the cutting tool to swing along an arc away from the work surface on its return stroke. A rule of thumb is that the top of the clapper box should be swiveled away from the surface to be machined.

The details of the tool head are illustrated in Fig. 3.29.

In the front of the slide, an apron is provided which carries the clapper box at its lower end. The clapper block or tool block fits snugly in the clapper box and is hinged about a pin passing through the clapper box and clapper block near its top ends as shown. The apron is secured to the slide with the help of a bolt with its back abutting against the slide. The bolt-passes through a slot (in the shape of a sector) in the upper end of the apron. By unscrewing the nut on the bolt, the apron and hence the tool can be swivelled as shown in the Fig. After swivelling to the desired position, the nut is tightened.

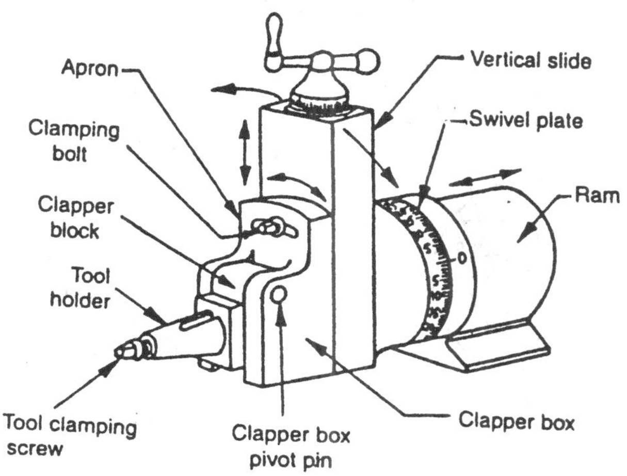

Fig. 3.29. Tool Head.

The slide is mounted on a swivel plate at its back. The swivel plate is bolted to the front of the ram. The bolt can be loosened and the slide and hence the tool head can be swivelled as shown, for machining inclined surfaces. The angle of inclination with the vertical can be read from the graduations on the swivel plate.

The above shaper in which the cutting takes place in the forward stroke is called the "Push cut Shaper", and is the most widely used. "Pull or draw cut" shapers cut in the direction corresponding to the return stroke of an ordinary shaper. The cutting force is directed towards the column and its advantage is that the chips are thrown away from the operator. These shapers are built with a long stroke and are designed to take heavy cuts. Such machines are generally found in railroad shops.

Depending upon the design of the table, the horizontal shapers are classified as : Plain Shapers and Universal shapers.

Plain Shapers. In these machines, the table has only two motions : Crosswise in the horizontal plane and vertical movement (up and down). The table is not-provided with a swiveling motion.

Universal Shapers. These machines are similar to plain shapers except that the table can be tilted to various angles, making it possible to machine inclined surfaces. The table can be swiveled through 360° about a central axis parallel to the cutting stroke direction and also perpendicular to it, that is, around two horizontal axis. The table also has movements in the horizontal plane and vertical direction (up and down) as in a plain shaper.

SHAPER DRIVE

1. Mechanical.

(*a*) **Ram drive.** The mechanical ram drive is a slotted arm quick return motion mechanism, (Fig. 3.30). The slotted arm is pivoted at its lower end to the column and its upper end is connected by a link to an adjustable ram block which is secured to the ram. Power from an electric motor is transmitted to the bull gear through gears and bullgear pinion. A pin fits freely into a hole in the bullgear block which can slide in the slot of the rocker arm. As the bull gear rotates, it carries the pin in a circular path with the help of the crank. This makes the rocker arm to swing back and forth about its stationary pivot C. Thus the rotary motion of the bull gear is converted into the reciprocating motion of the ram with the cutting tool. The length of the ram stroke depends upon the radius of the crankpin on the bull gear, that is, AB which can be adjusted. The larger the crank radius, the longer the ram stroke will be and vice versa.

Fig. 3.30. Mechanical Ram Drive.

$$\frac{\text{Cutting Stroke time}}{\text{Return Stroke time}} = \frac{\angle B_2 K B_1}{\angle B_1 L B_2}$$

It is clear that the return stroke is quick so as to minimise the return idle stroke time. The bull gear is massive in size so that it also performs the function of a flywheel. This makes the reciprocating movement of the ram smooth and without any jerks.

(*b*) **Feed Mechanism.** We know that the table feed is intermittent and is accomplished on the return (non cutting) stroke when the tool has cleared the workpiece. The cross feed is given to the table with the help of a cross feed screw, (Fig. 3.31), which is actuated by a pawl which engages a notched wheel (ratchet) keyed to the screw [Fig. 3.31(*b*)]. The spring loaded reversible pawl gets its motion from a slotted disc through a connecting rod and pawl arm or rocker arm. The slotted disc is driven from the bull gear through a spur gear drive and rotates at the same speed as the bull gear. Because the pawl is straight on one side and slanted on the other, it moves the ratchet wheel

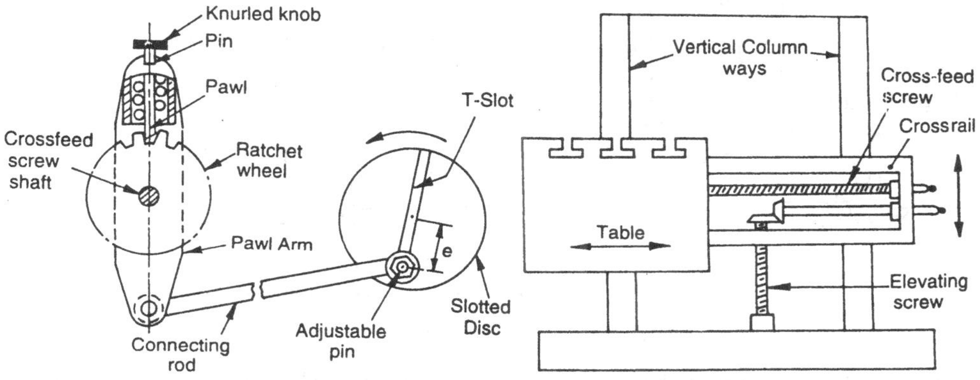

(*a*) (*b*)

Fig. 3.31. Feed Mechanism.

in one direction only. A spring pushes against the pawl to keep it in contact with the ratchet wheel. As the slotted disc rotates, the connecting rod oscillates the pawl about the cross feed screw, so that it moves the ratchet wheel by engaging with it and hence rotates the cross feed screw which moves the table, (Fig. 3.31 (*a*)). The straight side of the pawl pulls or pushes the gear for positive movement, while the slanted side merely slides over the teeth preparatory to taking another bite. A knob at the top of the pawl enables the operator to rotate it 180° to reverse the direction of feed or 90° to stop it altogether. The rate of feed is controlled by adjusting the eccentricity or offset of the driving pin in the slotted disc.

2. Hydraulic Drive. A typical hydraulic drive for a horizontal shaper is shown in Fig. 3.32. A constant speed motor drives a hydraulic pump which delivers oil at a constant pressure to the line. A regulating valve admits oil under pressure to each end on the piston alternately, at the same time allowing oil from the opposite end of the piston to return to the reservoir. The piston is pushed by the oil and, being connected to the ram by the piston rod, pushes the ram carrying the tool. The admission of oil to each end of the piston, alternately, is accomplished with the help of trip dogs and pilot valve. As the ram moves and completes its stroke (forward or return) a trip dog will trip the pilot valve which operates the regulating valve. The regulating valve will admit the oil to the other side of the piston and the motion of the ram will get reversed. It is clear that the length of the ram stroke will depend upon the position of the trip dogs. The length of the ram stroke can be changed by unclamping and moving the trip dogs to the desired positions.

The above system is a constant pressure system. Hence, the ram will travel to slightly different speeds under varying load conditions. The velocity of the ram travel will be directly proportional to the oil pressure and the piston area to which it is applied. The return stroke is quicker, since the piston area on which the oil pressure acts is greater as compared to the other end for which it gets reduced because of the piston rod.

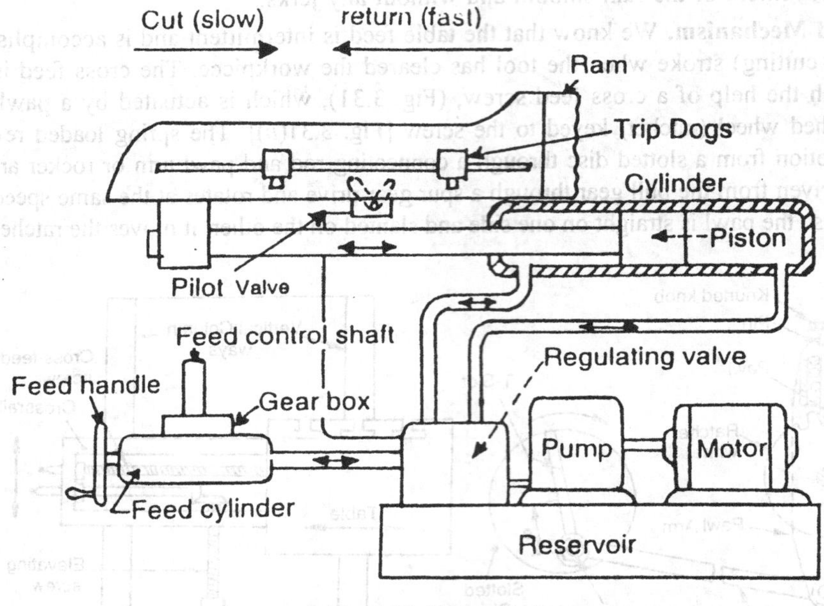

Fig. 3.32. Hydraulic Drive.

Another oil line is connected to a smaller feed cylinder to change the hydraulic power to mechanical power for feeding the work past the tool.

Advantages of Hydraulic drive :

1. Offers greater flexibility of speed.
2. Smoother operation.
3. Ability to slip when the tool is over loaded.
4. Ability to stall against an obstruction without damage to the tool or the machine.
5. Ability to change length and position of stroke or speed while the machine is running.
6. The hydraulic drive shows a very nearly constant velocity as compared with a mechanical drive, which has a constantly changing velocity because the horizontal component of the crankpin moving about its circle is constantly changing, Fig. 3.33.

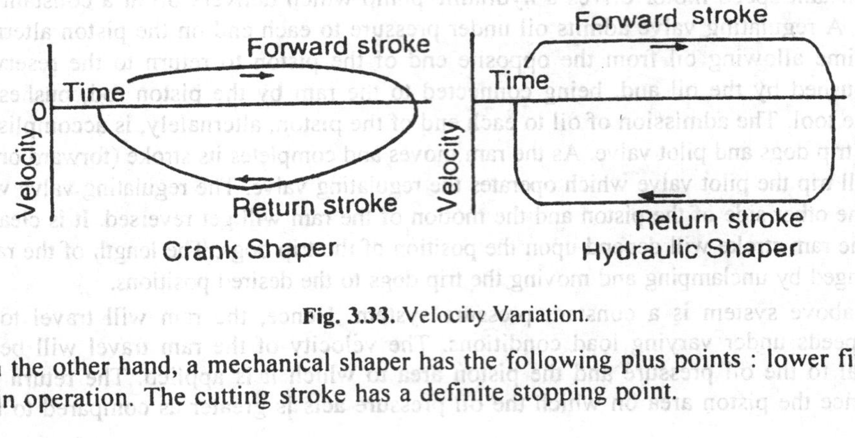

Fig. 3.33. Velocity Variation.

On the other hand, a mechanical shaper has the following plus points : lower first cost and simpler in operation. The cutting stroke has a definite stopping point.

Shaper Size. Shaper size is designated by its longest nominal cutting stroke. Thus a 600 mm shaper has ram travel enough to drive a tool across a 600 mm surface. On most horizontal shapers, the table can be fed cross-wise a distance at least as large as the stroke. That means, a 600 mm × 600 mm square.

Horizontal shapers range in size from small bench models with stroke of 175 or 200 mm to heavy duty models with strokes as much as 900 mm. Shaping machines are commonly provided with power feeds ranging from 0.2 to 0.5 mm/stroke.

Advantages of Shapers. The shapers have got the following advantages :

1. The single point cutting tools used in shapers are inexpensive. these tools can be easily grounded to any desired shape.
2. The simplicity and ease of holding work, its easy adjustment, and the simple tool give the shaper its great flexibility.
3. Shaper set up is very quick and easy and can be readily changed from one job to another.
4. Thin or fragile jobs can be conveniently machined on shapers because of lower cutting forces.

However, a shaper is by nature a slow machine, because of its straight line, forward and return (idle) stroke. The single point tool requires many strokes to complete a work. Also, cutting speeds in these types of machining operations (shaping, planing and slotting) are not usually very high since difficulties are encountered in designing machine tools with high speeds of reciprocating motion due to high inertia forces developed in the motion of the units and components of the machine. For all these reasons, the shaper does not find ready adaptability for assembly and production line. But because of its advantages discussed above, the shaper finds best use in toolroom, jobbing shops and die or jig making shops (for small or medium-sized work), where its flexibility is important and the relative slowness unimportant because so few identical pieces are made.

3.2.2. Vertical Shapers / Slotters. Vertical shapers and slotters are similar to the horizontal shaper except that the ram reciprocates in a vertical direction, cutting on the downward stroke, the upward stroke (return stroke) being idle. The terms vertical shaper and slotter are often used interchangeably by shopmen, although there is a slight difference between them. Whereas the vertical shaper has the facility of allowing the adjustment of ram nearly upto 10° from the vertical position, the slotting machine has only the vertical straight line movement of the ram. Therefore, on vertical shapers, inclined surfaces can also be cut. A slotter is, however, a more robust machine than a vertical shaper.

Most of these machines are provided with a rotary table. The rotary table can be indexed accurately for various operations or rotated continuously about the vertical axis by hand or power, to produce an arc. The table moves on a saddle that slides on the bed and has horizontal adjustments and feeds in two perpendicular directions, (Fig. 3.34). The rotary plus the two straight line motions of the table permit easy machining of circular, convex, concave and other curved surfaces. Slots and grooves can be accurately spaced around a workpiece. On the vertical shaper and slotter, the work is almost always clamped to the table.

Uses of Vertical Shaper / Slotter.

1. Internal machining of blind holes.
2. Work requiring machining on internal sections such as splines, keyways, various slots and grooves, and teeth.
3. Cutting of teeth on ratchet or gear rings which require primarily rotary feed.

4. Machining of die, punchet, straight and curved slots.

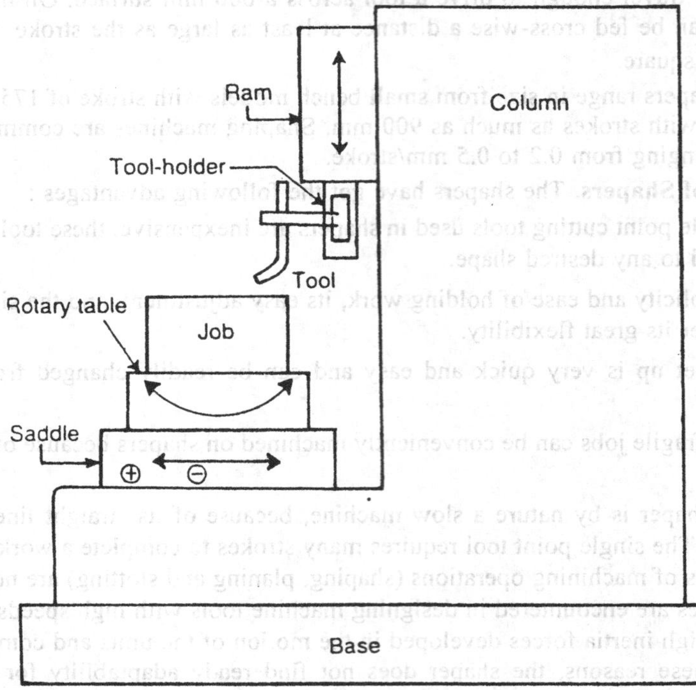

Fig. 3.34. Block Diagram of a Slotter.

Types of Slotters

1. **Puncher Slotter.** This type is used for machining heavier jobs produced by forging or stamping.

2. **Production Slotter.** Here, the upper portion of the machine frame and hence the ram, can be tilted for machining tapered jobs. This machine is of very heavy construction.

3. **Tool room Slotter.** This machine is a refined and more accurate form of production slotter. It is fitted with various sophisticated arrangements for accurate machining.

4. **Keyseater.** It is a special purpose slotter which is designed for cutting keyways in gears, pulleys, cams and similar parts.

Size of vertical Shaper / slotter. The size of these machines is given as : Maximum length of stroke (mm) × Diameter of work table (mm). The stroke of these machines ranges from 300 to 1800 mm. Feed ranges from 0.05 to 2.5 mm/stroke and the ram speed usually ranges from 2 to 40 m/min.

Ram drive Mechanism. The common drives for vertical shapers and slotters are :

1. **Hydraulic Drive.** This drive is similar to the one explained for horizontal shapers.

2. **Variable Speed Reversible Motor Drive.** This machine incor- porates a reversible motor with a multi-speed gear box to give a range of speeds to the ram.

3. **Slotted disc mechanism.** A slotted disc mechanism is shown in Fig. 3.35. The slotted disc is rotated with the help of a pinion and gear drive, from an electric motor. The rotary motion of the slotted disc is converted into the reciprocating motion of the ram with the help of connecting rod. The length of stroke of the ram can be changed by varying the eccentricity of the pin (its distance from the centre of the slotted disc).

Fig. 3.36 shows some typical surfaces machined on shapers and Slotters.

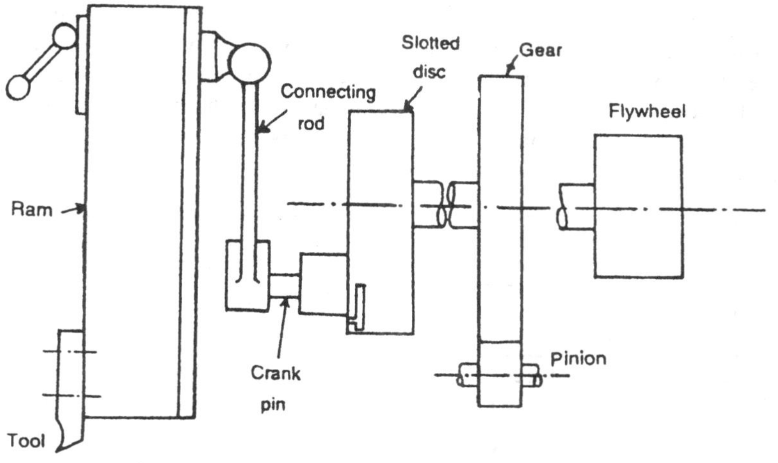

Fig. 3.35. Slotted Disc Mechanism.

3.2.3. Planers. In conventional planers, the work held on the table reciprocates past the tool/tools. However, in unconventional planers (pit planer and edge planer), the table is stationary and the tool head reciprocates. The planer is the largest of the reciprocating machines (shapers and millers). It has got the following advantages :

1. Larger work can be handled as compared to shapers and millers.
2. Capable of taking much heavier cuts as compared to shapers and millers.
3. There are no overhanging parts such as a ram. So, there is no work or tool deflection or distortion.
4. The work is mounted on a table which is supported throughout its entire movement. So, a maximum support is obtained.

Principal parts of a Planer. Fig. 3.37 is the block diagram of a double housing (column) planer. Its principal parts are discussed below :

1. Bed. The bed is a large, heavy casting strengthened with cross ribs to provide a stiff, heavy support for this heavy machine. Very large beds consist of two or more castings, carefully machined and bolted together. The length of the bed is about twice the length of the table. The bed provides the ways for the table to run on. It also houses the drive mechanism for the table.

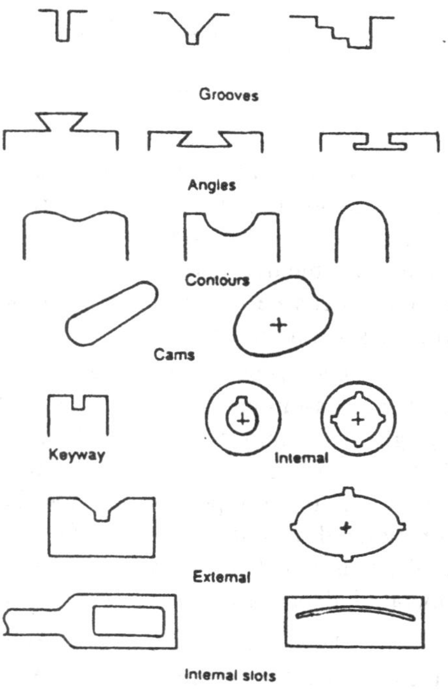

Fig. 3.36. Typical Surfaces Machined on Shapers and Slotters.

2. Table. The table on which the work is clamped, reciprocates along the ways of the bed and has T-slots in its upper surface for clamping the workpiece. The table is a single casting except in the case of two tables bolted together end to end. The table is powered from a variable speed motor through a reducing gear and a rack and pinion arrangement. The other drives can be : hydraulic drive and fast and loose pulley drive.

3. Housings. At the sides of the bed, two vertical housings, columns or uprights are arranged. The front face of each column is machined as a set of ways on which travel the cross rail which carries two tool heads. Most of these planers also carry two side tool heads (one on each housing). The housings enclose the various mechanisms which transmit power to the upper parts of the machine from the main motor drive. This includes : the vertical feed shaft, rail-elevating screw, feed bar for cross feed etc.

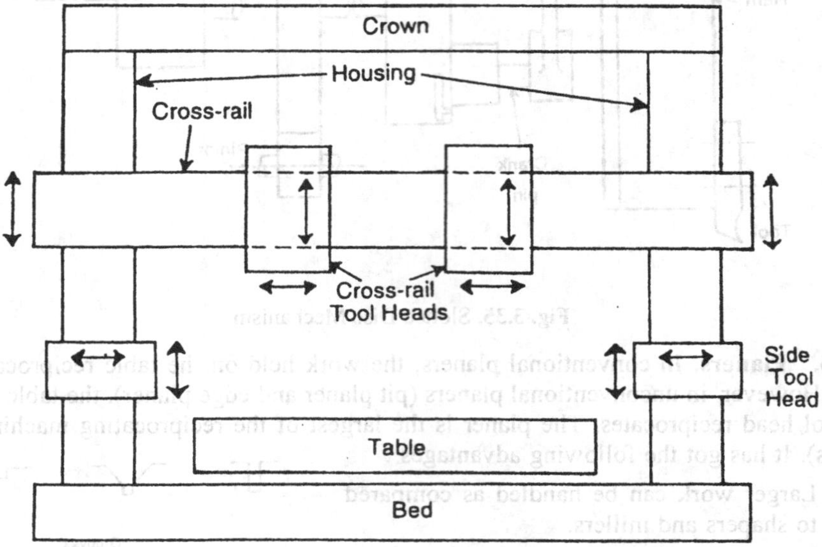

Fig. 3.37. Double House Planer.

4. Cross-rail. The cross-rail is a horizontal member of heavy cross-section connecting the two housings. It contributes to the rigidity of the machine. Its face is machined to provide ways for the two cross-rail tool heads. On many planers, the housings are connected together at the top by a cross member called a crown, to provide additional strength and rigidity to the structure. Cross-rail can be traversed vertically along ways on the housings by means of a drive powered from a separate motor.

5. Tool Heads. There are a maximum of four tool heads on a double housing planer, two on the cross-rail and two on the uprights (side tool heads). They are mounted on the rail and the uprights by means of saddles. The cross-rail tool heads are used for planing horizontal surfaces. Both tool heads have cross feed along the cross rail ways powered by their motors. The tool-slides with the tool posts on the tool heads can be traversed vertically to feed the tools down for a new depth of cut. Vertical surfaces of the workpiece are planed by tools mounted on the side tool heads which travel vertically along the two housings. The four tool heads of the planer operate independently of each other, or these may be controlled together.

A planer tool head can be normally swiveled on its base (saddle), upto about 60° of each side of its neutral position. The plate that swivels carries a slide that has a clapper box on the end for mounting the tool. The clapper allows the tool to swing upward from the workpiece on the re turn stroke as on a shaper. Many planers have air or hydraulic tool lifters (due to heavy tool head block of planers) that automatically swing the clapper box to raise the tool on the return stroke. The clapper box can usually be swivelled about 20° in each direction to place the tool in advantageous position as needed.

Types of Planers

1. Double Housing Planer. The double housing planer discussed above is the most commonly used planer. This design has one drawback. Because of the two housings, one on each side of the bed, it limits the width of the work that can be machined.

2. Open Side Planer. This design has only one column, (Fig. 3.38), which supports a cross-rail, with the cross rail cantilevered from it. Thus the maximum number of tool heads on this planer is 3. Its advantage is that it can accommodate jobs which are too wide to fit between

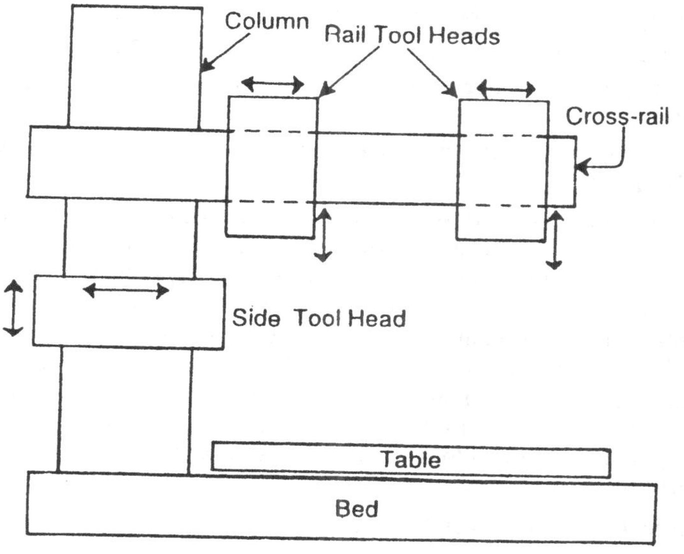

Fig. 3.38. Open Slide Planer.

columns of an equivalent double housing planer. The portion of the hob extending beyond the table side must be supported properly. They are not as strong and rigid as a double housing planer.

3. Planer Miller or Grinder. It is essentially the same as a double housing planer except that it has a milling cutter or a grinding head in place of one or more of the conventional tool heads on the cross rail.

4. Pit type Planer. When the work becomes extremely large and heavy, its reciprocation becomes very difficult. In such cases, the table is kept stationary and the tool reciprocates. Massive rails are mounted on both sides of the table. Tool heads (usually two) are mounted on the cross rails for two way planning. This design (Pit planer) is used to plane the largest kind of work (bases and slabs) which no other machine could handle. A pit planer is one of the largest machine tools. Because of the size of the parts machined on these planers, it is convenient and economical to move the machine than the work.

5. Plate Planer. In plate planer, the job/table is stationary while the tool and operator more bark and forth along the work for feed and cut. It is a special purpose (single purpose) machine tool, designed specifically for squaring or leveling the edges of the heavy steel plates (armour plate for squaring or leveling the edges of the heavy steel plates (armour plate, plate for ships and for vessels upto about 20 cm thick and preparing plates for pipe lines). The plate is clamped to a bed and the side mounted carriage supporting the cutting tool is moved back and forth along the edge.

6. Divided or Latching table Planer. In planers, lot of time is consumed in setting up the work. To reduce it and particularly when many identical pieces are machined at the same time, any type of planer may be provided with a divided table. The work on one table is set up while that on the other is machined. The two sections of the table may be coupled together for long work.

Planer Size. For a double housing planer, the size is given as : width, height and length of the largest workpiece which can be accommodated between the columns, under the cross rail in its highest position and on the length of the table, that is, distance between vertical housings, the height from the top of the table to the rail in its upper most position and the maximum length of the table travel.

Double housing planers have been built upto Size : 4.8 m square by 18 m.

The largest size of a pit planer is about : 12.5 m × 5.4 m × 22.8 m.

The size of an openside planer is given by the distance between table and maximum rail height and by table length.

Planer Drive. The two popular drives for reciprocating the table are : with d.c. reversible motor and the hydraulic drive.

1. Variable speed, reversible d.c. motor Drive. This drive has the advantage that the speed of the motor can be varied according to the field current supplied. The full speed forward to full speed return can be changed so quickly that it is hardly detectable. The common return spoke of a planer is three or four times as fast as the cutting stroke. The power from the motor goes through a clutch and a gear train to a rack mounted on the underside of the table. Adjustable trip dogs at the side of the table determines the length and the position of the stroke.

2. Hydraulic Drive. The hydraulic drive is quite similar to that used for a horizontal shaper. More than one hydraulic cylinder may be used to give a wide range of speeds. The chief drawback of the hydraulic drive on long planers is spasmodic movement of the table due to the compressibility of the hydraulic fluid.

The mechanical drive for cross feed movement of tool heads is similar to that used on shapers.

Product applications. The work customarily machined on a planer include the following: bases and tables of all kinds of machine tools, frames for printing processes, textile machines, locomotive frames and engines, forging hammer die blocks, large fixtures and moulds, rolling mill parts, parts for large hydraulic presses and altering rolled structural parts.

3.2.4. Cutting Speed, Feed and Depth of Cut. The cutting speed, V, in planing, shaping and slotting operations is the speed of the tool (or the work) in the direction of the primary cutting motion during the working stroke when the metal is being cut. The cutting speed may be either cons-tant or variable depending upon the design of the machine tool.

Shapers with crank and slotted arm mechanisms have a variable cutting speed and a variable return speed (Fig. 3.33) of the tool. The speed is minimum at the beginning and end of the stroke and maximum in the middle of the stroke. The cutting speed can be determined by the formula :

$$V = \frac{LN(1+K)}{1000} \text{ m/min.}$$

where L = Length of ram stroke, mm

N = number of full strokes / min,

that is, a full stroke comprises working and return strokes.

K = Ratio of return time to cutting time.

Planers with a rack and pinion drive for the table have a constant cutting speed, which can be determined from the formula :

$$V = \frac{2NL}{1000}, \text{ m/min.}$$

The feed, f, in planing, shaping or slotting (Fig. 3.39) is the relative movement of the tool or work in a direction perpendicular to the primary cutting motion per full stroke and is expressed in mm / stroke.

The depth of cut, d, is the thickness of layer of metal removed in one cut or pass. It is measured in a direction perpendicular to the machined surface.

The cross-sectional area of the undeformed chip is the product of the thickness, t, and the width, b, of the undeformed chip :

Area of undeformed chip $= t \times b = f \times d \text{ mm}^2$.

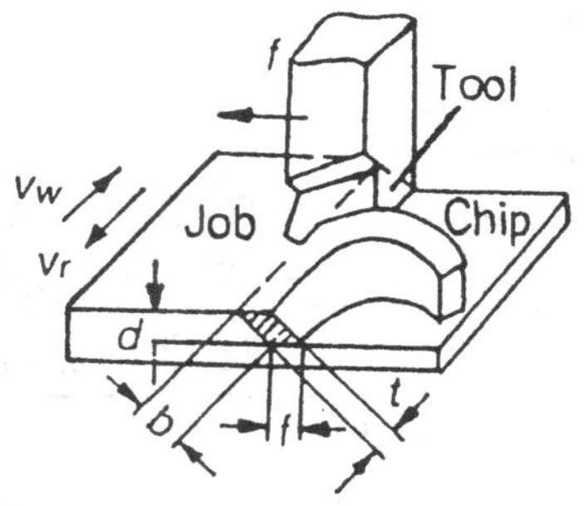

Fig. 3.39. Cutting Action in a Planer.

Cutting speeds in these kinds of machining operations are not usually very high, since difficulties are encountered in designing machine tools with high speed of reciprocating motion due to high inertia forces developed in the motion of the units and components of machine. Thus in planing cast iron, the recommended cutting speed is 15 to 20 m/min for roughing and 4 to 12 m/min for finishing. The corresponding depths of cut are 0.5 to 0.8 mm, and 0.08 mm respectively, with feed ranging from 1.50 to 2.00 mm / stroke.

3.2.5. Planing and Shaping Tools. The tools used in planing and shaping are similar to single- point lathe tools, consisting of the shank and the point. Planing and shaping tools are made of H.S.S. or tipped with cemented carbides. In the latter case, the shanks are made of carbon tool steel. It is a good practice to use planing and shaping tools with an offset, or gooseneck shank to reduce jamming of the tool during operation. When a straight shank tool is used, backward spring of the tool about point O, owing to the cutting forces acting on the tool, leads to gouging of the

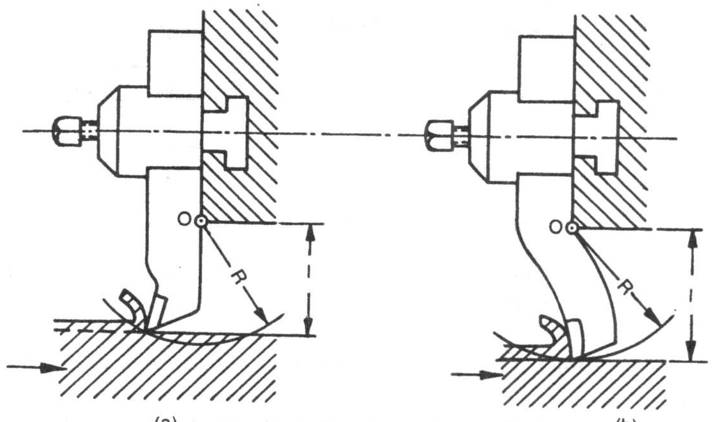

(a) **Fig. 3.40.** Shanks of Planing Tools. (b)

tool nose into the work, [Fig. 3.40 (*a*)]. The tendency for the tool to dig into the work will be less for a gooseneck shank, and will be completely eliminated if the tool nose is lined up with the base of the tool shank [Fig. 3.40 (*b*)] since the radius R along with the tool nose springs is equal to the overhang L.

3.2.6. Work-holding Devices for Shapers, Slotters and Planers. The commonly used work holding devices on these machine tools are given below :

1. T-Bolts and Clamps. The tables of machine tools such as shapers, slotters, planers, drilling machines and milling machines are provided with T-slots. Most jobs can be held on the machine table by means of T-bolts, blocks and clamp plates, Fig. 3.41 (*a, b, c*). A simple block or a step block (to accommodate jobs of different heights) can be used as the fulcrum block. For

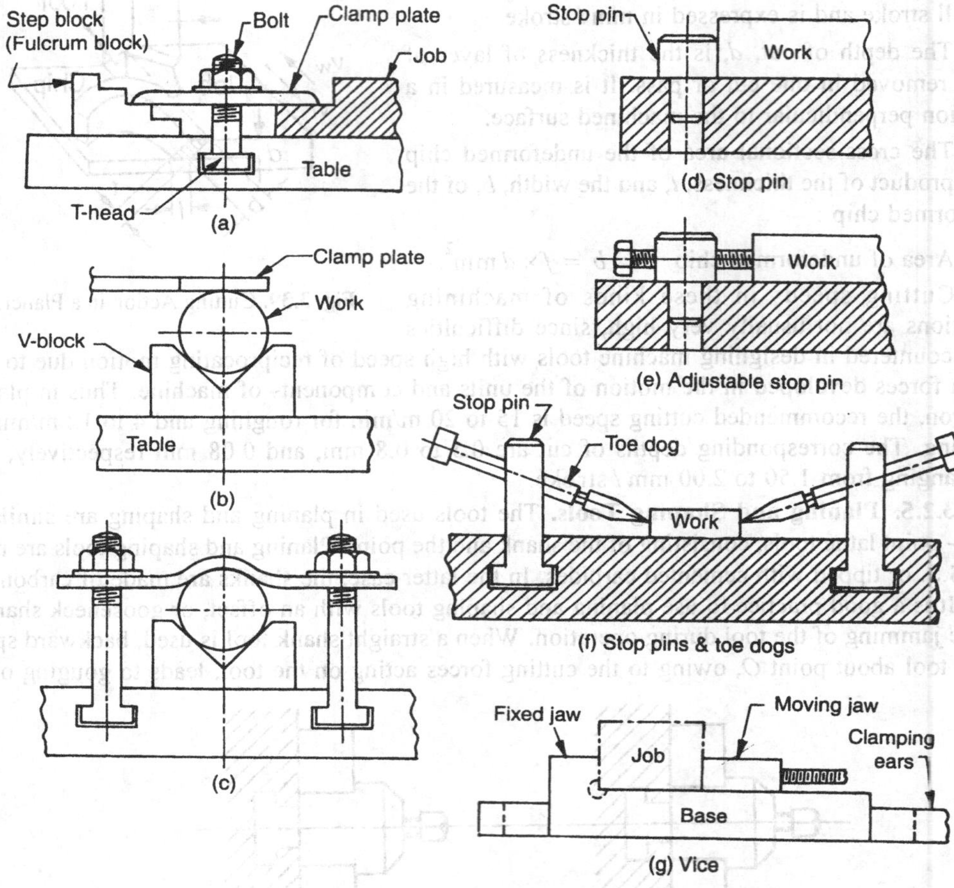

Fig. 3.41. Work-Holding Devices.

round jobs, V-blocks are used for the proper location of the jobs, Fig. 3.41 (*b, c*). T-bolt should be nearer to the job than to the fulcrum.

2. Stop pins. Fig. 3.41 (*d, e*).

3. Stop pins and Toe-dogs. Fig 3.41 (*f*). This arrangement is used for holding thin jobs (particularly on planers).

4. Vises. A vise is a common work holding device. It has two hardened jaws, one fixed and the other movable with the help of a handle. The job is held between the two jaws. The base of the vise carries slots or clamping ears to fix the vise to the table of the machine tool, with the jaws either parallel or perpendicular to T-slots. Special jaws can be attached (with the help of set screws) to the plain jaws of the vise to suit the different contours of jobs. For round jobs, Vees may also be fitted instead of standard plain jaws.

The 'Swivel vise' is made in two parts. It consists of a round swivel base over which the vise body is mounted. The base is graduated into 360°. The body is clamped to the base by means of clamping bolts. The body can be adjusted at any angle relative to the base and so the jaws can be set at any angle. If the vise, apart being swivelled in the horizontal plane, can also be tilted in the vertical plane, it is called 'Universal Vise'.

5. Fixtures. If a job cannot be held in the general purpose work holding devices mentioned above, then a special purpose work holding device, known as fixture, will have to be designed and fabricated.

PROBLEMS

1. What is a Shaper used for ?
2. Draw the block diagram of a horizontal Shaper and write about its important parts.
3. How the jobs are held on the table of a horizontal Shaper ?
4. Sketch the Tool Head of a Shaper and write how it functions.
5. What is the difference between a Push cut Shaper and a Pull cut Shaper ?
6. What is the difference between Plain Shapers and Universal Shapers ?
7. What is the value of having a universal table on a shaper ?
8. Sketch and explain the mechanical drive for the ram of a horizontal shaper.
9. Sketch and explain the mechanical feed drive of a horizontal shaper.
10. Sketch and explain the hydraulic drive of a horizontal shaper.
11. Write the advantage of hydraulic drive over mechanical drive.
12. How will you give the size of a horizontal shaper ?
13. Write the applications of shapers.
14. What is the difference between a vertical shaper and a slotter ?
15. Write the uses of a slotter.
16. Sketch and explain the working of a slotter.
17. Write about the various types of slotters.
18. Write about the various ram drive mechanisms of a slotter.
19. How will you give the size of a vertical shaper or a slotting machine ?
20. Describe the difference in operation between a planer and a shaper.
21. Write the advantages of planers.
22. Sketch and write about the principal parts of a double housing planer.
23. Write about the various types of planers.
24. How will you give the size of a planer ?
25. Write about the various drives for planers.
26. Write the product applications of planers.
27. List the advantages of Shapers.
28. List the Limitations of Shapers, Planers and Slotters.
29. Why the shanks of cutting tools used on shapers, and Planers have gooseneck ?
30. Discuss the common work-holding devices used on shapers, slotters and planers.

3.3. DRILLING MACHINES, REAMERS AND TAP TOOLS

3.3.1. General. Hole machining, constituting a considerable percentage of all machining operations, is accomplished as a rule, be rotary and axial feed motions of the cutting tool or work. Round holes are machined in drilling machines, boring machines, engine lathes, vertical boring mills, turret lathes, semi- automatic and automatic lathes, broaching machines and grinders. Whereas, drilling machines and machine tools of lathe type are used to originate or cut a hole where none previously existed, the boring machines, broaching machines and grinders are mainly used to enlarge and finish the existing holes (produced by machining, casting, forging or punching). Of course, holes can also be enlarged and finished on drilling machines and lathe type machine tools.

Holes upto 80 mm in diameter can be machined from the solid on drilling machines and machine tools of lathe type, holes over 80 mm in diameter are usually cut by special drilling heads and, as a rule, this operation is performed in boring machines.

Depending upon their purposes, holes should meet the following requirements :

1. The diameter size must be held within given limits.
2. The straightness of hole axis or its cylindrical surface must be as specified.
3. The hole must be of true cylindrical shape, that is, absence of taper, ovality and lobed-form.
4. The hole must be square to the faces of the part.

The machining of accurate holes always requires more machine time and more expenditure on cutting tools than the machining of an external surface of revolution of the same size and to the same grade of accuracy. The reason is that the hole making tools do not possess the same rigidity, especially as to the design of their mounting elements and clamping facilities, as turning tools. To attain the required accuracy in hole making, an increased number of passes is resorted to. This enables the error of the initial machining operation to be gradually reduced to the permissible limits.

3.3.2. Drilling process. The drilling process is an extensively used machining operation by which through or blind holes are cut or originated in a workpiece. The drilling tool is called a "drill" which is a multi-point cutting tool. The hole is produced by axially feeding the rotating drill into the workpiece which is held on the table of the drilling machine.

3.3.3. Drilling tools. The most widely employed drilling tool is the twist drill with two cutting edges or lips, (Fig. 3.42). A twist drill consists of the following main parts : the point, the body, the neck, and the shank. The point comprises the cutting elements while the body guides the drill in operation. The body of the twist drill has two helical grooves called "flutes" cut into its surface. The flutes form the cutting surface and also assists in removing the chips out of the drilled hole and making the coolant to get to the cutting edges. The drill is held and rotated by its shank. The cutting edges are straight and are separated by web thickness of the drill. To help provide adequate strength and rigidity to the drill, the web becomes progressively thicker towards the shank.

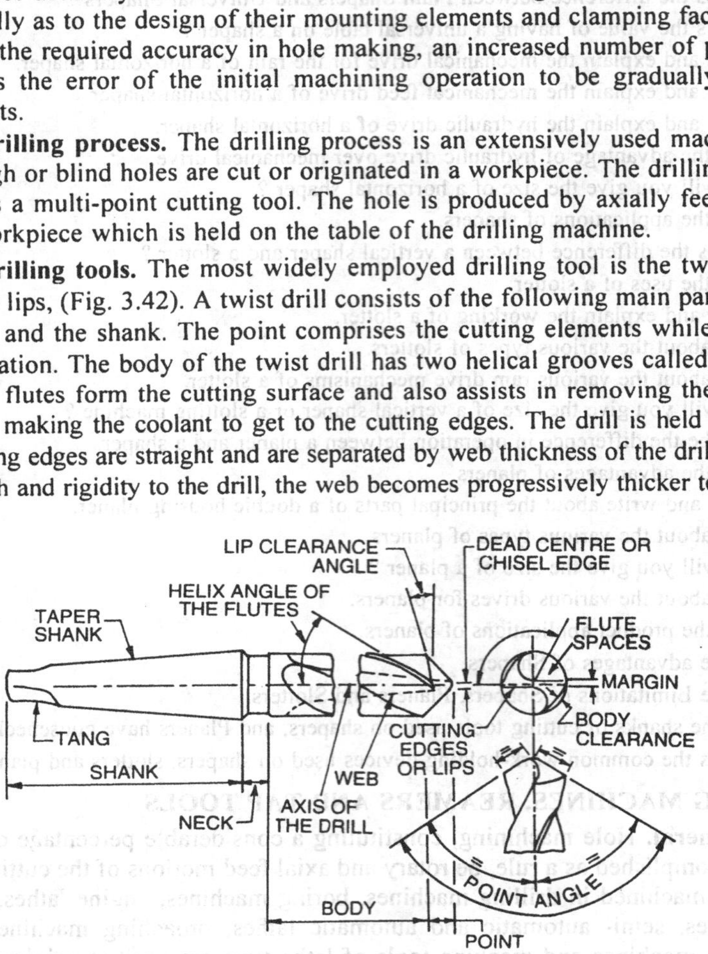

Fig. 3.42. A Twist Drill.

The body of the drill is made slightly less in diameter leaving a narrow "margin" at full nominal diameter along the edge of each flute. This reduces the rubbing action between the drill and the hole wall and allows the cutting fluid to reach the point of the drill. The two margins facilitate the removal of heat generated in drilling and properly guide and locate the drill. To help further in reducing the rubbing action, drill bodies are given a slight back taper (about 0.0075 mm per cm of length).

Lip relief (Clearance) angle. The heel of the drill point is backed off when ground to give relief behind the cutting lips. This will allow the cutting edges to cut without interference. This is equivalent to end relief angle of a single point cutting tool. It is kept 12° to 15°.

Point angle. The point angle is selected to suit the hardness and brittleness of the material being drilled. It is 116° to 118° for medium hard steel and cast iron, 125° for hardened steel and 130° to 140° for brass and bronze. It is only 60° for wood and fibre. This angle (half) refers to side cutting edge angle of a single point tool.

Helix angle. This angle is equivalent to back rake angle of a single point cutting tool. It is 24° to 30° for most drills.

Dead centre. The intersection of the two ground surfaces of the point with the web at the centre of the drill, produces a sharp raised line, known as the "dead centre" or "chisel edge.

The angle which the raised line at the dead centre makes with the cutting edges is called the "chisel-edge angle". This angle is an indication of the clearance on the cutting edge near the chisel edge. This angle generally varies from 120° to 135°.

The chisel edge does not do any cutting action. It only plastically displaces the material to be subsequently removed by the cutting edges. The chisel edge tends to wander or walk over the surface when starting a hole. The drill must be kept in place by creating an indentation on the job where the hole is to start, with a centre punch or a centre drill. The drill can also be kept in place by using drill bushings to guide it.

Materials. Carbon tool steels drills have a low first cost, but these should be used occasionally and at slow speeds. High speed steel drills are the most popular and have good strength. Drills tipped with cemented carbide are economical for high production but are expensive and must be handled carefully to avoid breakage. These are used mainly for drilling of malleable iron castings and nonferrous metals and alloys such as copper, brass, aluminium, Magnesium, zinc and also plastics, hard rubber etc. These are not used for steel components as there is likelihood of their breakage due to high tip pressure.

Drill Sizes. Standard drills are available in four size series, the size indicating the diameter of the drill body :-

(*i*) **Fractional Size.** Size range is $\frac{1}{64''}$ and $\frac{1}{4''}$ with increments of $\frac{1}{64''}$.

(*ii*) **Millimeter Size.** Size range is 0.5 to 10 mm with increments 0.1 mm.

(*iii*) **Numbered Size** (80 to 10). Size range is 0.0135″ to 0.228″ with very slight increments.

(*iv*) **Lettered Size (A to Z).** Size range is 0.234″ to 0.413″. with very slight increments.

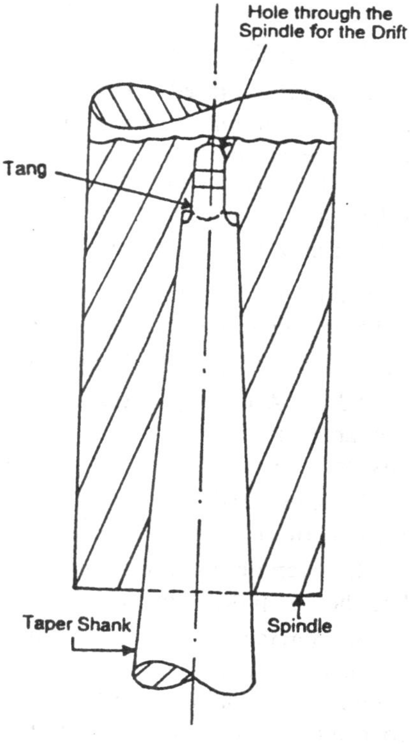

Fig. 3.43. Holding of a Taper Shank Drill.

How Drills are held. As written above, a drill is held and rotated by its shank. The drill shank can be tapered or straight. The straight or cylindrical shank is provided for smaller sized drills (upto 12.7 mm diameter). Taper shank drills are inserted directly into the taper hole of the drilling spindle (the taper is standard Morse taper, 1 : 20). To ensure against any possible slippage, the tang at the end of the taper shank engages a slot at the end of the tapered hole, Fig. 3.43. When the sizes of the tool shank and spindle hole tapers do not coincide, adapting devices such as tapered sockets or sleeves are employed. To remove the drill from the spindle hole, a wedge shaped drift which fits the hole through the spindle is used. Straight shank drills are held in drill chucks and clamped either by hand or by means of a chuck key, depending upon the design of the chuck. These chucks have taper shanks that enable them to be held in the tapered hole of the drilling machine spindle.

Advantages of two cutting edges Twist Drill

1. Two cutting edges are more efficient.
2. Cutting forces are balanced.
3. Helical flutes allow access of cutting fluid and help to dispose off the chips.
4. The small margins left on the cylindrical surface provide guidance.

Core Drills. The body of a core drill has three or four helical flutes. These drills can not start holes but are used to enlarge holes previously drilled or for enlarging punched, cast or forged holes. The helix angle varies from 10° to 30°. These drills have got the following advantages over a standard twist drill :

1. The three or four margins (locating and guiding elements of the drill) ensure higher machining accuracy by preventing deviation of the tool to one side in operation.
2. More cutting edges allow greater feed per revolution resulting in greater productivity.

These drills are sometimes called spiral reamers or core reamers.

Straight Fluted Drills. These drills are commonly used for drilling soft metals like brass or copper. The standard twist drill shows a tendency to grate or dig in the soft metal. A straight or flat drill consists of the head, the neck and shank, (Fig. 3.44). The head may have either parallel sides or sides with a certain amount of back taper. A back taper of 2° to 3° may be provided to reduce the friction in drilling. The drills with parallel sides are more accurate in operation and have a higher tool life. The point angle of a flat drill ranges from 90° to 120°.

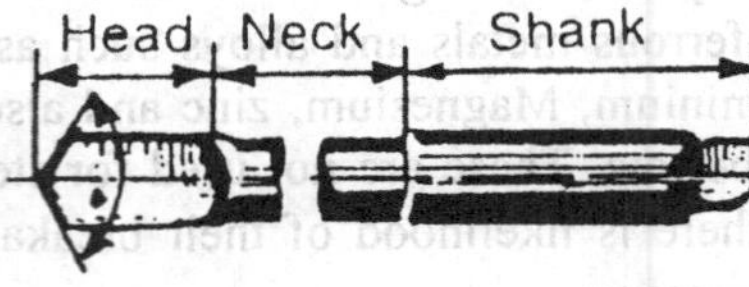

Fig. 3.44. Straight Fluted Drill.

3.3.4. Allied Operations. Holes drilled from the solid with one drill are further machined to : enlarge and finish the holes, produce seats for bolt head, nut, head of a screw or rivet and for a washer. These allied operations are shown in Fig. 3.45 and defined below :

1. **Core drilling.** Core drilling of holes is done with a core drill to enlarge and improve the geometric shape of the previously drilled hole.
2. **Step drilling.** It is the operation of cutting a hole of two or more diameters by one drill called the "Step drill" or "combination drill".
3. **Boring.** Boring is the operation of enlarging and finishing the complete hole with a single point cutting tool. This operation is normally done on boring machines but can also be accomplished on lathe group of machine tools. Boring corrects the position of the hole and makes it concentric with the axis of rotation of the spindle.
4. **Reaming.** Reaming usually follows drilling or core drilling and its purpose is to

remove all coarse traces of previous machining operations with the help of a multipoint cutting tool called "reamer". A reamed hole is of accurate size and good surface finish. Reaming can not correct the position of the hole since reamer merely follows the previously drilled hole.

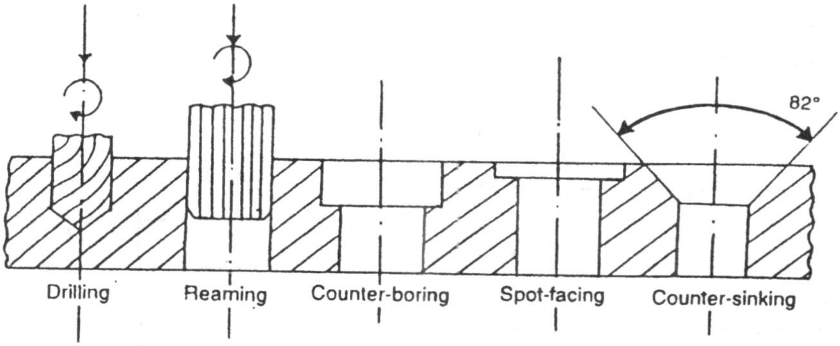

Fig. 3.45. Drilling and Allied Operations.

5. Counter boring. Consists in enlarging a portion of an existing hole to a larger diameter and making the surface at the bottom of the larger diameter flat and square. This is done to accommodate a screw head, bolt head or nut below the upper surface of the hole. The operation is accomplished by a special tool called a "Counterbore" or with a regular boring tool.

6. Counter sinking. is done after drilling a hole to chamfer the entrance to a hole or to make a conical recess or a seat for a flat headed screw or rivet so that the heads are flush or below the main surface. Standard countersinks are 60, 82, or 90° included angle. The tool, a countersink, is a short tool similar to a reamer.

7. Spot facing. is the machining of a boss or pad at the face of a hole. It is normally done to produce a square bearing surface for a washer and nut or the head of a bolt. Both spot facers and counterborers are pilot guided.

3.3.5. Drilling Machines. Drilling machines are classified into : general-purpose, single-purpose and specialised drilling machines.

General purpose Machines. Any hole making operation can be performed in the general purpose machines. This class includes : bench type drill press, Single- and multiple-spindle upright drill presses, radial drilling machines and others.

Single purpose Machines. This class includes : Centre-drilling and deep hole drilling machines (horizontal and vertical types) ; semi- automatic and automatic drilling machines.

Specialised Machines. Special unit-built multiple-spindle drilling machines, intended for performing a single definite operation, are used in mass production.

1. General purpose Drilling Machines. The block diagram of a single-spindle general purpose machine (drill press) is shown in Fig. 3.46. Its main parts are : the base, column, table and head. The upright column is supported by a heavy base. The table is suspended from the column and may be moved up and down and clamped at the desired height. The drill head is mounted on the top of the column and consists of the main operating parts. The cutting tools (held in spindle) are power driven from the motor by a pair of cone pulleys-V belt drive or through a speed gear box to obtain various speeds. The axial feed of the tool is controlled by hand (on small machines) but larger machines provide power feed for the drill spindle. The spindle has a keyway or spline, so that it may be moved up and down along its axis for feeding, but, still retains its drive at any point, (Fig. 3.47). The rack pinion is rotated through a worm gearing (in power feeding) and hand feed lever (in manual feeding). The pinion meshes with a rack secured to a non-revolving sleeve or quill in which the spindle freely rotates. The spindle is fed in the axial (vertical) direction together with the sleeve. Tables and often bases, are provided with T slots for clamping work or workholding devices.

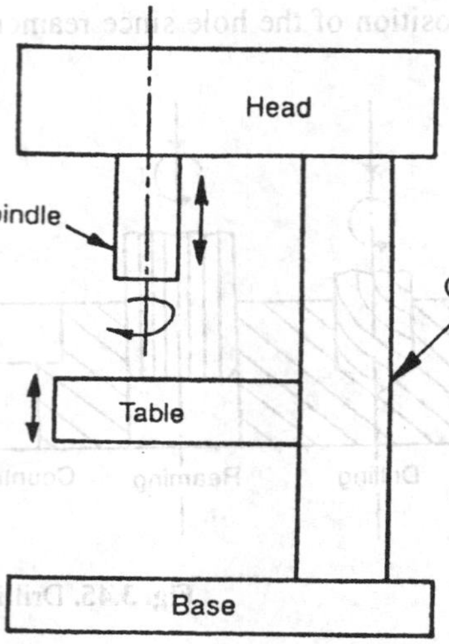

Fig. 3.46. Block Diagram of a Drill Press.

Size of Drill Press. The size of a drill press is expressed in one or more of the following ways :

(*i*) By the "swing" which is twice the distance from the nearest face of the column to the centre of the spindle. This indirectly refers to the diameter of the largest disc that can be drilled through the centre.

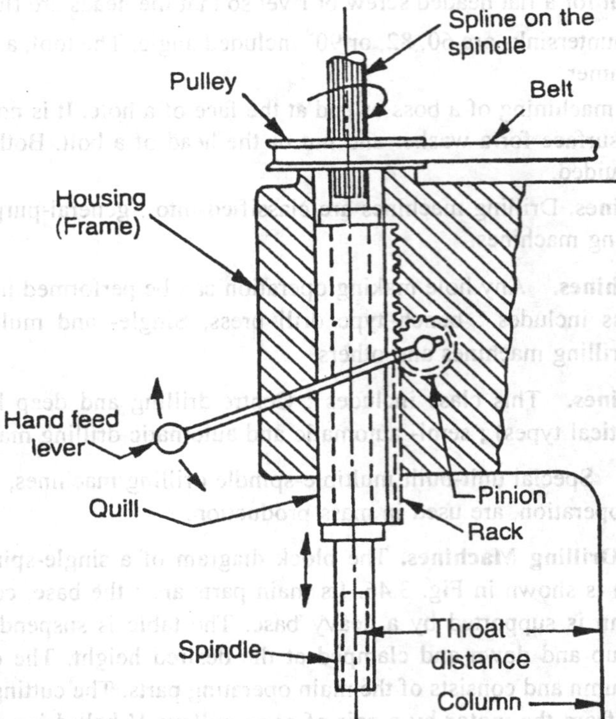

Fig. 3.47. Feeding Arrangement.

(*ii*) By the maximum diameter of the drill that can be used on steel.

(*iii*) By the maximum distance between the spindle and the table, that is, the maximum height of the job that can be accommo- dated with the table in its lowest position.

(*iv*) By the distance the spindle moves up and down, that is, by the length of the spindle feed.

Types of Drill Presses

(*a*) **Bench type.** This is the simplest type used in industry. They are also called "sensitive" drill presses because the feed lever is operated by hand. This allows the operator to "feel" how the drill is cutting. If the drill gets into trouble, the pressure on it can be released. This design is used for light work using small drills (upto about 9.5 mm to 12.5 mm diameter drills). The controls are light and delicate speeds from 800 to 9000 rev/min are a typical range.

(*b*) **Upright Drill Presses.** These are medium heavy duty machines and stand on floor. This machine usually has a gear- driven mechanism for different spindle speeds and an automatic or power feed device. This machine will take larger drills and cutting tools than the standard bench type machines. Single- spindle upright drilling machines can drill holes upto 75 mm in diameter and upto 350 mm deep. These machines have a wide range of spindle speeds and feeds and are employed, therefore, not only for drilling from the solid, but also for core drilling, reaming and tapping operations.

There are two constructions of the column :

(*i*) **Box - Column arrangement (Upright Machine).** The column is of box section. The box column is bolted to the base. The work table is incorporated with a bracket which slides on ways at the front of the machine. Support and elevating movement of the table is provided by a telescopic screw underneath its centre. The table on these machines can be swung.

(*ii*) **Round - Column type (Pillar Machine).** Here, the table instead of being carried on vertical slides is carried on the round pillar. This helps in swinging the table to one side so that tall work pieces can be mounted on the base. Here, the table support is less rigid than the upright machine and places a restriction on its width. A box-column machine being more rigid than a round column machine, is adopted for heavier work.

(*c*) **Multiple - spindle Drilling Press.** These machines also called as "cluster drilling machines" are mass production machines with the several spindles (from 4 to 48 or more) driven by a single power head and fed simultaneously into the work. These machines are very useful when a number of parallel holes must be drilled in a part. These machines are subdivided into two group (according to the design of their spindles) : those with nonadjustable (fixed position) spindles and those with adjustable spindles.

(*i*) **Non - adjustable Spindle Machine.** (Fig. 3.48). The main drive spindle is rotated by an electric motor through a change- speed gearbox. From the main drive spindle, the motion is transmitted to the drilling spindles, through appropriate gearing. The multiple-spindle head is moved up and down by the feed drive which has facilities for rapid traverse and feed motions. The worktable can be raised or lowered along the ways on the column. The drilling spindles have a fixed position in the head designed for drilling the holes in a definite work piece. Their arrangement may be either symmetrical or asymmetrical, as the case may be. The spindle head can be replaced by another head designed for some other workpiece or to perform some other operations. Different spindles in such a head may run at different speeds but they all have the same feed.

(*ii*) **Adjustable Spindle Machine.** The drilling head of a multiple- spindle machine with adjustable spindles is shown in Fig. 3.49. Here, the main drive spindle travels up and down the column ways with the drilling head, being actuated by the feed drive. The drill spindles are driven from the main drive spindle through gears Z_1 and Z_2 and through the universal joints and telescopic shaft. This arrangement enables the offset spindles to be driven. The drill spindles are mounted in a plate which has slots in various directions along which the spindles can be adjusted. Consequently, drill spindles can be positioned in the plate as required for a certain workpiece. They can be readjusted to other positions for drilling some other work piece.

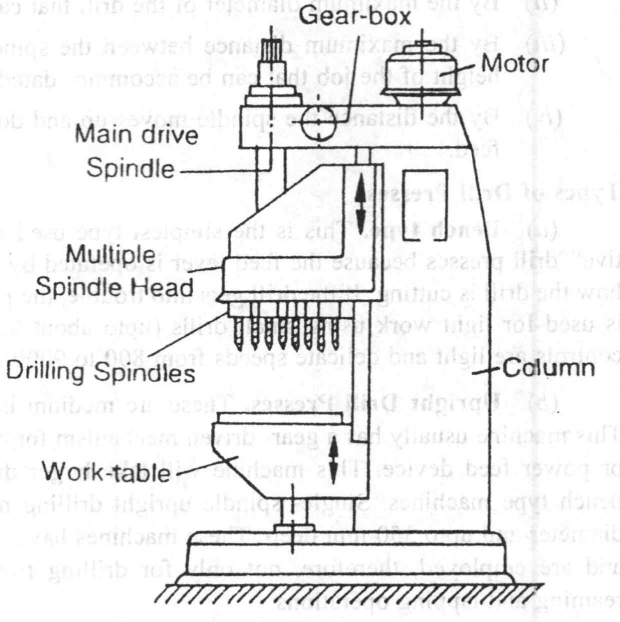

Fig. 3.48. Non-Adjustable Multi-Spindle Machine.

(*d*) **Gang Drilling Press.** A gang drilling press is equivalent to 2, 3, or 4 or more (upto 8 or 10) upright drill presses in a row with a single common long table or base. This machine is very convenient because of its compactness and is used in mass production. The machine can be set up so that work can be transferred from spindle to spindle to undergo two or more operations which may be similar (drilling holes of different diameters) or different (drilling, reaming, counter- boring etc.). In another set up, the same operations may be performed at all spindles. With automatic feed control, two or more operations may be going on *simultaneously* attended only by one operator.

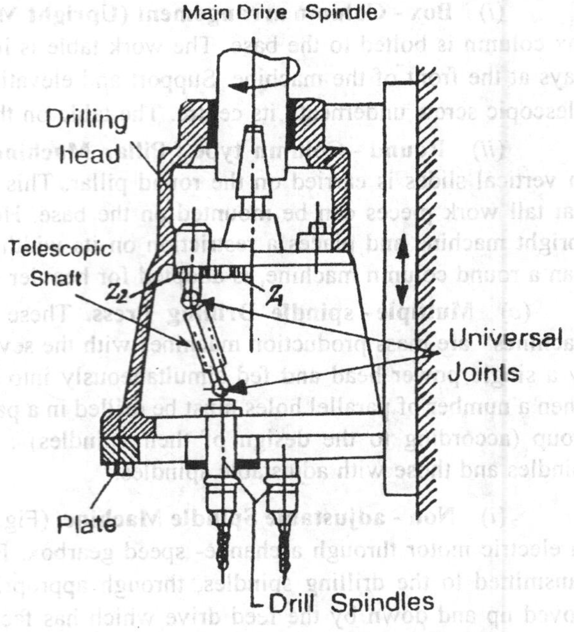

Fig. 3.49. Adjustable Multi-Spindle Machine.

(*e*) **Turret Drilling Press.** A turret drilling press, (Fig. 3.50), over- comes the floor space restriction caused by a gang drill press. Numerical control is also available. Two fixtures can be located on the worktable. This enables loading and unloading of one part while the other part is being machined. This reduces the cycle time.

(*f*) **Radial Drilling Machine.** These machines, (Fig. 3.51), are used for drilling holes at different locations on heavy and bulky workpieces which are either inconvenient or impossible to mount on the table of an upright drilling machine and can not be moved easily. A large arm extends out from the column. It can be raised or lowered along the column with the help of an elevating screw and it also swings in a complete circle around the column. The drilling head moves back and forth on this arm. On most radial drilling machines, the movements of the arm, the drill head, and the spindle are controlled by power feeds. Due to the back and forth movement of the drilling head along the arm and the swinging motion of the arm around the column, holes can be drilled in many different locations without moving the workpiece.

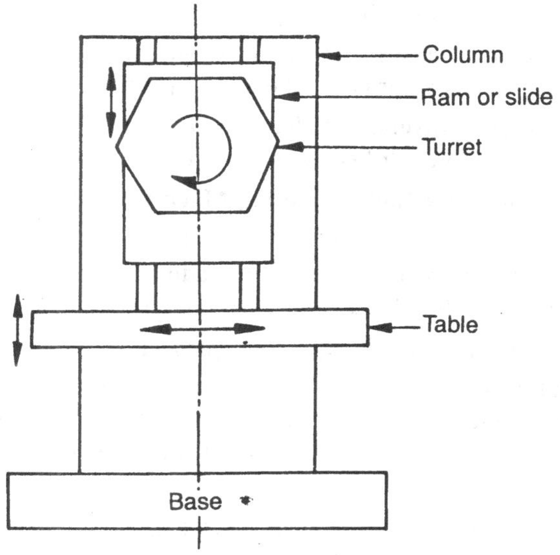

Fig. 3.50. Turret Drilling Press.

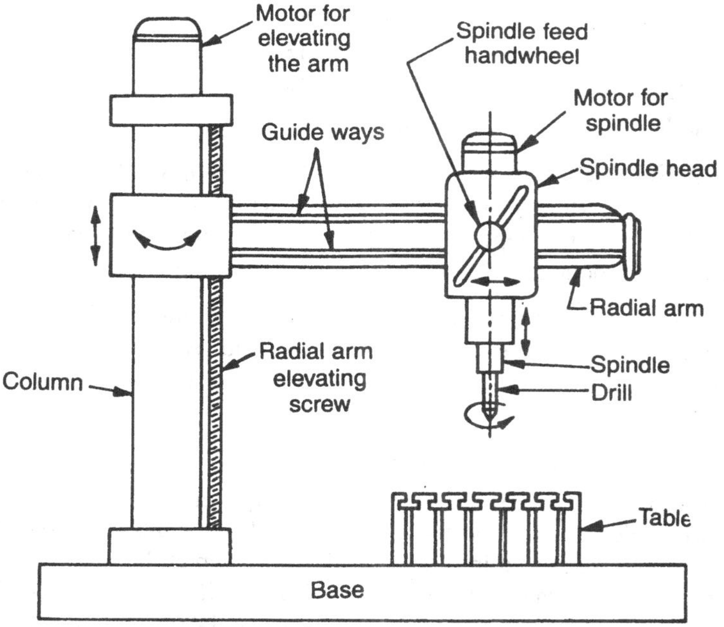

Fig. 3.51. Radial Drilling Machine.

A radial drilling machine is available in three designs :

(*i*) **Plain.** This design has only vertical spindle motion.

(*ii*) **Semi-universal.** In this machine, the spindle head can be swung about a horizontal axis normal to the arm to permit the drilling of holes at an angle in a vertical plane.

(*iii*) **Universal.** Radial arm can also be rotated about a horizontal axis. This enables holes to be drilled at any desired angle.

Size of a Radial Drilling Machine. The size of a radial drilling machine is given as :

1. Diameter of column in cm;
2. In metres which is the radius of the largest disk in which a centre hole can be drilled with the drilling head at its outermost position on the arm.

3.3.6. Cutting Speeds and Feeds. As already discussed, the cutting speed depends upon: the type of material being drilled, cutting tool material, the quality of hole desired, the efficient use of cutting fluid, the way in which the work is set up or held and the size and type of drilling machines. The suggested cutting speeds for drilling process on drill presses are given in Table 3.1.

Feed of the drill is the axial distance the drill advances into the work piece for each complete revolution of the drill and is given in mm/rev. The correct feeds for different sizes of drill are given below :

Table 3.1

Drill size, mm	feed, mm/rev
3.2 and less	0.025 - 0.050
3.2 to 6.4	0.050 - 0.10
6.4 to 12.7	0.10 - 0.18
12.7 to 25.4	0.18 - 0.38
25.4 and large	0.38 - 0.64

3.3.7. Work Holding Devices. Workpieces must be set up correctly and held securely for drilling. If the workpiece springs or moves, the drill usually breaks. The type of work holding device used on drilling machines depends upon the shape and size of the workpiece, the required accuracy and the rate of production. Some of the work holding devices are listed below :

1. **Machine Vise.** A machine vise is widely used for holding small work of regular shape such as flat, square or round pieces. The vise may be held by hand or clamped to the table, depending upon the size of the work. Many vises have open slots or ears on the sides to receive T-bolts see Fig. 3.41.

2. **V - blocks.** are used for holding round workpieces.

3. **Strap clamps and T-bolts.** are used to clamp large and oddshaped work pieces directly to the table. The work piece is first located on the table. The T-bolts are inserted in the T- slots of the table. Strap clamps are placed in position and blocked up at the outer end with step blocks or pieces of scrap metal. This makes the clamp level or parallel with the top of the workpiece. The workpiece is then clamped by tightening the nut of the T-bolt. The T-bolts must always be as close to the work as possible. The step blocks or scrap pieces that support the end of the clamp must be as far from the T bolt as possible, see Fig. 3.41.

4. A three jaw chuck may be fastened to the table and used for drilling the ends of round or hexagonal stock.

5. **Drilling jigs.** These are special workholding devices designed to hold *a particular workpiece* and to guide the cutting tools. *Jigs enable drilling to be done without previously laying out the workpiece.* These are used for duplicating the parts in *mass production.*

3.3.8. Deep Hole Drilling. Holes with a length to diameter ratio over 5 are called deep holes, for example, holes through machine tool spindles, hollow shafts etc. In drilling deep holes with the use of twist drills in ordinary drill presses, it is impossible to obtain accurately straight holes. The straightness of the hole depends upon the rigidity of the twist drill and the guiding action of the margins, adjacent to the flutes, which slide on the surface of the drilled portion of the hole. If the drill lips are of unequal length after sharpening, or if they become dull unequally, the drill will begin to run off the centre of the hole. The amount of drill run off also depends upon drill operation at the initial stage of drilling when only the chisel edge, square to the drill axis, cuts the metal. Other factors, such as large elastic deformation of the drill in operation (buckling), excess play in the spindle bearings, and nonuniform adhering of chips on the lips of the drill may lead to run off. Other difficulties encountered with deep hole drilling are the supply of coolant to the cutting edge and removal of chips.

To avoid drill run off and enable straight holes to be obtained in deep-hole drilling, the following measures are taken :

(1) Holes are drilled with low rates of feed ; the drill is carefully sharpened so that both lips are inclined at the same angle ; excess wear of the drill or adherence of chips to the lips is avoided ; and proper and ample cutting fluid is provided.

(2) The hole is first spotted with a short starting twist drill of large diameter and with a point angle of 90°. This is especially bene- ficial in drilling holes with small- diameter drills in turret lathes and automatics.

(3) Twist drills should be guided by a jig bushing in drilling holes with a comparatively low length-to-diameter ratio.

(4) It is better to drill deep holes with workpiece, instead of drill, rotation. In this case, the drill seems to centre itself, and the tendency to run off centre is substantially reduced.

If the workpiece rotates in the deep-hole drilling, the drill may be either stationary or rotate in the opposite direction to the workpiece. Drill rotation, in this case, is a means for increasing the cutting speed. Hole drilling with rotation of the workpiece is used in engine and turret lathes as well as in deep-hole-drilling machines (for drilling the bores of gun barrels, machine tool spindles, hollow shafts, etc.) which are most frequently of the horizontal type.

Special Deep Hole Drills. Where length-to-diameter ratio of the hole is 10 or more, the following special types of drills are used :

Half round drills, gun drills and trepanning drills. These special drills require the provision of a sufficiently reliable guidance for axial motion of the drill along the surface of the portion of the hole that has already been drilled. Such drills may have a single or several cutting edges (lips):

1. **Half-round Drills.** This is the simplest type of deep-hole drill, (Fig. 3.52 (*a*)). These drills have a single lip and a long bearing surface of the hole over an arc of an angle slightly larger than 180°. To reduce friction between the bearing surface and the hole walls and possible jamming of the drill, the head of the drill has a back taper (0.03 to 0.05 mm per 100 mm of length), the diameter being smaller towards the shank.

To start the drill with proper guidance at the beginning of the operation, it is necessary first to accurately bore the end of the hole to the diameter of the drill over a depth at least one half of the diameter. Half round and similar drills can be used either to drill holes from the solid or to enlarge holes previously drilled by twist drills of smaller diameters.

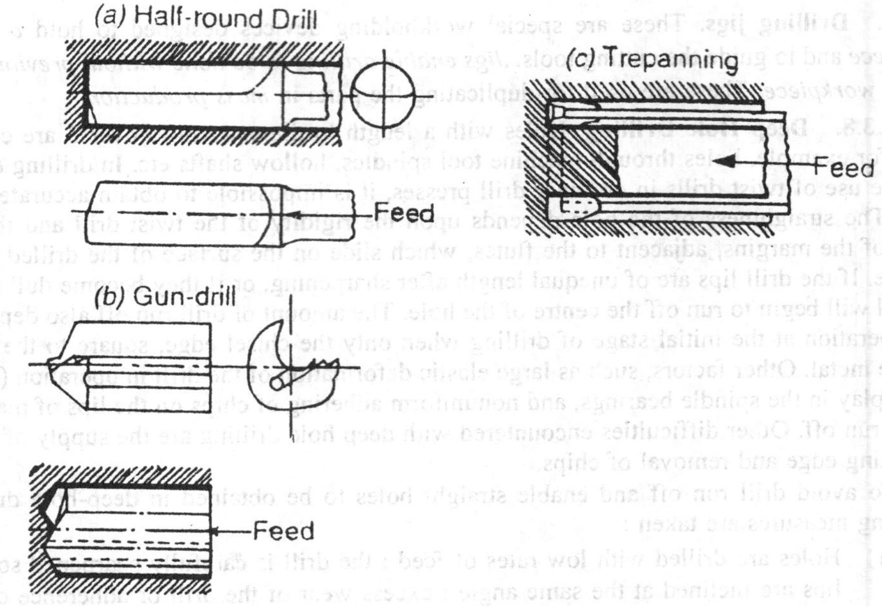

Fig. 3.52. Deep Hole Drills.

2. Gun Drill. A more efficient type of deep-hole drill is the gun drill, (Fig. 3.52 (*b*)). The bearing surface of these drills provides better guidance for the cutting element because it extends over an arc of the hole circumference of an angle from 250° to 260°. The bodies of these drills, like the half round design, have a back taper, the largest diameter being at the point. Flats are provided on the guiding member to reduce friction in the hole and improve cooling. The lip (cutting edge) of these drills is often designed in the form of a zig zag line to break up the chip.

Gun drilling was originally used to produced long, straight holes in gun barrels ; but the technique is now widely used for producing all sorts of true, straight holes. A hydraulically operated head withdraws the drill each time it drills a depth about equal to its diameter ; this clears chips and assures penetration of coolant. Gun drills can drill holes from the solid in addition to enlarging previously drilled holes.

3. Trepanning Drill. The most efficient design of special drill for large-diameter deep hole drilling is the "trepanning drill", Fig. 3.52 (*c*). This drill cuts an annular recess through the hole, leaving a core that is removed at the end of the operation in the form of a cylindrical rod. These drills consist of a head, in which cutting blades or bits are secured to cut the annular recess from the solid, and a hollow drive tube, joined to the head by thread. Such drills are used in deep-hole-drilling machines for making large hollow shafts, long machine tool spindles, etc.

The rates of feed used in deep-hole drilling range from 0.02 to 0.5 mm per revolution, depending upon the hole diameter, workpiece material, required finish and specified accuracy of the hole. Speeds average 75 to 105 m/min for carbide-tipped drills in steel. In most designs of special deep-hole drills, the coolant is delivered at high pressure to the cutting edge through the clearance between the periphery of the hole and the shank and the drive tube. The coolant flushes the broken-up chips out through a hole in the centre of the drill body, shank and drive tube.

As already mentioned, the deep hole drilling machines are most frequently of the horizontal type. The workpiece is rotated by a chuck a~d supported (stabilised) by steady rests, while the drill is fed axially. This procedure reduces the amount by which the drill departs from the hole axis in drilling.

3.3.9. Specialised Drilling Machines. In mass production, special unit-built multiple-spindle drilling machines are employed for drilling housing-type and other parts. These machines are assembled from standard units and mechanisms. Such machines perform drilling, reaming, tapping and other operations that are commonly done in drill presses. In many cases, such multiple-spindle machines are built into an automatic transfer line. As a rule, the only special units of these machines are the spindle head and the jig or fixture for holding the workpiece. To changeover to the production of a new part, it is usually necessary to design, manufacture and install these two special units only. Various arrangements of the standard units to obtain different designs of unit-built machines are illustrated in Fig. 3.53.

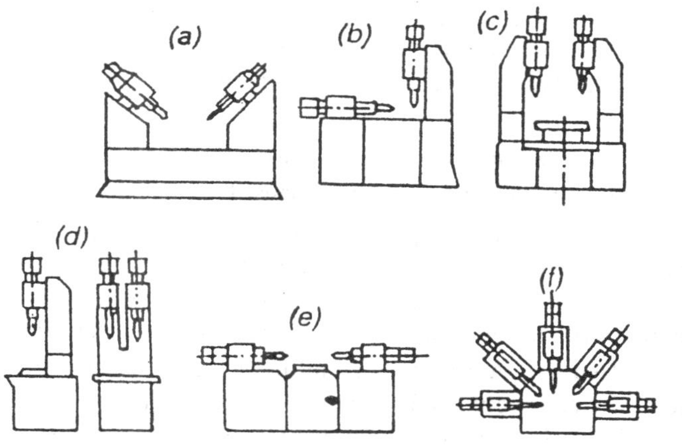

Fig. 3.53. Specialised unit built drilling machines.

PROBLEMS

1. Define drilling operation.
2. Write the requirements which must be met by a drilled hole.
3. Sketch a Twist Drill. Write its main parts and their functions.
4. Why the body of a Drill is made slightly less in diameter than the full nominal diameter ?
5. Why the Drill bodies are given a slight back taper ?
6. Write about the materials for Drills.
7. What are the four systems of drill sizes ?
8. How drills are held in a drilling machine ?
9. What is a core drill ? When it is used ? Write its advantages over a 'standard twist drill'.
10. What is a 'flat drill' ? Sketch it.
11. What is a "Step drill" or "Combination drill" ?
12. Sketch and explain the following allied operations : reaming, counterboring, countersinking and spot facing.
13. How the different drilling machines are classified ?
14. Draw the block diagram of a drill press. Name its four major parts.
15. How is the size or capacity of a drill press determined ?
16. How is the feed controlled on most bench and small floor-type drill presses ?
17. What is the difference between Box-Column type and Round-Column type upright drill presse ?
18. What is a 'multispindle drilling press' ?
19. Sketch and describe the two types of multiple-spindle drill presses.
20. What is the difference between a gang type and a multiple-spindle-type drilling machine ?
21. Sketch and describe a Turret Drilling Press.

22. What is a radial drilling machine ? Sketch and describe it.
23. How is the size or capacity of a radial drilling machine is specified ?
24. What factors affect the cutting speed for drilling ?
25. Name and describe five devices or methods for holding work to be drilled.
26. What is 'deep hole drilling' ? What difficulties are encountered while drilling deep holes with the use of twist drills on conventional drill presses ?
27. What precautions are taken to obtain accurate deep holes ?
28. Name the special deep hole drills.
29. Sketch and describe : Half-round drills, Gun drill and Trepanning drill.
30. Sketch and describe specialised drilling machines.

3.3.10. Reaming. Reaming is the operation of finishing a hole very smoothly and accurately in size. A drill will not produce a hole having sufficiently good qualities of finish and accuracy for many purposes. Therefore, when a very accurate, smooth hole is required, the hole is first drilled a little undersize. Then it is reamed to the correct size.

Reamer. A reamer is a multipoint rotary cutting tool, generally of cylindrical shape, which removes relatively small amounts of material as it is rotated and fed into a previously drilled or bored cylindrical hole. It imparts to the hole the necessary smoothness, parallelism, roundness and accuracy in size. It will not correct any error in the hole, because it follows merely the previously drilled hole.

A reamer consists of three main parts : fluted section, neck and shank, (Fig. 3.54). The fluted part consists of chamfer l_1, starting taper l_2, sizing section l_3 and the back taper l_4. Chamfer length or bevel lead length l_1 ensures proper and easy entry of the reamer into the hole. The main cutting action of reamer is done by starting taper l_2. The sizing section serves to guide the reamer and also smooths or sizes the hole. The back taper l_4 (with a difference of from 0.01 to 0.08 mm between the maximum and minimum diameters) reduces friction between the reamer and the hole surface.

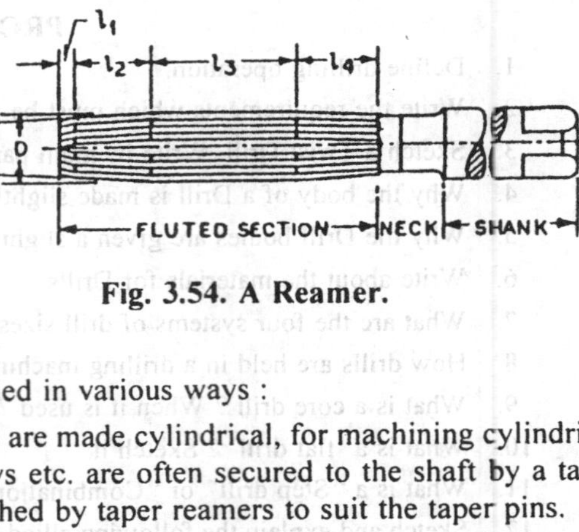

Fig. 3.54. A Reamer.

Types of Reamers. Reamers can be classified in various ways :

(*a*) **According to shape.** In shape, reamers are made cylindrical, for machining cylindrical holes and taper, for tapered holes. Collars, pulleys etc. are often secured to the shaft by a taper pin. Taper holes for this are drilled and then finished by taper reamers to suit the taper pins.

(*b*) **According to the method of application.** By the method of application, they may be hand-type and machine-type.

1. Hand Reamers. These reamers are operated by hand with a tap wrench fitted on the square end of the reamer, the work being held in the vise or vice versa. These are solid straight reamers and have straight shanks with square tang for a wrench. The reamers cut on the flutes which may be straight or helical. They may be made of carbon steel or H.S.S. The depths of cut for these reamers are about 0.13 mm. These reamers produce the most accurate diameters and are used only for sizing holes.

2. Machine Reamers. A machine reamer is similar to a hand reamer except that it has a taper shank, and reaming can be done in a drill press, in a lathe or in a milling machine. There are two designs of machine reamers :

(*i*) **Rose Reamers.** A rose reamer does not cut on the circum- ference of its flutes, but is bevelled off or chamfered at usually 45° to cut on its ends. The chamfered ends form the cutting edges. Rose reamers are used for greater depths of cut (upto about 1.6 mm) and as roughing reamers particularly for clearing out cored holes. These reamers are also called as "Rose chucking reamers".

(*ii*) **Chucking Reamers.** Chucking reamers or "Fluted chucking reamers" have the ends of their teeth ground on a slight chamfer for end cutting like a rose reamer. In addition, cutting is also done from the side edges of the teeth along the circumference of the reamer just like a fluted reamer. These reamers are considered precision machine reamers and can produce more accurate holes than rose reamers.

(*c*) **By Construction.** By construction, the reamers can be classified as :

Solid type reamers and shell type reamers of one diameter and adjustable with inserted blades. Reamers are also classified as "floating reamers", and "expansion reamers".

1. Solid Reamers. Solid-type reamers are made from one piece of material, such as H.S.S.

2. Shell type Reamers. Solid-type reamers in large diameters become very costly. To reduce the cost of such reamers, the cutting portions are made as separate shells made of H.S.S. These shells are mounted on standard shanks made of a lower-cost steel and which can be used interchangeably with other reamer shells. Shell type reamers are made with diameters greater than 19 mm. Large diameter shell type reamers are quite cheap as compared to equivalent solid type reamers. However, they are not very rigid and accurate and may not ream tougher materials with sufficient accuracy.

3. Adjustable Reamers. For very large diameter reamers, the reamer cost can be further reduced by using inserted teeth or blades in the shells. Only the blades are made of H.S.S. For large quantity production, the teeth or blades with tungsten carbide tips can be used. The blades can be moved up or down to increase or decrease the diameter. Also, the blades can be adjusted for wear when regrinding. The blades may be replaced individually. The drawbacks of such reamers are that the blades are difficult to adjust and they do not have the rigidity of solid reamers.

4. Expansion Reamers. The body of an expansion reamer is bored taper and is slitted to permit a slight expansion. A taper plug which runs through the hole is the expander, which is operated by a screw. This permits adjustment of the reamer diameter within a relatively small range as compared to adjustable-blade reamer. As the reamer wears or is resharpened, it can be adjusted back to its original size. Also, if a slightly larger hole than standard size is required, this reamer can be slightly enlarged to meet the need.

5. Floating Reamers. Here, the holders are not rigid but are floating. This permits the reamer to follow the previously made hole naturally and without restraint resulting in a better hole.

(*d*) **Left hand and Right hand Reamers.** Reamers are classified as left-hand or right hand depending upon whether they are rotated to the left or to the right while looking from the shank toward the cutting end of the reamer.

(*e*) **According to the Type of Flute.** The reamers can be made either with straight flutes or with helical flutes. Helical flutes have got the following advantages :

1. Smoother surfaces are produced because the cutting action is more of a shearing action is this case.
2. Reamers with helical flutes should be used when the surfaces to be reamed have interruptions, such as a keyway parallel to the axis of the hole.

The helical flute can be "right-hand" or "left-hand". If the helical flute of a reamer rotates as it advances to the right toward the cutting end, it is known as "right-hand helical flute". If the opposite is true, it is known as "left-hand helical flute". Generally, for best results, a right-hand reamer with right-hand helical flutes is used for roughing cuts, and a right-hand reamer with left-hand helical flutes is used for finishing cuts.

Hints on Reaming Operation

1. Holes to be reamed must always be drilled undersize. Common practice is to drill the hole about 0.4 mm undersize. Depending upon the material and hole size, the following is usually allowed on diameters for machine reaming :

0.25 mm on a 6.35 mm hole

0.375 mm on a 12.7 mm hole

upto 0.625 mm on a 38 mm hole

For hand reaming, not more than 0.125 mm to 0.25 mm of stock should be left in the hole, owing to pressure required to force the tool through.

2. Never turn a reamer backward. It will ruin the delicate cutting edges.

3. Reamers are made with an even number of flutes.

4. Apply plenty of cutting oil or coolant to the reamer when reaming all metals except cast iron which contains free graphite.

5. Reamers are operated at cutting speeds slower than for drilling, to avoid early resharpening. A speed that is about half to two-thirds that for a drill of equal size is correct. Feeds may be large compared with those for drilling (upto about five times large), because depths of cut are small.

3.3.11. Tap Tools. Taps are the tools for cutting internal threads. A tap is similar to a threaded bolt, with one to four flutes cut parallel to its axis. The flutes perform three functions :

1. provide cutting edges.
2. conduct the cutting fluid to the cutting region, and
3. act as channels to carry away the chips formed by the cutting action.

The flutes can be either straight, spiral, helical or spiral pointed. Taps with straight flutes are most commonly used, since it is easier to cut and sharpen these flutes. Tapping can be done manually or on drilling machines, tapping machines, turret lathes and automatics. A hole of diameter slightly larger than the minor diameter of the thread to be cut, must already exist, for thread tapping. The hole can be made by drilling, boring or casting. The two main types of taps are : solid taps and collapsing taps.

(*a*) **Solid taps.** Solid taps are of one piece construction. These taps are usually worked manually but can also be used on machine tools, such as lathes, drill presses and special tapping machines. Taps are made of high carbon or high speed steel. The shank of the taps is kept plain and the end is squared. To operate the tap by hand (Hand taps), it is held at the squared end with the help of a 'tap wrench', which is used to screw the tap into the hole. To cut any particular size, hand taps are available in sets of three : taper, plug and bottoming. The three taps are identical in size and length, but differ in the amount of chamfer at the bottom end. The taper plug has about 8 to 10 threads chamfer at the bottom end, the plug tap has 2 or 3 threads chamfered, whereas, a bottoming tap has no taper threads at its bottom end. The tapered are cut to the full depth gradually, so less effort is required. If a hole is open at both ends, then, after the taper tap, plug is used for finishing the threads as deep into the hole as its shape will permit. Lastly, the bottoming tap is used to finish the entire thread portion. So, the three taps should be used in the order mentioned above. The bottoming tap is the only tap which would nearly reach the bottom of a blind hole. The three taps are shown in Fig. 3.55.

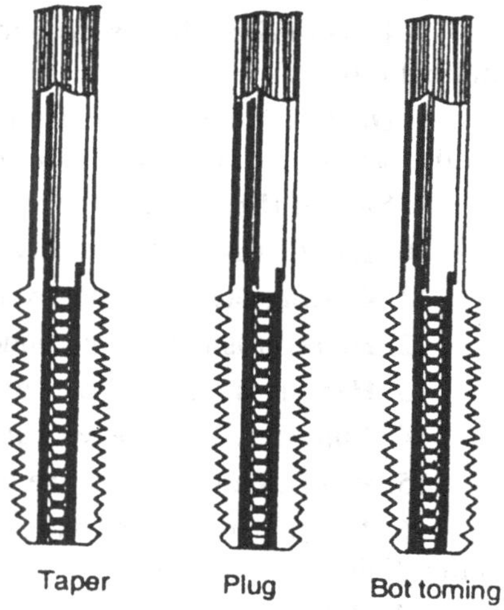

Fig. 3.55. Solid thread taps.

While threading, a combined rotary and axial motion is given to the tap. When using a solid tap on a drill press, a special tapping attachment is used. This makes the tap to rotate slowly as it is fed downward into the job. At the end of tapping, when the spindle is raised, the tap automatically starts rotating in the reverse direction at a higher speed to back the tap out of the hole in a shorter time. On screw machine or turret lathe, a special holder is used for the tap, in which a pin prevents the tap from rotating while it is fed into the job. At the end of travel, the tap pulls the pin so that it is free to rotate with the work. The machine spindle is then reversed in motion and the pin again stops the tap from rotating while it is being backed out of the hole.

(*b*) **Collapsing taps.** For better results, a tap (or a die) should not be backed off the thread it has just produced, because, during backing off, they catch tiny chips which can do damage to the product. So, for good finish and to speed up operations, collapsing taps are used, which collapse inward automatically when the thread is completed. This makes it possible to withdraw the tap from the hole without reversing the machine spindle.

Nomenclaure : Refer to Fig. 3.56

1. **Axis.** It is the longitudinal centre line through the tap.

2. **Body.** The body of a tap is the threaded and fluted part of the tap.

3. **Thread.** It is the cutting tooth of the tap which produces the thread in a hole.

4. **Angle of thread.** It is the angle included between the sides of the thread, measured in the axial plane.

5. **Crest.** It is the top surface joining the two sides of a thread.

6. **Root.** It is the bottom surface joining the sides of two adjacent threads.

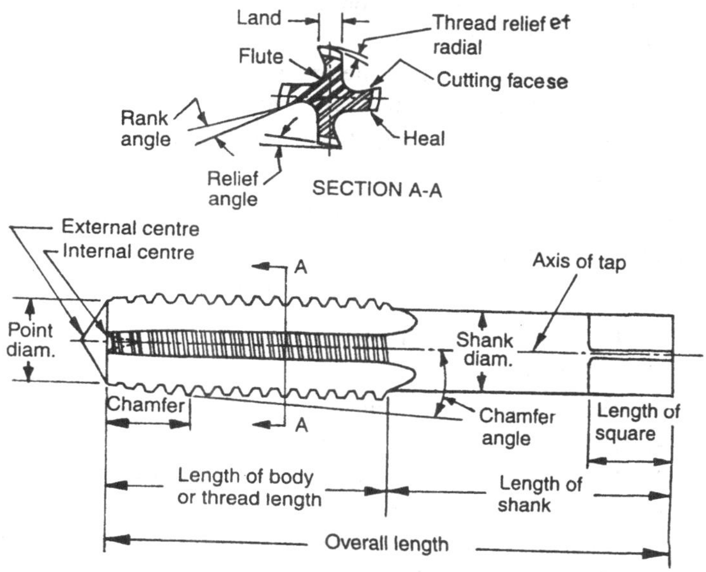

Fig. 3.56. A Solid Tap.

7. **Base of thread.** It is the bottom section of a thread; the greatest section between the two adjacent roots.

8. **Depth of thread.** The depth of the thread profile is the distance between the top of crest and the base or root of thread measured perpendicular to the axis of the tap.

9. **Side of thread.** It is the surface of the thread which connects the crest with the root.

10. **Land.** It is the threaded web between flutes.

11. **Cutting face.** It is the front part of the threaded section of the land.

12. **Hook.** It is the curved undercut of the cutting face of the land.

13. **Heel.** It is the back part of the threaded section of the land.

14. **Chamfer.** The tapered outside diameter at the front end of the threaded section.

15. **Point diameter.** It is the outside diameter at the front end of the chamfered protion.

16. **Flute.** It is the groove providing for the cutting faces of the teeth, chip passage and cutting fluid.

17. **Helix.** It is the curve of an ordinary screw thread.

18. **Helix angle.** It is the angle made by the helix of the thread at the pitch diameter with a plane perpendicular to the axis.

19. **Shank.** It is the part of the tap behind the threaded and fluted section of the tap. The tap is held or located and driven by the shank.

20. **Square.** It is the squared end of the tap.

21. **Radial rake angle.** It is the angle formed in a diametral plane between the face and a radial line from the cutting edge at the crest of the thread from.

22. **Chamfer angle.** It is the angle formed by the tapered outside diameter at the front end with the top axis.

23. **Web.** The central portion of the tap situated between the roots of the flutes and extending along the fluted section of the tap. Its thickness increases from the front and towards the shank end of the flutes.

24. **Back taper.** The reduction in diameter of the tap body of the threaded portion from the front end towards the shank end.

25. **External Centre.** It is the one-shaped end of the tap. It is provided only for manufacturing purposes and only for small taps and usually at the thread end.

26. **Internal Centre.** A small drilled and countersunk hole at the end of the tap, necessary for manufacturing purposes.

27. **Thread Relief.** It is the radial clearance providing a gradual decline in the major, pitch, and minor diameters of the lands, back of the cutting face.

PROBLEMS

1. Define the reaming operation.
2. Define a Reamer.
3. Sketch a Reamer and explain the functions of its different parts.
4. Discuss the various types of Reamers.
5. What is the difference between Rose Reamers and Chucking Reamers ?
6. Write a brief note on reaming operation.
7. Write briefly on :- Solid Reamers, Shell type Reamers, Adjustable Reamers, Expansion reamers and Floating reamers.
8. What are Left handed and Right handed reamers ?

9. What are left-hand flute and Right-hand flute reamers ?
10. What is the function of a tap ?
11. Give the functions of flutes on taps.
12. What are : Taper tap, Plug tap, Bottoming tap and Collapsing tap ?

3.4. BORING MACHINES

When the internal surface of a hollow part is turned, that is, a single point tool is used for enlarging a hole, the operation is called as "boring", (Fig. 3.57). The single point cutting tool for this purpose is mounted on a bar called "boring bar". Even though a boring bar may on occasion use more than one tool bit, at is still spoken of as "Single point boring".

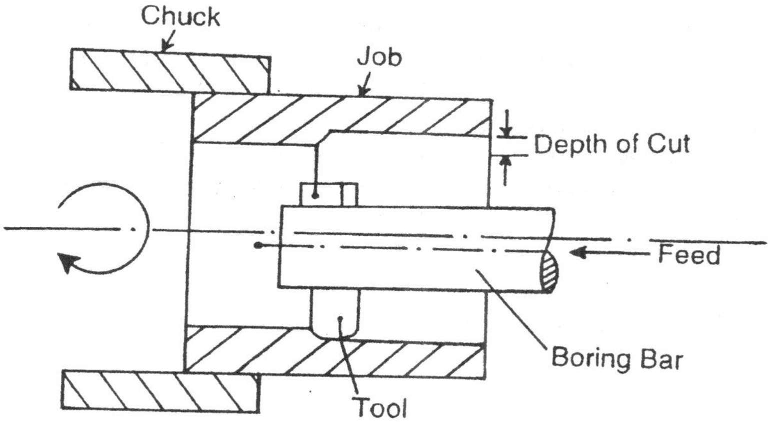

Fig. 3.57. Boring operation.

Boring tools usually operate with a large overhang from the tool holder. This factor does not allow shank type boring tools to cut a heavy chip. For this reason, deep holes are commonly bored by bar-type tools (often called bits) held is a holder or a bar (see Fig. 3.57).

While reaming is widely used to size holes more accurately than is possible with drills, boring as a rule must be resorted to in order to obtain the maximum in precision sizing, location and surface finish accuracy. Boring operation is chiefly used for machining holes with precise centre to centre distance in housing type parts (engine blocks, gear boxes etc.). Boring is also used for machining large diameter holes, since drills in larger sizes are relatively expensive.

Boring Vs Reaming

1. Boring can correct hole location, size, or alignment and can produce a good finish if a fine feed and a correct tool are used. The reamer follows the hole already in the workpiece and so can not correct location.
2. Reaming is faster than boring but boring operation is often preferred because of location correction advantage.
3. Reaming involves the use of a tool of fixed size, which is different for each size of hole and a large hole would require an expensive reamer, while a boring tool can make a hole of any size.

3.4.1. Boring Machines. Boring can be done on most machine tools that do turning (workpiece rotating with tool stationary), for example, lathes. However, boring also can be done using a rotating tool with the workpiece remaining stationary (drill presses, milling machine). Also, specialized machine tools (boring machines) have been developed that will do boring operation.

General purpose boring machines can be subdivided into machines for ordinary and precise boring operations. The latter include jig-borers, used to produce holes with especially high precision centre-to-centre distances. Boring machines can be of either horizontal or vertical design.

Horizontal Boring Machines. Horizontal boring machines can be used to drill, bore and ream holes and to mill flat surfaces. Such machines are widely used in piece and small-lot production for machining housing type parts (gear boxes, machine tool headstocks etc.). In mass production, holes of larger diameter are bored in multiple-spindle boring machines, using boring cutter heads as tools

In horizontal boring machines, the tool revolves and the work is stationary.

The workpiece is mounted on the table and is clamped with ordinary strap clamps, T - slot bolts and nuts, or it is held in a special boring fixture if so required.

Various types of rotary and universal swivelling attachments can be installed on the boring machine table to bore holes at various angles in horizontal and vertical planes.

In piece and small-lot production, holes are bored to layout lines ; in medium-and large-lot, as well as mass production, devices for coordinate positioning are used, as well as jigs for guiding the cutting tools.

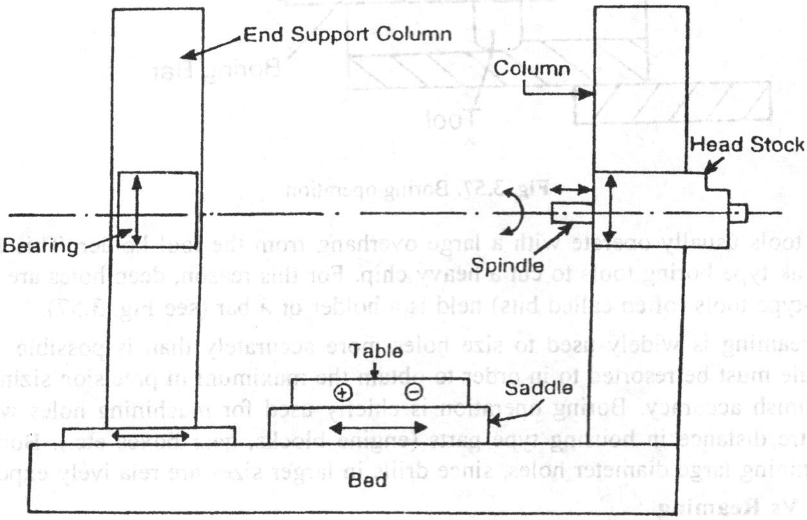

Fig. 3.58. Table Type HBM.

Types of Horizontal Boring Machines. Horizontal boring machines are of four principal types : Table, planer, floor and multispindle head type.

1. **Table type HBM.** (Fig. 3.58). The work is held stationary on a coordinate work table having in-and-out as well as back-and-forth movements, that is, perpendicular and parallel to the spindle axis. The head stock carrying the spindle can be moved vertically along a column for adjustment to varying heights of work. At the opposite end of the bed is a support column which carries a bearing for supporting the end of the boring bar. The spindle carrying the tool can be fed axially. Alternatively, the table travels parallel to the spindle axis (longitudinal feed). This method of boring with longitudinal feed of the table is employed when the holes are of consider-

able length and bending of the boring bar is possible. Machines of this type built for high spindle running accuracy and high positioning accuracy along the two axes of traverse which are at right angle to the spindle axis, are known as horizontal jig boring machines.

2. Planer type HBM. This machine, (Fig. 3.59), is similar to the table type HBM except that the work table has only in-and-out movements, that is, perpendicular to the spindle axis. Other features and applications of this machine are similar to the table type HBM.

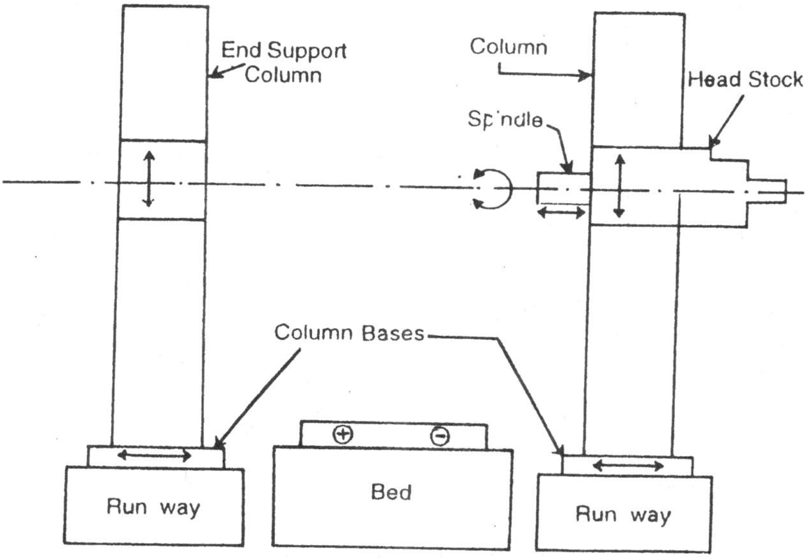

Fig. 3.59. Planer type HBM.

3. Floor type HBM. Here, there is no table and the job is mounted on a stationary T-slotted floor plate, (Fig. 3.60). This design is used when very large and heavy jobs can not be

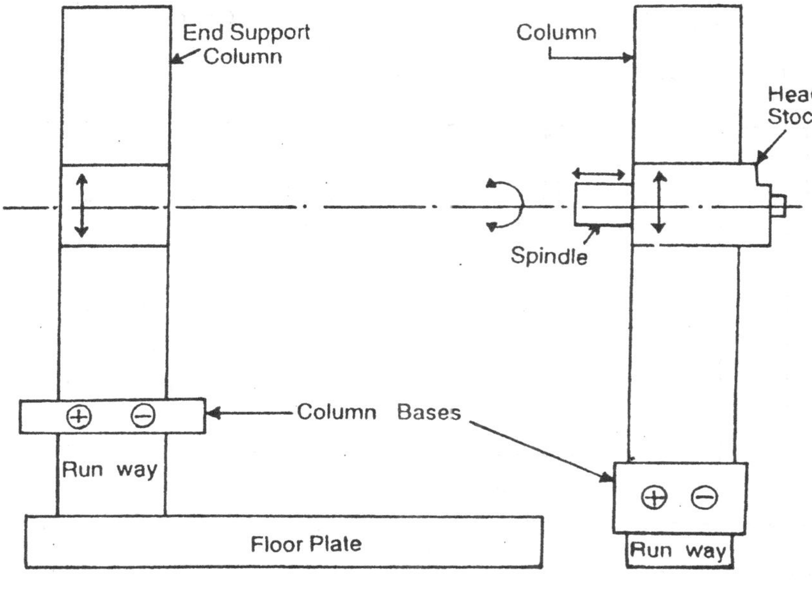

Fig. 3.60. Floor type HBM.

mounted and adjusted on the worktable. Horizontal movement perpendicular to the spindle axis is obtained by traversing the column carrying the headstock, on guide ways.

4. Multiple - spindle HBM. This machine, (Fig. 3.61), is just like a planer and is used for boring holes of larger diameter in mass production.

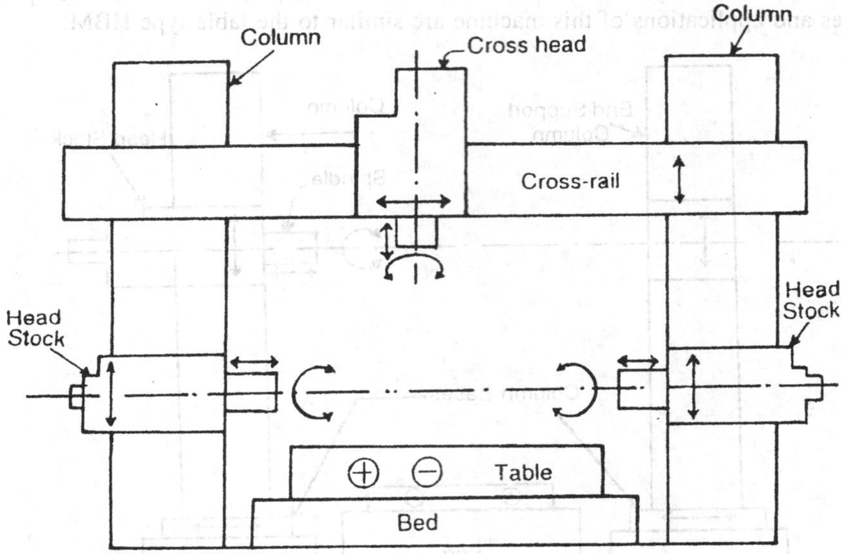

Fig. 3.61. Multi-spindle HBM.

Vertical Boring Machines. For convenience, parts whose length or height is less than the diameter are machined on vertical boring machines. On a VBM, the work is fastened on a

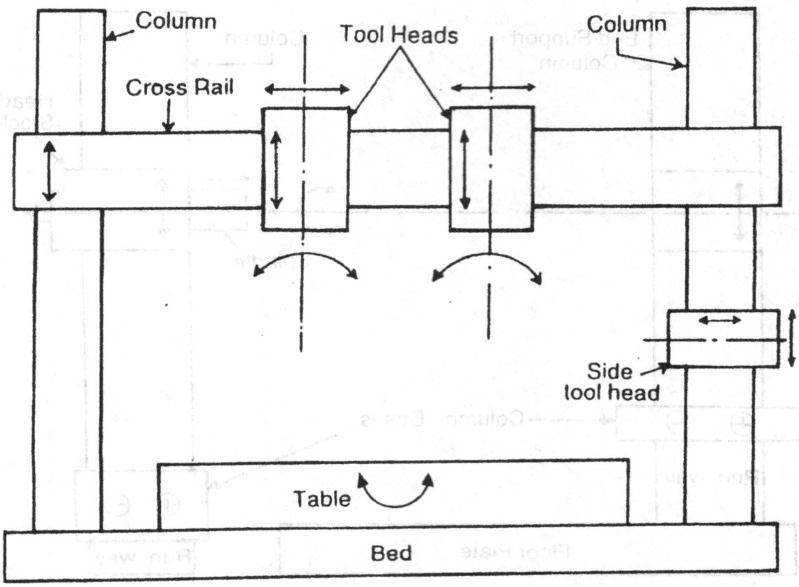

Fig. 3.62. Double Column VBM.

horizontal revolving table, and the cutting tool or tools, which are stationary, advance vertically into it as the table revolves.

There are two designs of a VBM : Single column VBM and double column VBM. The single column VBM looks like a drill press or a knee-type vertical milling machine. Guide ways are employed on the column to support the spindle head in the vertical direction. A double column VBM is shown in Fig. 3.62. The work is accommodated on the horizontal revolving table at the front of the machine. The circular work can be clamped on to the table with the help of jaw chucks whereas the T-slots can be used with bolts and clamps for setting up and holding irregular work. A horizontal cross rail is carried on vertical slide ways and carries the tool holder slide or slides. On machines designed for working on large batches of similar articles, a single slide with turret may be employed. Most machines also have a side tool head.

3.4.2. Size of Boring Machines. The size of boring machines is given by the diameter of the spindle (75 to 350 mm). In the case of VBM, one also has to give : dimensions of column height and table size (diameter), 1200 mm to 3600 mm.

3.4.3. Jig Borers. A jig borer or jig boring machine is a very precise vertical type boring machine. The spindle and spindle bearings are constructed with very high precision. The table can be moved precisely in two mutually perpedicular directions in a plane normal to the spindle axis, (Fig. 3.63). The co-ordinate method for locating holes is employed. Holes can be located to within tolerances of 0.0025 mm. Jig boring machines are relatively costlier. Hence, they are found only in the larger machine shops, where a sufficient amount of accurate hole locating is done. Jig borers are basically designed for use in the making of jigs, fixtures and other special tooling.

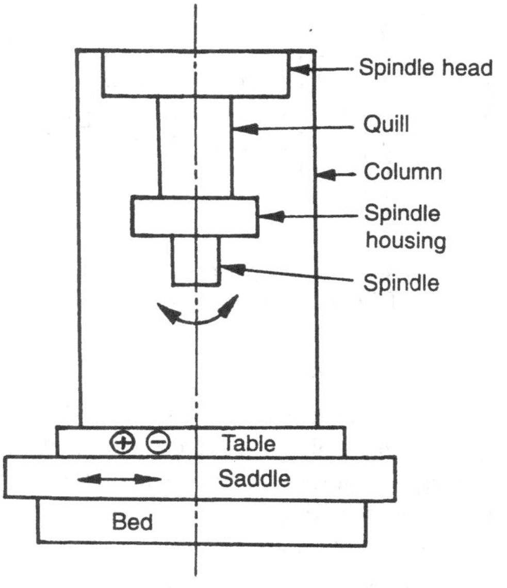

Fig. 3.63. A Jig-borer.

3.4.4. Boring with coordinate positioning. In boring with co- ordinate positioning, the axes of the holes are specified by two co-ordinate dimensions from the basic datum surfaces of the part. The axis of the boring machine spindle is set to the co-ordinate position, that is to the co-ordinate dimensions, by means of scales and vernier reading devices, installed on the machine, or by means of inserting the required end measures in accordance with the co-ordinate dimensions of each hole axis. Before beginning to bore the holes, the work is properly set up on the machine table to a dial indicator mounted in the boring spindle. Then the spindle is set to the "Zero" co-ordinate position by means of a locating tool inserted into the tapered hole. These tools may be of three types : locating dial indicator, locating microscope or a precise cylindrical bar of a definite size with a taper shank. The last of these is used in conjunction with clearance gauges. Then, boring can be begun after first positioning the spindle vertically and the table horizontally (in the transverse direction) so that the spindle axis coincides with the specified axis of the hole to be bored. This method of boring increases the accuracy of the centre-to-centre distances between the holes, simplifies machining and may in some cases, increase the output.

3.4.5. Boring Tools. As already noted, a boring tool consists of a single point cutting tool (boring bit) held in a tool holder known as boring bar, Fig. 3.64, where the boring bit is held in a cross-hole at the end of a boring bar. The boring bit is adjusted and held in position with the help of set screws. The material of the boring bit can be: Solid H.S.S., Solid carbide, Brazed Carbide, Disposable carbide tips, Ceramic tips or diamond tips. Boring tools are of two types:

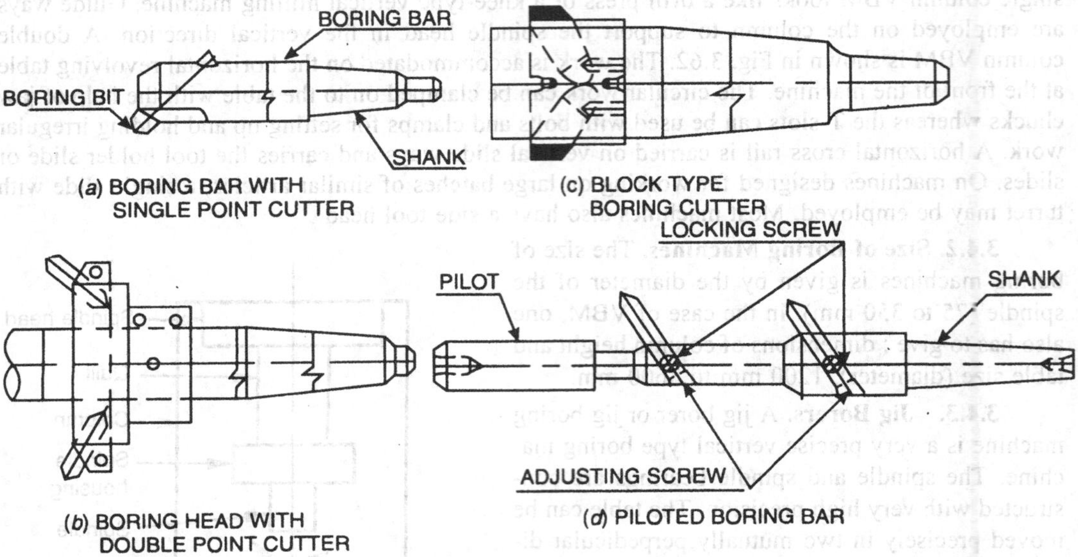

Fig. 3.64. Boring Tools.

1. Fixed type 2. Rotating type

Fixed type boring tools are used on work rotating machines, such as lathes, whereas rotating type boring tools are used on tool rotating machines such as drill presses, milling machines and Boring machines. Lathes, drill presses and milling machines can be used only for small jobs. However, for large jobs boring machines or mills are used. The boring bar or the boring head in the case of rotating type tools, is held in the main spindle of the machine.

In the case of centre lathe, fixed type tools are mounted in the tools post and the longitudinal feed to it is given with the help of compound rest slide or carriage. In such tools, there is no in built arrangement for tool adjustment for bore sizes. However, for fixed type tool ho ders held in turret of turret lathes and rotating type tools used in boring machines, there is in-built size adjusting facility. Such adjustable boring bars can be preset away from the machine, or a presetting device. Boring bars are also built with cartridges which hold throw away type inserts. Built in adjustment facility (set serews) is provided both for radial and axial adjustment. For precision boring or on jig boring machines, micrometer adjustment facility is built in, to ensure the true boring of the hole. For multi-point boring, the boring bars are built with more number of cartridges.

Since the boring bar is to reach the whole length of the job and it is overhanging, the problem of its deflection, chatter and vibration, is there. Due to this, maintaining dimensional accuracy can be a significant problem. The boring bar, therefore, must be sufficiently stiff, that is, made of a material with high modulus of elasticity, for example, WC, to minimise ceflections and avoid vibration and chatter. However, boring bars have been designed with capabilities for damping vibrations.

PROBLEMS

1. Define boring operation.
2. Compare 'boring' and 'reaming' operations.

3. Name the various types of HBM. Discuss them with the help of neat sketches.
4. Sketch a double column VBM and explain its working.
5. How the size of boring machines is specified ?
6. What is the difference between a VBM and a Jig borer ?
7. Sketch a Jig borer and explain its working.
8. Discuss : Boring with co-ordinate positioning.
9. Write a short note on "Boring Tools".

3.5. MILLING

Milling may be defined as a machining process for removing excess material from a work piece with a rotating cutting tool. The rotating cutting tool called the "Milling cutter" is a multiple-point tool having the shape of a solid of revolution with cutting teeth arranged (equally spaced) either on the periphery or on end face or on both. The work may be held in a vise, a 3-jaw chuck, an index head, a rotary table, between centres, in a special fixture or bolted to the machine table. Milling process involves simultaneous rotary cutter and usually linear (sometimes rotary) motion of the work, with the work fed against the cutter. Milling process is used for producing flat, contoured or helical surfaces, for cutting threads and toothed gears and for making helical grooves.

In general, all milling operations can be grouped into the following two types :

1. Peripheral Milling. Here, the finished surface is parallel to the axis of the cutter and is machined by cutter teeth located on the periphery of the cutter, (Fig. 3.65 (*a*)).

2. Face Milling. In face milling, the finished surface is at right angle to the cutter axis and it is obtained by the teeth on the periphery and the flat end of the cutter, (Fig. 3.65 (*b*)).

Surface Generation Methods. There are two methods of milling flat surfaces with plain milling cutters (these cutters have teeth only on the periphery of the cutter) depending upon the relative direction of feed of the worktable and the rotation of the cutter:

1. Conventional (Up) Milling. Here, the direction of feed of the worktable is opposite to the direction of rotation of the cutter, Fig. 3.66 (*a*).

2. Climb (Down) Milling. Here, the direction of the worktable feed and the rotation of the cutter is the same, Fig. 3.66 (*b*).

In either case, the individual chips cut in the milling process are of variable thickness, having a cross section resembling a comma. In conventional milling, the thickness gradually increases during the cut ; in climb milling the cutter tooth takes a chip of maximum thickness at the beginning of cut and of zero thickness at the end.

The advantages of conventional milling are that the load on each cutter tooth is gradually increased and the teeth start cutting under the metal skin of the workpiece. The initial tooth contact usually occurs in clean metal and ends by lifting or peeling off the rough surface scale. Due to this reason, this method of milling is preferred for machining sand castings, forgings and metals that have a rough or hard abrasive surface scale. The disadvantages are that the cutter tends to lift the workpiece from the worktable and as the teeth tend to dull, each tooth slides across the workpiece surface a minute distance before it begins to cut, producing a characteristic wavy surface. There is tendency of lifting the work table also, alongwith lifting the job from the table or fixture (since the cutting forces are directed upwards). This increases the clearance between the table and bed or saddle ways. In making heavy cuts, such clearances lead to vibrations which impair the surface finish produced on the job. Again, chips accumulate ahead of the cutting zone where they can be picked up by the teeth and carried around to impair the finish. Also, it is difficult to machine thin sheets and workpieces where clamping is inadequate. Since, the table is to be fed against the cutting force, the feed motor and main motor powers consumed are comparitively more. The favourable

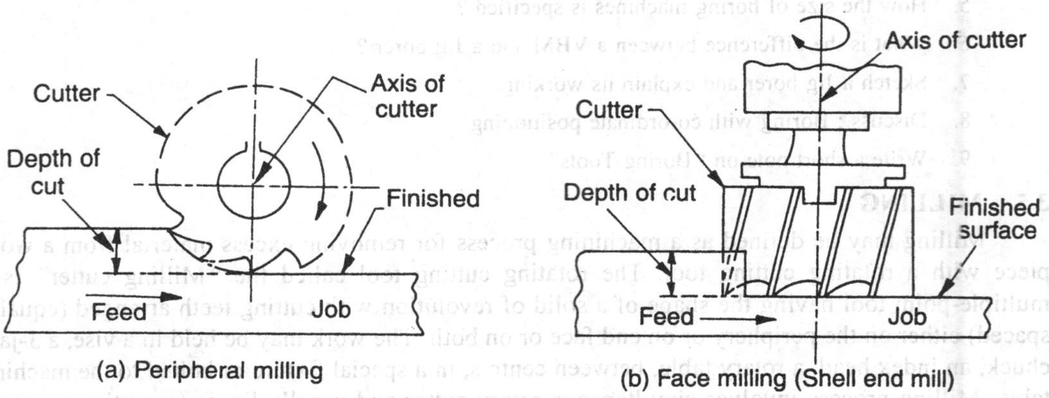

Fig. 3.65. Type of milling operations.

conditions for conventional milling are that the work must be rigidly held to offset the tendency of the cutting forces to lift or pull the work out of the vise or fixture.

The advantages of climb milling are : job is forced against the table and the table against the ways. This eliminates excess clearance in jointing surfaces and the resulting vibrations. Also, the chips are deposited behind the cutter and out of its way. These factors give a better surface finish than in conventional milling. Again, less power is required in climb milling, because, less power is consumed in the feed motion. Increased cutting speeds and feeds are possible.

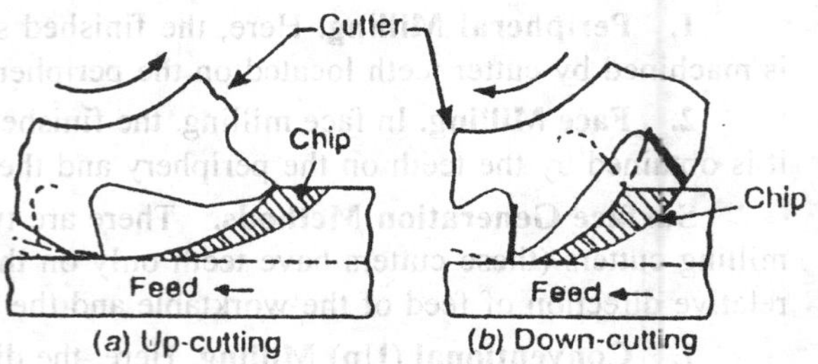

Fig. 3.66. Up and down milling.

Also, thin sheets can be easily machined. The favourable conditions for climb milling are : the workpiece has no hard skin, the milling machine is in good condition and there is no excessive backlash in the table screw and nut, since any looseness will allow the cutter to draw the workpiece ahead and take bites that are too large. The method is usually used for milling operations on slender and intricate parts.

3.5.1. Milling Machines. Milling machines were basically develop- ed to machine flat surfaces. But, the present milling machines can machine flat, contoured and helical surfaces, cut gears and do various other jobs. Due to all this, a milling machine is one of the most useful and necessary machine tools found in the shop and it ranks next to the lathe in importance.

Milling machines are designed to hold and rotate milling cutter or cutters, hold the workpiece and feed the workpiece to the milling cutter in one of several directions.

Types of Milling Machines. Milling machines can be classified in different ways :

(*a*) According to the axis of the spindle of the machine, we have : Horizontal milling machines with horizontal spindle.

(*b*) Vertical milling machines with vertical spindle.

According to their purpose, the milling machines may be classified as :

(*i*) General purpose milling machines

(*ii*) Production milling machines

(*iii*) Special purpose milling machines

1. General purpose Milling Machines. The most common kinds of milling machines in this category are the column and knee type models which are all single spindle machines. The various models under this category are :

(*a*) Plain (Horizontal spindle).

(*b*) Universal (Horizontal spindle with swivel table).

(*c*) Omniversal (Horizontal spindle with swivel table and swivel knee).

(*d*) Vertical spindle.

(*a*) Plain Column and Knee Type Milling Machines. This machine is called by this name because the spindle (the part that rotates) is fixed in the column. A block diagram of the machine is shown in Fig. 3.67. The structure of the machine is mounted on the base or bed of the machine. It provides rigidity and strength to the machine. It may also serve as a reservoir for

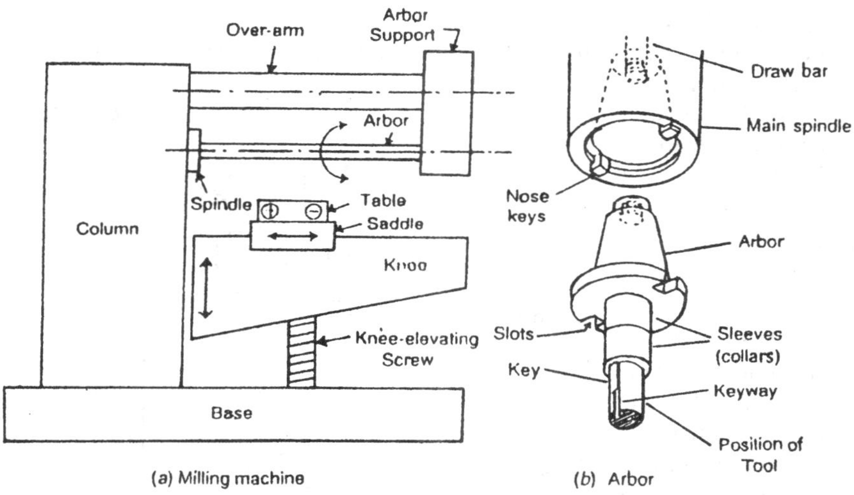

Fig. 3.67. Plain Column and Knee type Milling Machine.

cutting fluid. The column is the main supporting frame. It is a heavy box like structure. Inside this is space for motor and much of the driving mechanism. The face of the column is a very accurate slide. The "knee" slides vertically (up and down) on the column face. A feed, or elevating screw extends from the base to the knee. On top of the knee is a "saddle". The saddle can move transversely (cross movement toward and away from the column). On top of the saddle is the "table". The table moves longitudinally (back and forth) in front of the column. The surface of the table is cut with T slots to which work holding devices such as the vise can be clamped. The machine provides for three movements of the workpiece :

(*i*) Vertical (up and down) movement of the knee.

(*ii*) Transverse or cross (in-and-out) movement of the saddle.

(*iii*) Longitudinal movement (back-and-forth) movement of the table.

Along any one of these coordinate directions, feeding may be accomplished. Movements along the other two coordinate directions are then used for locating the cut, which includes obtaining the depth of cut.

The driving power for all the cutting tools comes from the rotation of the spindle. This is a hollow shaft that rotates in bearings. The front (end near the table) has a tapered hole. The arbor is an accurately machined shaft for holding and driving the arbor-type cutter (Fig. 3.67 (*b*)). It is tapered at one end to fit the spindle nose and also has two slots to fit the nose keys for locating and driving it. When it is in place, a draw in- bar goes through the spindle and is threaded on one end to screw into the threaded taper end of the arbor, holding the arbor tightly in place. The outer end of the arbor is supported by an overarm support in a proper bearing. The overarm extends out from the top of the column above the spindle. It can be moved in or out and clamped at different distances from the column.

(*b*) **Universal Milling Machine.** The universal horizontal milling machine differs from the plain horizontal type in that its table can be swivelled to enable helical grooves to be milled (*e.g.* the helical flutes of twist drills or the teeth of helical gears). The saddle is in two parts so that the table can be horizontally rotated.

(*c*) **Omniversal Milling Machine.** An omniversal milling machine has an additional movement as compared to universal milling machine. The knee can be rotated about the column face on an axis perpendicular to it. This enables to machine tapered spiral grooves in reamers, teeth on bevel gears and angular holes etc.

(*d*) **Vertical Milling Machines.** The vertical milling machine is very similar to the plain horizontal milling machine except that the spindle is held in a vertical position instead of horizontal. Vertical milling machines are of two types :

Fixed bed type, and Column - and - knee type, (Fig. 3.68).

The column-and-knee type vertical machine has a vertical head and spindle, but the same feeds and adjustments as the plain horizontal machine. The overarm provides a strong support to the spindle. On some models, the head can be moved up and down to perform operations like grooving, slotting, die forming, facing, drilling and boring of accurate holes, with their exact locations. The machine can be with a fixed head and of swivelling head type with the head swivelling around a horizontal axis.

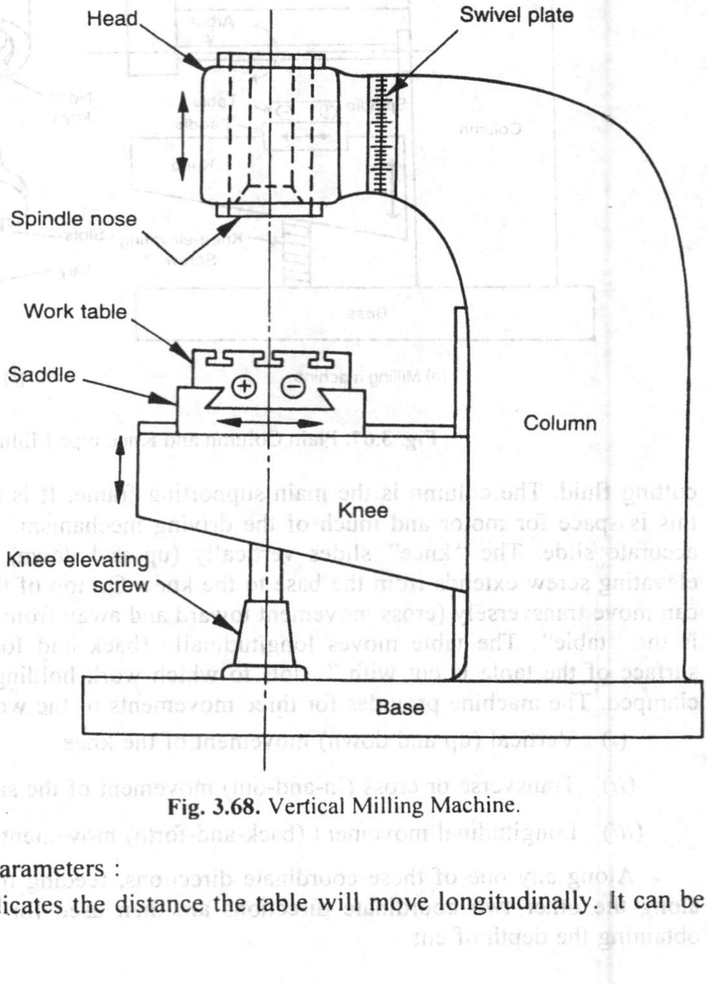

Fig. 3.68. Vertical Milling Machine.

Size of Milling Machines. The sizes of column-and-knee milling machines are given in terms of the following parameters :

(*a*) *Table Travel.* This indicates the distance the table will move longitudinally. It can be upto the range of 1.5 m.

(*b*) In terms of the power of the machine.

(*c*) In terms of the type of the machine : plain, universal, omni- versal, vertical etc.

2. Production Milling Machines. Production or Manufacturing milling machines are designed to remove metal rapidly and to require a minimum amount of attention from the operator. The various models in this category are discussed below :

(*i*) **Planer Type Milling Machines.** A planer type milling machine will look like a double column planer, but has milling heads mounted in various planes, vertical milling heads on the cross rail and horizontal heads at the sides (on columns). This enables it to machine a workpiece on several sides simultaneously. The planer type milling machines are primarily intended for producing long straight surfaces on large and heavy machine parts.

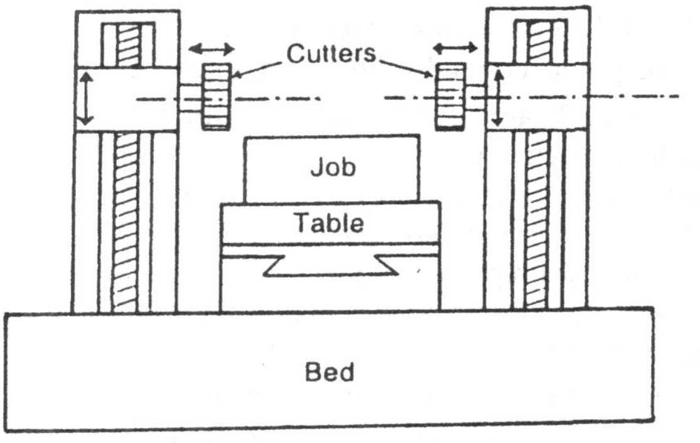

Fig. 3.69. Duplex Bed type Milling Machine.

(*ii*) **Bed Type Milling Machines.** The smaller versions of planer type milling machines, having one horizontal spindle on one side or two spindles, one on each side, are called simplex and Duplex fixed bed milling machine, (Fig. 3.69). These machines are larger, heavier and have greater rigidity than the column-and-knee type and are not adapted to tool room work. The in-and-out adjustment of the cutter is made possible because the spindle is mounted in a quill. The fixed bed provides for controlled longitudinal travel only.

(*iii*) **Drum Type Milling Machines.** These machines are of the continuous-operation type. They are mostly found in large-lot and mass production shops for production of large parts such as motor blocks, gear cases, and clutch housings. Two flat surfaces of the workpiece can be milled simultaneously. A drum type milling machine is shown in Fig. 3.70. A square drum (sometimes it may be a regular pentagon or hexagon), is mounted on a shaft passing through the frame. Parts are carried in fixtures mounted on the drum faces. The drum rotates continuously, carrying the parts between face mills. The milling heads can be adjusted along the housing and clamped as required for the set up. In addition to rotation, the milling spindles also have axial adjustment to set the cutters to the depth of cut. The output of such machines depends upon the number of simultaneously machined parts and the speed of rotation of the drum (rate of feed).

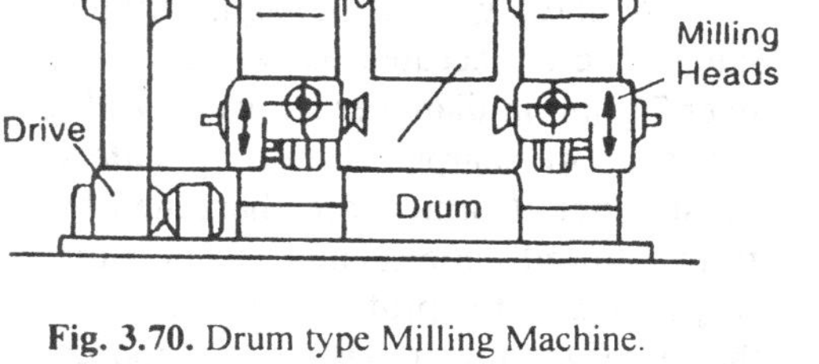

Fig. 3.70. Drum type Milling Machine.

(*iv*) **Hand Operated Millers.** These machines are usually of the column-and-knee type and are common among the smaller production machines. Table movements are hand controlled through a handwheel or lever (which rotates a pinion that engages a rack on the underside of the table), providing high sensitivity and making them ideal for delicate operations (light cuts) on small parts, *e.g.*, slotting the heads of screws. The cutter is driven by a motor.

(*v*) **Continuous or Rotary Millers.** These machines are usually of the vertical spindle type and have a rotary table for feeding jobs under the cutter continuously and are normally used for facing operations. It is used for facing automobile cylinder heads and crank pins etc.

An unusual departure from the conventional rotary machines is the rotary-head machine. Suited primarily for tool, die, and small- quantity production, this miller adds a bit to the versatility of the rotary-table machine but is designed for much smaller work. Intricate radial cam work can be produced readily from drawings without the use of templates or patterns.

(*vi*) **Tracer Controlled Production Milling Machines.** The vertical spindle rotary-head machine discussed above eliminates the need for templates or patterns in the production of small quantities of complicated forms, cams etc. However, when mass production is required an automatic tracer controlled miller is indicated. A variety of vertical and horizontal tracers are available and represent the ideal machine for the production milling of odd shaped parts such as connecting rods, complicated gear housings, and extremely large or small cam tracks of intricate shape. Intricate machine parts can be profiled in stacks or groups where high accuracy for interchangeability is necessary but quantity does not warrant expensive tooling.

A roller on a tracer control valve rides on a template bolted to the rear of the table and causes the cutter to cut in a path corresponding to that of the template.

3. Special purpose Milling Machines. The number and design of special purpose millers is probably unlimited as the design of machine parts themselves. Where quantity is sufficiently large, a special machine can offset its cost by virtue of increased output and automaticity. Thus we have : die sinking machines, cam millers, thread millers and keyway millers etc. Die-sinking machines are duplicating machines used to make forging dies, moulds, forming dies etc. "Profilers" are capable of reproducing external and internal profiles from templates in two dimensions, "Duplicators" do the job in three dimensions. Some machines are hand operated, others are power fed.

"Cam millers" produce disk cams. The profile of the cam is cut on a slowly revolving workpiece by an end mill positioned by a master cam revolving in unison with the workpiece.

3.5.2. Milling Cutters. A milling cutter is a multiple-edged rotary cutting tool having the shape of a solid of revolution with cutting teeth arranged either on the periphery or on end face or on both. Usually, the cutter is held in a fixed (but rotating) position and the workpiece moves past the cutter during the machining operation.

Types of Milling Cutters. There are very many different kinds of milling cutters :

According to the design, milling cutters may be of the solid or inserted blade type. The latter have replaceable blades that are mechanically retained and are usually adjustable. Cutters over 75 mm in diameter are usually of inserted blade design to save tool steel. According to how the teeth are sharpened, milling cutters may be either "profile sharpened" or "form relieved". Most cutters (former) are sharpened by grinding a narrow land back of the cutting edges ; On form-relieved cutters, only the face of each tooth is ground to sharpen it without changing the shape. Hence, the profile of the form-relieved cutters is maintained after resharpening throughout the life of the cutter. This is of importance for form cutters which are very difficult to sharpen if they are of profile-sharpened design.

Some of the more common kinds of cutters are discussed below :

1. Plain Milling Cutters. These cutters are cylindrical in shape and have cutting teeth on the periphery (around the outside) only. They are used for milling plain or flat surfaces. A plain milling cutter may have either straight (Fig. 3.71 *a*) or helical teeth (Fig. 3.71 *b*). In the operation of a cutter with straight teeth each tooth enters the cut simultaneously along its full length. This leads to intermittent loads acting on the machine and impairs the quality of the surface machined. Cutters with helical teeth operate more smoothly since the teeth enter the cut gradually and the

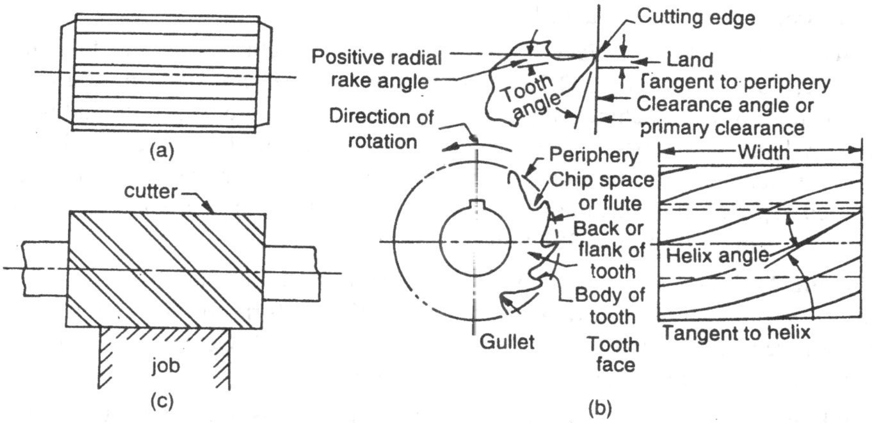

Fig. 3.71. Plain Milling Machine.

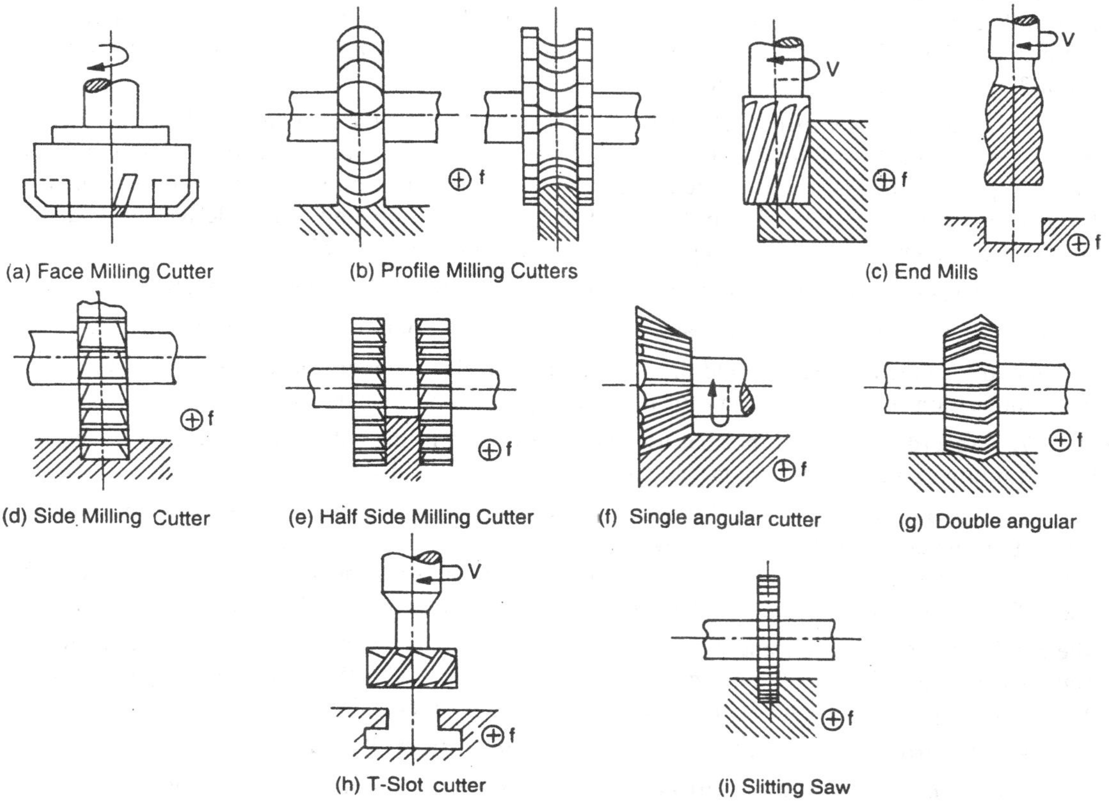

Fig. 3.72. Milling Cutters.

load on the machine is more uniform. When the width of the cutter is more than the width of the workpiece, it is called "Slab cutter", (Fig. 3.71 (*c*)). If a set of two plain cutters operate side-by-side to mill a wide surface, one has right hand helix and the other has left-hand helical flutes so that the axial forces are counterbalanced.

2. Face Milling Cutters. Face milling cutters, with teeth on the end face are used for machining large, flat surfaces, (Fig. 3.72 *a*). The cutter may be mounted on an arbor or rigidly clamped on the nose of the machine spindle. Face milling cutters are similar to end mills but they are considerably larger in diameter. They are designed to machine flat surface which are perpendicular to the rotational axis of the spindle. These cutters are used on vertical milling, planer type milling and bed type milling machines. Facing milling is more of a production operation done with cutters from about 140 mm to 380 mm in diameter.

3. Profile Milling Cutters. Cutters with a curved tooth outline of the same shape as the profile of the workpiece, (Fig. 3.72 (*b*)), are called "profile or form milling cutters". some of the common shapes are concave, convex and corner rounding cutters". Gear milling cutters are a kind of form milling cutters to machine any standard gear tooth.

4. End Mills. End mills have cutting teeth on the end as well as on the periphery. End mills may be made as solid cutters in which the cutter body is part of the shank, (Fig. 3.72 (*c*)). They may also be made as "Shell end mills" in which the cutters have a centre hole for mounting them on a short arbor. End mills are used in tracer-controlled profile milling operations. They are employed for making deep grooves in base parts, profile recesses, steps, perpendicular planes etc. End mills can be used on a horizontal milling machine, but it is better to use them on a vertical milling machine. Their diameter varies from about 3 mm to 50 mm. Shell end mill vary from about 30 mm to 150 mm in diameter.

5. Side Milling Cutters. These cutters have teeth around the periphery and also on one or both sides. Plain side milling cutters have straight teeth around the outside and side teeth on both sides [Fig. 3.72 (*d*)]. These cutters are made in width from about 5 mm to 25 mm and in diameters from 50 to 200 mm. "Half-side milling cutters" have teeth on only one side and around the circumference. They are made as right-and left-hand cutters, [Fig. 3.72 (*e*)].

6. Angular Milling Cutters. are made as single-(Fig. 3.72 *f*) or double-angle (Fig. 3.72 *g*) cutters and are used to machine angles other than 90°. Some of the common cutters are the 45° and 60° single angle cutter and the 45°, 60° and 90° double-angle cutters. They find the most wide application in tool production for milling chip flutes on various cutting tools.

7. According to the purpose or use, we have T-slot milling cutters, (Fig. 3.72 *h*), Woodruff keyslot milling cutters and dove tail cutters for milling dovetail guides.

8. Metal - slitting Saws. For cutting splines and narrow grooves on workpieces and also for cutting materials, metal slitting saws are used, (Fig. 3.72 *i*). Large diameter (over 120 mm) slitting mills are often called disk saws. Such mills are either made solid or with inserted teeth made from H.S.S. or hard metal. To promote uniform work, the end faces of these mills are alternately chamfered at an angle of 45°. Usually, the chamfers cut ⅕ – ⅓ the length of the cutting edge, therefore each tooth cuts chips of a smaller width than than of the groove. This allows the chips to freely move in the tooth space and facilitates their removal.

9. According to the method of mounting the Cutters. We have

(*a*) **Arbor Type Milling Cutters.** These are made to be mounted on a machine arbor. For this, they have an accurately ground centre hole. The arbor is supplied with a variety of collars or spacers which enable exact location of cutters on the arbor. Very thin shims may be used to position the cutters most accurately. Arbor type cutters are used in plain horizontal and universal horizontal milling machines.

A "stub arbor" is a short arbor on which the cutter is mounted on one end and held in position with a nut. The stub arbor has no outer support and is intended for facing work.

(*b*) **Shank Type Cutters.** These cutters have shanks (like drills and reamers) by which they are held and located. The shank may be straight or tapered. The taper shanks are inserted into the taper hole of the spindle and are clamped by a draw bolt passing through the hollow spindle. The cutter is retained from rotation by a key and is clamped by a screw. For cutters which have shanks that do not fit the spindle nose, adapters are available. Larger shank mounted cutters may have removable shanks and are known as "Shell-type cutters". These shanks need not be of expensive tool material. Collet holders are used to clamp straight shank cutters. Shank mounted cutters are used in horizontal and vertical spindle milling machines.

(*c*) **Facing Type Cutters.** When the diameter of a milling cutter, similar to an end mill, is sufficiently large, provisions are made to mount such cutters directly on the milling machine spindle nose, with bolts ; or may be mounted at the end of a stub arbor or used with an adaptor.

3.5.3. Construction of Milling Cutters. As already written, the milling cutters can be solid, carbide-tipped solid cutters and inserted blade cutters with H.S.S. or carbide tipped blades, the body being of constructional steel.

The advantages of carbide-tipped cutters (either solid or inserted-blade type) are :

(*i*) Their high production capacity.

(*ii*) The high quality of the surfaces they produce.

(*iii*) Elimination of grinding operation in some cases, the possibility of machining hardened steels and the reduction in machining costs that their use leads to.

Due to these advantages, they have been successfully applied in metal cutting industry where they have replaced many solid cutters of tool steels. Alongwith the especially popular carbide-tipped face milling cutters, carbide-tipped side and form milling cutters and various end mills are used in industry.

3.5.4. Cutter Materials. General purpose solid milling cutters are commonly made of H.S.S. Some cutters are tungsten-carbide teeth which may be brazed on the tips of the teeth or individually inserted and held in the body of the cutter by some mechanical means. Carbide-tipped cutters are especially adapted to heavy cuts and increased cutting speeds.

3.5.5. Hand of Milling Cutters. The hand of any cutter may be determined by looking at the cutter end of the spindle (that is viewing from the front end as mounted on the spindle). If the cutter rotates counter clockwise, it is right handed ; if it rotates clockwise, it is left handed. Among all the types of milling cutters, only face mills and end mills can be classified as to direction of rotation. Other cutters are changed from one hand to the other by reversing their mounting on the arbor.

3.5.6. Cutting Speeds, Feeds and Depths of Cut. Since the milling cutter is a multi-point cutter, the feed may be given as :

Feed per rev., mm/rev, f = Feed per tooth, mm × number of cutter teeth

$\therefore$ Feed per min. (table feed), f_t = Feed per rev × cutter speed (r.p.m.)

or $F\,(\text{mm/min}) = f_t \times n \times N$

where f_t = Feed rate per tooth, mm

n = number of cutterteeth

For H.S.S. plain milling cutters,

f_t p= 0.05 to 0.6 mm/tooth for milling steel = 0.1 to 0.8 mm/tooth for milling C.I.

The r.p.m. of the cutter is obtained as,

$N = \frac{1000 \times V}{\pi D}$, D is cutter diameter in mm. V is the cutting speed in m/min.

The average cutting speed is given in Table 3.2

Table 3.2 Cutting speed (H.S.S. Cutter)

Material being cut	*Brass*	*C.I.*	*Bronze*	*Mild Steel*	*Hard C- Steel*	*Hard alloy Steel*	*Al*
V, mpm	45 - 60	21 - 30	24 - 45	21 - 30	15 - 18	9 - 18	150 - 300

The cutting speed with carbide tools can be as high as five times the cutting speeds for H.S.S. Tools.

Depth of cut = 3 to 8 mm for roughing = 0.5 to 1.5 mm for finishing.

Example. A slot is to be milled by a side and face milling cutter with 10 teeth and of diameter 150 mm. The cutting speed is 50 m/min. and feed is 0.25 mm/tooth. Determine the table feed in mm/min.

Solution. Given. $n = 10$, $V = 50$ m/min

Diameter of cutter = 150 mm

Feed, $f_t = 0.25$ mm/tooth

Now Table feed, F, mm/min= $f_t \times n \times N$; N = RPM of the cutter

$$= \frac{1000 \times V}{\pi \times D} = \frac{1000 \times 50}{\pi \times 150} = 106 \text{ r.p.m}$$

$\therefore$ F = 0.25 × 10 × 106 = 265 mm/min.

Example. A steel workpiece is to be milled. Metal removal rate is 30 cm^3/min. Depth of cut is 5 mm and width of cut is 100 mm. Find the Table feed.

Solution. Now, Metal removal rate, MRR is given as,

MRR = Depth of cut × width of cut × rate of feed

$\therefore$ $30 \times 10^3 = 5 \times 100 \times F$

$\therefore$ Rate of feed (Table feed), F = **60 mm/min.**

3.5.7. Work - holding Devices. There are many different devices for holding the workpiece to be machined :

(*i*) The work may be clamped on the table by means of T-bolts, strap clamps and pads.

(*ii*) **Plain Vise.** This is the most common work holding device used on a milling machine. It can be fastened to the table by means of T-bolts with the jaws either parallel or at right angles to the T-slots.

(*iii*) **Swivel Vise.** The vise is made in two parts. The top part can be turned in a complete circle. The base is divided into 360°. The jaws can be set to any angle.

(*iv*) **Universal Vise.** If the vise, apart from being swivelled in the horizontal plane, can also be tilted in the vertical plane, it is called as "Universal Vise". It is used for milling compound angles.

(*v*) **Universal Chuck.** It is used to hold round workpieces and is used mostly on the dividing head (discussed ahead).

(*vi*) **Rotary Table.** The rotary table is made in two parts and is fastened to the machine table by means of T-bolts. The base is divided into 360° allowing the workpiece to be rotated in a complete circle. It is used to hold workpieces for accurate spacing, dividing and radius milling operations.

(*vii*) **Dividing Head.** The dividing head is a device for holding and turning the workpiece so that a number of equally spaced divisions or cuts can be taken around it. It consists of a headstock and a tailstock. The job may be held between centres or it may be held in a chuck mounted on the headstock spindle.

(*viii*) Work may also be held in various types of milling fixtures.

3.5.8. Principal Types of Milling Operations. Milling machines can be employed to produce a large variety of surfaces :

1. Milling Flat Surfaces. Horizontal Flat surfaces may be milled on :

(*i*) Plain horizontal and Universal horizontal machines with plain milling cutters as shown in Fig. 3.71 (*c*).

(*ii*) Vertical-spindle milling machines with face milling cutters, as shown in Fig. 3.65 (*b*).

Vertical flat surface are milled on :

(*i*) Plain horizontal and Universal horizontal machines by using side milling cutters.

(*ii*) Vertical-spindle milling machines using end mills.

(*iii*) Vertical flat surfaces may also be milled on planer-type machines with face milling cutters, (Fig. 3.73 *a*).

Inclined flat surfaces are milled on :

(*i*) Plain milling machines with single angle cutters (See Fig. 3.72 *f*).

(*ii*) On vertical spindle machines with the spindle swivelled to the required angle, (Fig. 3.73 *b*). This arrangement is possible only on swivel-head vertical spindle milling machine in which the head can be swivelled in a vertical plane parallel to the column face.

"Recesses" are milled on vertical spindle machines with end mills (See Fig. 3.72 *c*). In this case, the two perpendicular surfaces are machined simultaneously.

2. Milling Slots, Grooves and Splines. Rectangular, T-and dovetail slots are milled, as a rule, on vertical-spindle machines by means of suitable shank-type milling cutters, (see Fig. 3.72 *h*). Rectangular slots can also be machined on plain horizontal spindle milling machines with side milling cutters.

Keyways of rectangular cross section are milled on vertical- spindle machines or on keyway milling machines by using end mills and special cutters. Keyways can also be milled on horizontal spindle machines by using side milling cutters.

Splines may be milled on horizontal spindle machines by using single-and double-angle cutters.

3. Milling Contoured Surfaces. Contoured surfaces of relatively small width are milled on plain milling machines, using formed milling cutters. The profile of the cutter coincides with that of the workpiece. Surfaces of more complex shape are milled by a gang of simple cutters, (Fig. 3.73 *c*). The operation is called as "Gang milling". Complex shaped surfaces can also be milled

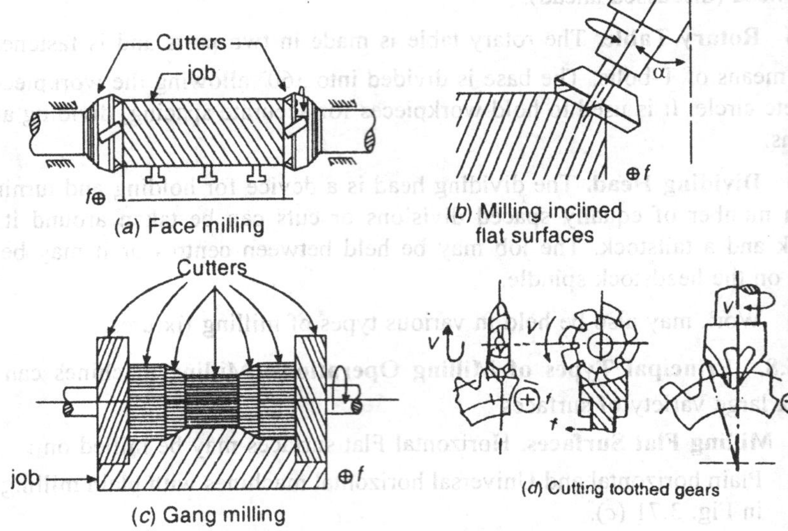

Fig. 3.73. (*a*, *b*, *c*, and *d*). Milling Operations.

using a template on a tracer-controlled machine. Three dimensional contoured surfaces (for example on dies etc.) are milled on tracer-controlled contouring machines.

4. Cutting Toothed Gears and Helical Grooves. Spur and helical gears are cut by form milling cutters on plain milling machines or by end-mill type cutters on vertical-spindle machines, (Fig. 3.73 *d*). The cutter profile must exactly coincide with the tooth space of the gear. Dividing head is used to index the gear blank through the required angle from one tooth space to the next.

5. Gang Milling, Straddle Milling, String Milling. When two or more milling cutters are mounted on an arbor so that each cutter will produce its own distinctive surface as the workpiece is fed to it, the operation is called "Gang milling", (Fig. 3.73 *c*).

"Straddle milling" is a special form of gang milling in which side milling cutters are used to machine both sides of a workpiece simultaneously (see Fig. 3.72 *e*).

In "String milling", two or more workpieces are mounted on the milling machine table in a line, so that they are successively fed to one or more milling cutters.

String milling results in a substantial reduction in handling time, as it is overlapped by the machining time.

Depending upon the job set-up and tool set-up, the other milling methods are given below :

1. Abreast Milling. This is the simultaneous milling of several workpieces, placed in a row, parallel to cutter axis or of several surfaces of the same workpiece. It can be accomplished by :

(*i*) Mounting the required number of plain, side and form cutters on an arbor, or

(*ii*) Mounting the required number of face milling cutters on different spindles

(*iii*) Using a face milling cutter of sufficiently large diameter, or

(*iv*) Using a plain milling cutter of sufficient length.

Abreast milling leads to a sharp reduction in labour input, as a result of combination of several operation elements and less handling time. Same or different operations can be performed on each piece.

2. Combined Abreast-and-String Milling. The simultaneous milling of several workpieces (or several surfaces of a single workpiece), placed in a single or several rows on the machine table, is combined with consecutive milling. This method, in addition to the reduction in labour input, enables the machine time to be considerably reduced, due to, the reduction in handling time.

3. Transfer-Base Milling. Here, two fixtures are mounted on opposite ends of a two-station index base, which is secured at one end of the machine table. After one workpiece has been milled, the table is returned to the initial position, the second fixture is indexed through 180° and the second work-piece is milled, while the first fixture is being unloaded and loaded. Economy is due to the combining of machine and handling times.

4. Index Milling. Here identical multiple operations are done on one or more pieces, which are indexed (rotated through required angle) each time to present a new position and repeating the same operation, for example, milling of splines on a shaft or cutting gear teeth (See Art. 3.5.10).

5. Reciprocal Milling. Fixtures are mounted on the left and right-hand ends of the machine table. While milling the workpiece in one fixture, the other fixture is being unloaded and loaded. Thus, the machine is performing the productive work during the handling time.

6. Progressive Milling. Here, two or more similar or different operations are performed either simultaneously or successively, on separate workpieces on the same machine. Workpieces are progressively moved from one fixture station to the next, until all the desired operations are performed [Fig. 3.74].

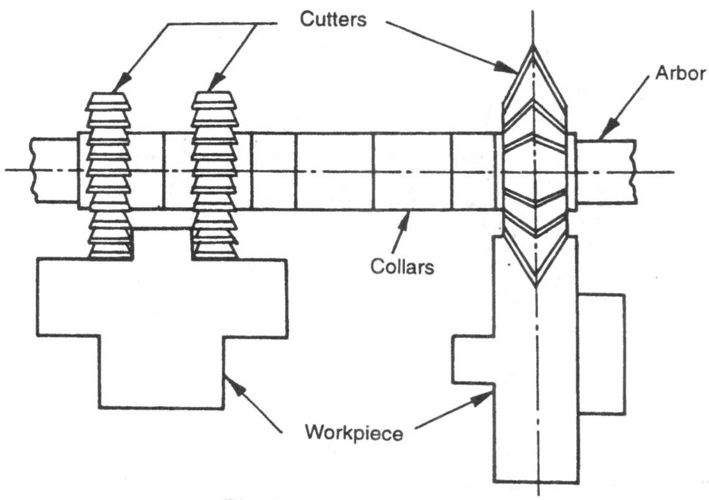

Fig. 3.74. Progressive Milling.

3.5.9. Cutter Angles. Refer to Fig. 3.71.

1. Radial rake angle = 10° to 20°.

   Larger values are used for milling soft materials and smaller values for harder materials.

   Carbide tipped cutters have negative rake angles = 10° to 15°.

2. Clearance of relief angle = 12° to 25°.

3. Helix angle : = 20′ to 30′ for plain milling cutters.
= 20′ to 15′ for side cutters and end mills.

3.5.10. Dividing Heads. The general construction and function of a dividing head has already been given under work holding devices.

An important function and use of milling machines is for cutting slots, grooves etc. which are to be equally spaced around the circumference of a blank, for example, gear cutting, ratchet wheels, milling cutter blanks, reamers etc. This necessitates holding of the blank (workpiece) and rotating it the exact amount for each groove or slot to be cut. This process is known as " Indexing". The dividing head is the device used for this purpose. It is lined and bolted to the machine table so that the axis passing through the headstock centre and tailstock centre is at right angle to the spindle axis of the machine. The headstock of the dividing head consists essentially of a spindle to which is keyed a 40-tooth wormwheel. A single threaded worm meshes with this wheel. The worm spindle projects from the front of the head and has a crank and handle attached. The head spindle is bored with a taper hole and is also screwed on its end. The workpiece is mounted between centres, one inserted into the dividing head spindle and the other into the tailstock of the head. The workpiece may also be mounted on a mandrel between these centres. A chuck may be mounted on the spindle nose for holding short work pieces having no centre holes.

The workpiece is rotated by turning the index crank by means of handle. Since the gear

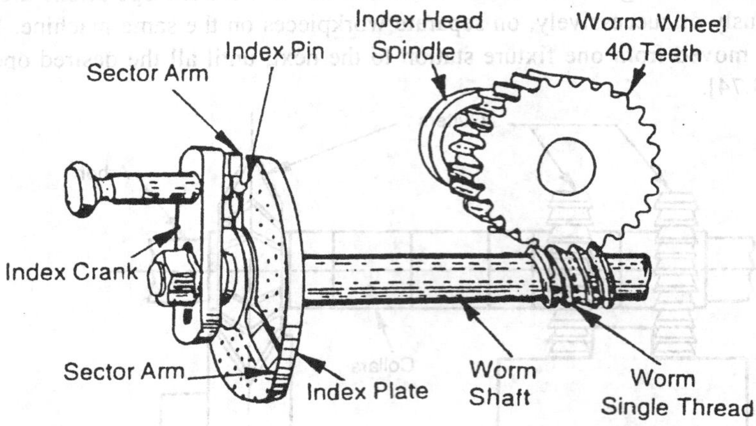

Fig. 3.75. Indexing Mechanism of a Dividing Head.

ratio of worm and worm wheel is 40 : 1, it takes 40 turns of the crank to rotate the spindle and hence the workpiece through one complete revolution. Thus one turn of the crank rotates the work $\frac{1}{40}$ th of a turn. If divisions other than factors of 40 are required "Index plates" are used. An index plate has several circles of holes (each circle containing a different number of holes) and is mounted on the worm shaft, (Fig. 3.75). A pin on the crank can be adjusted to a radius such that it will fit in any desired circle of holes. By using different circles of holes and index plates, any fractional part of a turn of the index crank can be obtained. The two sector arms shown on the front of the index plate, determine the angle through which the index crank is turned for indexing.

Types of Dividing heads. The various dividing heads used with milling machines are :

1. Plain. A plain dividing head has a fixed spindle axis and the spindle rotates only about a horizontal axis.

2. Universal. In these models, the spindle can be rotated at different angles in the vertical plane from horizontal to vertical. A universal dividing head performs the following functions : indexes the workpiece, imparts a continuous rotary motion to the workpiece for milling helical grooves (flutes of drills, reamers, milling cutters etc.) and setting the workpiece in a given inclined position in reference to the table.

3. Optical. These models are used for high precision angular setting of the work piece with respect to the cutter. For reading the angles, an optical system is built into the dividing head.

Methods of Indexing. The various methods of indexing are discussed below :

1. Direct Indexing. In direct indexing, the index plate is directly mounted on the dividing head spindle. The intermediate use of worm and worm wheel is avoided, (Fig. 3.76). For indexing, the index pain is pulled out on a hole, the work and the index plate are rotated the desired number of holes and the pin is engaged. Both plain and universal heads can be used in this manner. Direct indexing is the most rapid method of indexing, but fractions of a complete turn of the spindle are limi-ted to those available with the index plate.

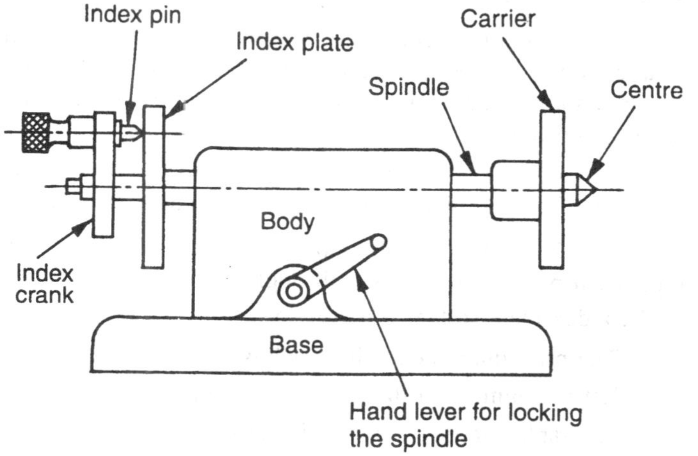

Fig. 3.76. Direct Indexing.

With a standard indexing plate having 24 holes, all factors of 24 can be indexed, that is, the work can be divided into 2, 3, 4, 6, 8, 12 and 24 parts.

2. Simple indexing. In simple or plain indexing, an index plate selected for the particular application, is fitted on the worm shaft and locked through a locking pin, (Fig. 3.75). To index the work through any required angle, the index crank pin is withdrawn from a hole in the index plate. The spindle and hence the work is indexed through the required angle by turning the index crank through a calculated number of whole revolutions and holes on one of the hole circles, after which the index pin is relocated in the required hole.

If the number of divisions on the job circumference (that is number of indexings) needed is z, then the number of turns that the crank must be rotated for each indexing can be found from the formula,

$$n = \frac{40}{z}$$

Dividing heads are usually furnished with a set of three index plates having six concentric hole circles with different numbers of equally spaced holes on each hole circle. A typical set has the following number of holes :

I 15, 16, 17, 18, 19, 20
II 21, 23, 27, 29, 31, 33
III 37, 39, 41, 43, 47, 49

Example. Let $z = 16$

$$\therefore \quad n = \frac{40}{16} = 2\frac{8}{16}$$

That is, for each indexing we need two complete rotations of the crank plus 8 more holes on the 16 hole circle of plate I.

Example. Let $z = 45$

$$\therefore \quad n = \frac{40}{45} = \frac{8}{9} = \frac{8}{9} \times \frac{2}{2} = \frac{16}{18}$$

We will employ index plate I and use hole circle with 18 holes. For each indexing, the crank will rotate through 16 holes.

3. Compound Indexing. When none of the index plates has a hole circle which would enable the work to be divided by simple indexing method, more involved methods are employed. One method is "compound indexing". The compound indexing is achieved in two stages, by using two different hole circles of one index plate :

(*i*) By a movement of the crank in the usual way as in simple indexing, say, n_1 holes in hole circle N_1, with the lockpin engaged in circle N_2 of the index plate.

(*ii*) By adding or subtracting a further movement by rotating the crank and the index plate together forward or back ward, through n_2 spaces in the N_2 circle. (by disengaging the locking pin of the index plate so that it is free to turn).

The procedure is explained below :

Let z = number of divisions needed on the work.

$\therefore$ Crank rotations for each indexing = $40 / z$

Since this can not be obtained by simple indexing, it is achieved by compound indexing. Therefore, addition or subtraction of the two movements given above should be equal to $40 / z$. This is done as explained below :

(*i*) Write z above and 40 below a straight line and factorise them.

(*ii*) Select two numbers representing two hole circles in the same plate. Wrte these numbers below 40 and factorise them. Write their difference above z and factorise it. These hole numbers are to be chosen in such a manner that all the factors above the line get cancelled out with factors below the line.

Let these hole numbers be N_1 and N_2.

(*iii*) Let n_1 be the number of holes to be indexed in N_1 hole circle and n_2 the number of holes to be indexed in N_2 hole circle.

$$\text{Then,} \quad \frac{n_1}{N_1} \pm \frac{n_2}{N_2} = \frac{40}{z}$$

From here, n_1 and n_2 are found out by trial and error. Then the total indexing will be :

n_1 holes in N_1 hole circle by rotation of the crank $\pm\, n_2$ holes in N_2 hole circle by rotating the crank and index plate together.

Example. Let $z = 87$

Steps.

(*i*) $87 = 29 \times 3$

$40 = 2 \times 2 \times 2 \times 5$

(*ii*) Let $N_1 = 29$ and $N_2 = 33$

$$\therefore \quad \frac{4 = 2 \times 2}{87 = 29 \times 3}$$

$$\frac{40 = 2 \times 2 \times 2 \times 5}{\begin{array}{l} 29 = 29 \times 1 \\ 33 = 11 \times 3 \end{array}}$$

Since all the factors above the line get cancelled out, therefore, selection of N_1 and N_2 is correct.

(*iii*) Now, indexing equation is :

$$\frac{n_1}{29} \pm \frac{n_2}{33} = \frac{40}{87}$$

or $$33n_1 \pm 29n_2 = 440$$

By trial and error $n_1 = 23$ and $n_2 = 11$ with minus sign.

that is, $$33 \times 23 - 29 \times 11 = 440$$

$\therefore$ Indexing equation will be :

$$\frac{23}{29} - \frac{11}{33} = \frac{40}{87}$$

that is, Movement of crank by 23 holes in 29 hole circle forwards and movement of crank and index plate both by 11 holes in 33 hole circle backwards.

Note. The method of compound indexing is little used to-day, as it has been replaced by differential indexing.

4. Differential Indexing. Differential indexing in reality is an automatic method of doing compound indexing. It is achieved in a single step as compared to two stages needed in compound indexing.

In differential indexing, the index plate is connected to the headstock spindle by means of a gear train. Fig. 3.77 shows one such design where z_1, z_2, z_3 and z_4 are interchangeable gears. During indexing, the index plate rotates in relation to the crankmovement. For this, the locking pin which kept the index plate locked while doing simple indexing, is disengaged. As the index crank is turned for indexing, rotating the spindle through worm and worm gear, the index plate will receive power through the change gears, equal bevel gear and the sleeve, and will rotate slowly. The index plate can be made to rotate either in the same direction or in the opposite direction to the index crank (by gear train design).

Indexing is performed in the same manner as that for simple indexing except that the location of the hole from which the index pin is turned will move slightly during indexing. The required movement of the index plate is calculated and taken care of by the gear train.

Differential indexing is thus more straight forward and so has wider applications as compared to compound indexing.

Procedure

Let z = number of divisions required to be indexed for one complete revolution of the spindle and hence the work-piece.

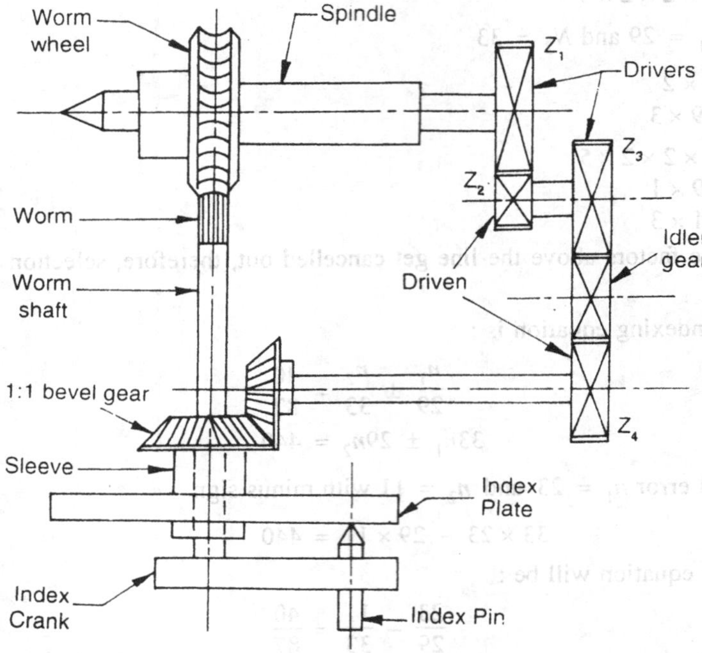

Fig. 3.77. Differential Indexing.

k = A number very nearly equal to z and which can be used in simple indexing method.

∴ number of crank turns for each simple indexing,

$$n = 40 / k$$

∴ Number of crank turns needed for z indexings,

$$N = \frac{40}{k} \times z$$

But, we know that the crank must make only 40 turns for the spindle (and hence the work) to turn through one complete circle. So,

(*i*) If $N > 40$, then $(N - 40)$ turns have to be subtracted. This is achieved through the change gears so that while the spindle makes one turn, the index plate makes $(N - 40)$ turns in the opposite direction to that of the crank.

(*ii*) If $N < 40$, then the index plate should rotate $(40 - N)$ turns in the same direction as that of the crank.

The gear ratio will be :

$$i = \frac{40}{k} (k - z).$$

Thus, the movement of the index handle (crank) operates according to the principle of simple indexing and the gear ratio makes it possible to find gears which take care of residual divisions.

Example. Do differential indexing for 93 divisions.

Solution. $z = 93$

$\therefore$ $$\text{Simple indexing} = \frac{40}{93}$$

It is clear from the available index plates, that 93 can not be simple indexed.

So, let $k = 90$, which can be simple indexed each indexing $= \frac{40}{90} = \frac{4}{9} = \frac{8}{18}$, that is, 8 holes in an 18 hole circle.

$\therefore$ for 93 indexings,

$$N = \frac{8}{18} \times 93 = 41\frac{1}{3} \text{ turns of the crank}$$

Since N > 40, the index plate must rotate $\frac{4}{3}$ turns backwards, that is, in the opposite direction.

$$i = \frac{40}{k}(k - z) = \frac{40}{90} \times 3 = \frac{4}{3}$$

In the Brown and Sharpe dividing head, the gears supplied are : 24(2), 28, 32, 40, 44, 48, 56, 64, 72, 86 and 100 teeth.

$$\therefore \quad i = \frac{4}{3} = \frac{32}{24} = \frac{\text{Drivers}}{\text{Driven}}$$

It is a simple gear train.

Example. $z = 127$

Let $k = 128$

$$\text{Simple indexing} = \frac{40}{128} = \frac{5}{16}$$

$$\therefore N = \frac{5}{16} \times 127 = 39\frac{11}{16} \text{ turns of the crank.}$$

since N is < 40, therefore, the index plate must rotate $\left(40 - 39\frac{11}{16}\right)$, that is, 5 /16 turns in the same direction as the crank, as the spindle completes one turn.

$$i = \frac{40}{k}(k - z) = \frac{40}{128} \times 1 = \frac{5}{16} = \frac{5 \times 1}{8 \times 2} = \frac{40}{64} \times \frac{24}{48} = \frac{\text{Drivers}}{\text{Driven}}$$

It is a compound gear train.

In Fig. 3.77, gears Z_1 and Z_3 are drivers and gears Z_2 and Z_4 are driven gears.

$\therefore$ With this gear train and an indexing of 5 holes in a 16 hole circle, the 127 divisions would be obtained.

Rule of Thumb.

(*i*) If $(k - z)$ is positive, the index plate must rotate in the same direction as that of the crank.

(*ii*) If $(k - z)$ is negative, the rotation of the index plate is in opposite direction to that of the crank.

Example. Let $z = 153$

(*i*) Let $k = 155$

$\therefore$ Simple indexing $= \dfrac{40}{155} = \dfrac{8}{31}$, that is, 8 holes in 31 hole circle.

$$\text{Change gears} = \frac{40}{k}(k - z) = \frac{40}{155} \times 2 = \frac{80}{155} = \frac{16}{31}$$

It is clear that this gear ratio can not be accommodated with the gears available. Therefore, choice of $k = 155$ is not correct.

(*ii*) Let $k = 150$

$\therefore$ For simple indexing $= \dfrac{40}{150} = \dfrac{4}{15}$, that is, four holes in 15 hole circle.

$$\text{Change gears} = \frac{40}{k}(k - z)$$

$$= \frac{40}{150} \times 3 = \frac{12}{15} = \frac{2 \times 6}{3 \times 5} = \frac{32}{48} \times \frac{48}{40} = \frac{\text{Drivers}}{\text{Driven}}$$

$\therefore$ Change gears : $Z_1 = 32$, $Z_2 = 48$, $Z_3 = 48$, $Z_4 = 40$

(*iii*) We can obtain another solution also.

Now indexing of 153 divisions by

simple indexing $= \dfrac{40}{153} = \dfrac{1}{3.825}$ which is approximately equal to $\dfrac{7}{27}$, that is, 7 holes in a hole circle 27 for each indexing.

Now number of crank turns for 153 indexings

$$= \frac{7}{27} \times 153 = 39\frac{18}{27}$$

which is less than 40 by $\dfrac{9}{27}$ turns. Therefore, the gear train must rotate the index plate in the same direction as the crank.

$$\text{Change gears} = \frac{\text{Drivers}}{\text{Driven}} = \frac{9}{27} = \frac{1}{3} = \frac{24}{72}, \text{ that is, a simple train.}$$

$\therefore$ For 153 divisions the indexing is : 7 holes in a 27 hole circle ; gear ratio 24 / 72 and the crank and the index plate rotating in the same direction.

Note. It is clear from above that for a given problem, many solutions are possible. Solution (*iii*) is better than (*ii*) as a simple gear train is needed. To avoid wastage of time if the gear ratio does not fit the available gears as in solution (*i*) above, the hole circle selected for approximate simple indexing should be a number whose factors are accommodated in the available gears, for example, use hole circles of 18, 20, 21, 27 etc.

Rules of Thumb for gear trains.

(*a*) If $k > z$ and the gear train is simple, then only one idler gear is to be used.

(*b*) If $k > z$ and the gear train is compound, no idler gear is to be used.

(*c*) If $k < z$ and the gear train is simple, then two idler gears are to be used.

(*d*) If $k < z$ and the gear train is compound, then no idler gear is to be used.

5. Angular indexing. Angular indexing is used when it is necessary to cut grooves or slots subtending a given angle at the centre of the circle upon which they are spaced.

We know that 40 turns of the index crank will rotate the head spindle and hence the workpiece through one revolution, that is, 360°.

$\therefore$ 1 turn of the crank = 9° of the spindle

$$\therefore \text{ Turns of crank to give any angle} = \frac{\text{Angle required}}{9}$$

Example. Do angular indexing for 38°.

$$\text{Indexing} = \frac{38}{9} = 4\frac{2}{9} = 4\frac{4}{18}$$

that is, four complete turns of the crank and 4 holes in a 18 hole circle.

Example. Do angular indexing for $51^\circ\ 37'$

$$\text{Indexing} = \frac{51^\circ\ 37'}{9} = 5 + \frac{6^\circ\ 37'}{9^\circ} = 5 + \frac{397'}{540'}$$

Now no index plate is available with 540 hole circle. So, the exact indexing is not possible. For approximate indexing, let us convert the residual turn to a continued fraction :

$$
\begin{array}{l}
397\,)\,540\,(\,1 \\
\quad\;\; 397 \\
\hline
\quad 143\,)\,397\,(\,2 \\
\qquad\;\; 286 \\
\hline
\qquad 111\,)\,143\,(\,1 \\
\qquad\quad\;\; 111 \\
\hline
\qquad\quad 32\,)\,111\,(\,3 \\
\qquad\qquad\;\; 96 \\
\hline
\qquad\qquad 15\,)\,32\,(\,2 \\
\qquad\qquad\quad\; 30 \\
\hline
\qquad\qquad\quad 2\,)\,15\,(\,7 \\
\qquad\qquad\qquad 14 \\
\hline
\qquad\qquad\qquad 1\,)\,2\,(\,2 \\
\qquad\qquad\qquad\quad\; 2 \\
\qquad\qquad\qquad\quad\; 0
\end{array}
$$

$\therefore$ We can write the residual fraction as :

$$\frac{397}{540} = \cfrac{1}{1 + \cfrac{1}{2 + \cfrac{1}{1 + \cfrac{1}{3 + \cfrac{1}{2 + \cfrac{1}{7 + \cfrac{1}{2}}}}}}}$$

$$\text{First fraction} = \frac{1}{1}$$

Second fraction $\dfrac{1}{1+\dfrac{1}{2}} = \dfrac{1}{\dfrac{3}{2}} = \dfrac{2}{3}$

Third fraction $= \dfrac{1}{1+\dfrac{1}{2+\dfrac{1}{1}}} = \dfrac{1}{1+\dfrac{1}{3}} = \dfrac{1}{\dfrac{4}{3}} = \dfrac{3}{4}$

Fourth fraction $= \dfrac{1}{1+\dfrac{1}{2+\dfrac{1}{1+\dfrac{1}{3}}}} = \dfrac{1}{1+\dfrac{1}{2+\dfrac{1}{\dfrac{4}{3}}}}$

$= \dfrac{1}{1+\dfrac{1}{2+\dfrac{3}{4}}} = \dfrac{1}{1+\dfrac{1}{\dfrac{11}{4}}} = \dfrac{1}{1+\dfrac{4}{11}}$

$= \dfrac{1}{\dfrac{15}{11}} = \dfrac{11}{15}$ and so on

All the fractions will be :

$\dfrac{1}{1}, \dfrac{2}{3}, \dfrac{3}{4}, \dfrac{11}{15}, \dfrac{25}{34}, \dfrac{186}{253}, \dfrac{397}{540}$

Now with the available index plates, the nearest indexing for the residual turn is $\dfrac{11}{15}$, that is, 11 holes in a 15- hole circle.

$\therefore$ Angle will be $= 5\dfrac{11}{15} \times 9 = 51^\circ, 36'$, that is, an error of $1'$.

Now Cincinnati and Parkinson Index Plates have the following hole circles :

First : 24, 25, 28, 30, 34, 37, 38, 39, 41, 42, 43

Second : 46, 47, 49, 51, 53, 54, 57, 58, 59, 62, 66

Using the cincinnati plate, the closest fraction to the residual fraction is 25/34

$\therefore$ Angle will be $= 5\dfrac{25}{34} \times 9 = 51^\circ, 37', 3\dfrac{9}{17}''$

That is an error of only $3\dfrac{9}{17}''$ in the angle. This is well within the accuracy of the ordinary dividing heads.

PROBLEMS

1. Define the "milling" process.
2. Define a "milling cutter".
3. What type of surfaces can be produced by a milling process ?
4. Sketch and explain the two basic types of milling operations.
5. Sketch and contrast the two milling methods of machining flat surfaces.
6. What is the basic type of milling machine ?
7. Sketch and explain the working of a plain column and knee type milling machine.
8. What are the three motions on a plain milling machine ?
9. What are universal and omniversal horizontal milling machines ?

10. Sketch and describe a vertical milling machine.
11. What is the difference between a fixed head and a swivelling head vertical milling machine ?
12. Describe the difference between a horizontal milling machine and a vertical milling machine.
13. What is feed on a milling machine ?
14. How is the size of a column and knee type milling machine given ?
15. Sketch and describe a Bed type milling machine.
16. Sketch and describe a Drum type milling machine.
17. Write on the following milling machines :
    (*a*) Planer type milling machine. (*b*) Hand operated millers.
    (*c*) Continuous or rotary millers.
    (*d*) Tracer controlled production milling machines.
    (*e*) Special purpose milling machines.
18. How the milling cutters are classified according to :
    (*i*) design. (*ii*) the way the teeth are sharpened.
19. What is a plain milling cutter ?
20. What is a "slab cutter" ?
21. What is face milling ?
22. Name four types of side-milling cutters.
23. What is the difference between face milling and end milling ?
24. What is a "shell end mill" ?
25. Why is end milling done best on a vertical milling machine ?
26. How can sawing be done on a milling machine ?
27. Describe the three types of milling cutters according to the method of mounting the cutters.
28. Describe the construction of milling cutters.
29. Write the advantages of carbide-tipped milling cutters.
30. Write on the materials for milling cutters.
31. What is meant by hand of milling cutters ?
32. Describe the various work holding devices used on milling machines.
33. How will you cut the following types of surfaces on milling machines :
    (*a*) flat surfaces. (*b*) Slots, grooves and splines.
    (*c*) Recesses. (*d*) Contoured surfaces.
    (*e*) Toothed gears and helical grooves.
34. Write on the following milling operations :
    (*i*) Straddle milling (*ii*) Gang milling (*iii*) String milling.
35. Describe a dividing head.
36. What are : plain, universal and optical dividing heads ?
37. Sketch and explain the direct method of indexing.
38. Describe Simple indexing method.
39. Describe compound indexing method.
40. Sketch and describe Differential indexing method.
41. What is angular indexing ?
42. Do the indexing for the following number of divisions :
    (*i*) 16 (*ii*) 45 (*iii*) 53 (*iv*) 18

43. What is the advantage of a helical-tooth cutter over a straight- tooth cutter for slab milling ?
44. How does a duplicator differ from a profiler ?
45. Write briefly on :- Abreast Milling, combined Abreast and String milling, Transfer-base milling, Index milling, Reciprocal Milling and Progressive milling.

3.6. GRINDING OPERATION

Grinding Operation is a method of machining workpieces by the use of a rotary abrasive tool, called "grinding wheel". Such wheels are made of fine grains of abrasive materials held together by a bonding material, called a 'bond'. Each individual and irregularly shaped grain acts as a cutting element (a single point cutting tool). The grinding operation can be : rough grinding or finishing (precision grinding). Rough grinding is a commonly used method for removing excess material from castings, forgings and weldments or as a method for removing or snagging thin fins, sharp corners, burrs or other unwanted projections from various shapes of workpieces. Precision grinding is the principal production method of cutting materials that are too hard to be machined by other conventional tools or for producing surfaces on parts to higher dimensional accuracy and a finer surface finish as compared to other manufacturing methods. Since cutting edges of the grits are extremely thin it is possible to remove much smaller chips and refine surfaces to a much greater accuracy of finish and dimension than with other machining methods.

3.6.1. Abrasives. Abrasives include fine crushed substances in the form of grains and powders obtained by crushing natural or synthetic materials. The main characteristics of abrasive materials are : grain size and shape, hardness, toughness, resistance to attrition and friability. After attrition takes place, a grain must be friable to break apart in chunks to form new sharp edges.

There are two types of abrasives used in metal machining : Natural and Artificial or Synthetic.

(*a*) **Natural abrasives.** The natural abrasive materials are : Sandstone or Solid quartz, emery (50 to 60% crystalline Al_2O_3 + Iron oxide), corundum (75 to 90% crystalline Al_2O_3 + Iron oxide) and natural diamond. In the case of natural diamonds, the abrasive grains called "borts" are produced by crushing unsuitable gem stones to the required particle size. Their use is limited because of the tendency of dulling and glazing of the cutting faces. Natural abrasives lack uniformity of properties and reliability and have been largely replaced by manufactured abrasives.

(*b*) **Artificial abrasives.** These abrasives include : Aluminium oxide, Silicon carbide, Diamond (a form of pure carbon), boron Carbide and Cubic Boron nitride. The manufactured abrasives have well defined and controlled properties of hardness, toughness and type of structure.

1. Aluminium Oxide. Aluminium oxide, Al_2O_3, is known by such trade names as "Alundum" and "Aloxite". Aluminium oxide abrasive is composed chiefly of crystalline aluminium oxide. It is obtained by melting materials rich with this oxide (bauxite ore which is mainly aluminium hydroxide) in an electric furnace. Iron chips and coke are added to combine with and remove impurities. The refined aluminium oxide comes out of the furnace in a large lump called a "pig". It is crushed and rolled into small grains, treated magnetically to remove ferrous impurities and washed. Aluminium oxide is softer than silicon carbide. So it is tough and is not easily fractured. Hence it is extensively used for grinding materials of high tensile strength, such as, most steels, H.S.S., ferrous alloys, non-ferrous cast alloys and annealed malleable and ductile iron.

White Aluminium Oxide :- It is the most refined form of Al_2O_3 and contains a very high percentage of it. This cutting tool material is recommended for cool and precision grinding, where heat is required to be kept at a minimum temperature. It is especially recommended for tool room jobs. This material is particularly suitable for surface grinding and tool and cutter grinding.

2. Silicon Carbide. Silicon carbide is a chemical compound of silicon and carbon. It is produced by fusing quartz sand and powdered coke in an electric furnace. Sawdust and salt are also added to the mix. At temperature around 2300°C, the silicon of sand combines with carbon of coke to form silicon carbide. The sawdust burns and leaves pores to let the gases escape. The salt helps remove impurities. After the process has run its course, the furnace is cooled. The core of loosely knit silicon carbide crystals is broken into individual grains. Two types of silicon carbide are available : Pure silicon carbide of green variety and black or grey. Green silicon Carbide is harder and so is more brittle and friable and is in general a better abrasive than black silicon carbide. Both the types of silicon carbide have a higher hardness than artificial Al_2O_3. Being extremely brittle, Si C is not recommended for grinding hard materials. Silicon carbide wheels are used for grinding non-ferrous metals, non-metallic elements and cast irons.

Green Silcon Carbide :- This variety of SiC catting tool material is used for grinding Tungsten Carbide tools. This abrasive material is especially prepared for this class of work and has long been recognised as the only efficient abrasive for carbide tipped tools.

3. Diamond. Synthetic diamonds (a form of pure carbon) are mainly used for truing and dressing other grinding wheels, for sharpening carbide tools, and for processing glass, ceramics and stone.

4. Boron Carbide. Boron carbide is a combination of boron and carbon (B_4C). It is manufactured as a powder of grayish-black colour with the grain size not over 120 μm. The grains are characterized by sharp faces and a higher abrasive ability. Boron Carbide is mainly used for grinding and lapping very hard metals, hard alloys, glass and jewels (ruby, topaz).

5. Cubic Boron Nitride. Cubic boron nitride is a combination of boron and nitrogen. Its manufacture and properties have already been discussed under chapter 7. It is used for grinding workpieces of tool and die steels as well as many types of hard high-alloy steels

Grain Size. For manufacturing abrasive tools and for application of abrasives in finishing and polishing operations, abrasive materials are crushed to obtain particles of desired size. These particles are known as "grains" or "grits". For some applications, the crushing is continued until the abrasive becomes a fine powder, often called a "flour". The metal removal rate and the quality of a ground workpiece depends largely on the grain size of the abrasive material of which the grinding wheel is made. Grain size is denoted by a number indicating the number of meshes per linear inch of the screen through which the grains pass when they are graded after crushing. A grain size of 60 means that the abrasive will pass through a screen having 60 openings per linear inch.

The screen or sieve number and the mean diameter of the abrasive grain are related approximately as,

$$\text{Mean diameter of abrasive grain} = \frac{15}{\text{grit size}}, \text{ mm}$$

The grain size may be broadly classified as given below : Generally, coarse-sized grains (upto 24) are normally used for roughing which involves heavy material removal. Finer grits (70 to 180) are used for finishing operations where material to be removed is less. For producing very high surface finish, the grain sizes ranging from 220 to 600 and finer (upto 1000) are used, for example, grinding threads etc. Medium size grains (30 to 60) are used in operations which require both stock removal and finish, that is, for ordinary grinding and tool grinding.

The standard grain sizes are :

1. Coarse : 10, 12, 14, 16, 20, 24
2. Medium : 30, 36, 46, 54, 60
3. Fine : 70, 80, 90, 100, 120, 150, 180
4. Very fine : 220, 240, 280, 320, 400, 500, 600, 700, 800, 900, 1000

3.6.2. Bonding Materials. A grinding wheel consists of abrasive grains that are supported or held together by a bond. All bonds must be sufficiently strong to with stand the stresses of the high speed rotating grinding wheel. They must be capable of holding the abrasive grains firmly, yet must not be so dense as to impede the cutting action. The choice of bond depends on operating conditions of the abrasive tool : grinding speed, pressure on the tool, heat formation in the grinding zone and conditions of cooling.

Bonds are classified as non-organic and organic. Metallic, Vitrified and silicate bonds are non-organic. Resinoid, Rubber, Shellac and Oxychloride bonds are organic. These bonds are discussed below :

1. **Vitrified bond.** The most widespread is the vitrified bond. It is made of refractory clay, feldspar and quartz. It hardens to a glass like structure on heating, (The literal meaning of " Vitrified" is : to convert into glass by the action of heat). The advantages of vitrified bond are : it is strong, highly porous, resistant to heat, highly chemically stable (not affected by water, oils and acids), has a high cutting capacity and properly removes heat. Vitrified-bonded wheels can be used with a peripheral speed upto 35 m/*s*. The disadvantage of vitrified-bonded wheels is their brittleness. Vitrified bonds are strongest and hardest.

2. **Silicate bond.** This bond consists of water glass ($NaSiO_3$) mixed with zinc oxide, lime and other fillers. It hardens on heating at about 260°C. It is not as strong as vitrified bond, because self-dressing (crum- bling-out of dull grains) occurs more intensively. The wheels become soft when exposed to moisture. Silicate bonded wheel is the softest. Silicate bond is used for face grinding with large area of wheel in contact and for grinding of items sensitive to heat, such as cutters and precision tools.

3. **Resinoid bond.** A resinoid bond is a synthetic resin or plastic. Grinding wheels with a resinoid bond are very strong and elastic, highly stable under variable loads, but their porosity is lower than that of vitrified-bonded wheels. Resinoid has a polishing effect on the surface, thus diminishing its roughness. Resinoid-bonded wheels work at the surface speed of 35 - 40 m / s and is used for high speed operations such as cutting off or fettling. It gives a very rapid stock removal rate. The disadvantages of these wheels are : loss of hardness when heated to a temperature of 200 – 250°C and insufficient resistance to alkaline coolants (it becomes soaked). These wheels can be readily reinforced with steel rings, or fibreglass or other fibres to increase their flexural strength.

4. **Rubber bond.** It basically consists of synthetic rubber with vulcanizing additions and fillers. In manufacturing the wheels, the abrasive material is mixed with rubber, then sulphur and other components are added. This mixture is pressed in heated moulds ; rubber becomes hard and retains elasticity when heated to 150°C. The resulting grinding wheels are stronger and may be rotated at higher rotational speeds. These wheels being slightly elastic are used in slitting and cutting-off operations (Elastic wheels less than 8 mm thick) and also in finish grinding. They are not softened by alkaline cooling liquids and are water resistant. However at temperatures above 150°C, the wheel becomes soft (dull) and requires cooling and dressing. These wheels are also used for centreless grinding control wheels.

5. **Shellac bond.** This is also a organic bond like resin and rubber. The abrasive material is mixed with shellac in a heated solution. It is then pressed and rolled into the desired shapes. Baking is done for several hours at about 150°C. This bond is primarily used for making strong, thin wheels having some elasticity. These wheels produce high surface finish and are used for grinding parts like camshafts and mill rolls. Occasionally, these wheels are used for slitting wheels (> 0.80 mm thick).

6. **Oxy-Chloride bond.** This organic bond is used particularly for disc grinding wheels. These wheels are less brittle and less sensitive to side loads as compared to vitrified-bond wheels. Also, the temperature during grinding is lower than with other bonds except the silicate bond. The drawbacks of these wheels are : affected by acidic solutions, dampness and sudden changes in

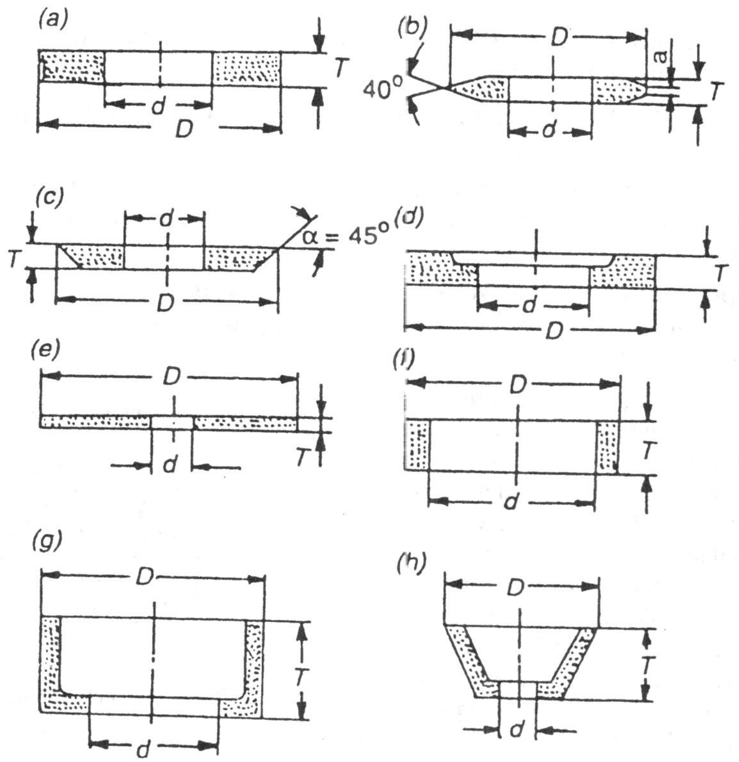

Fig. 3.78. Grinding Wheel Shapes.

temperature. Also, the strength of these wheels is lowered with long storage.

7. **Metallic bond.** Metallic bond is used for holding together the grains of diamond and cubic boron nitride. Copper, tin, aluminium and their alloys are used as bonding materials.

3.6.3. Grinding Wheels.

(*i*) **Manufacture.** The manufacture of the organic-bond grinding wheels have been explained in the previous article. To make vitrified grinding wheels, there are three methods : puddling, tamping and pressing. Firstly, the clay and abrasive material are thoroughly machine mixed. In the puddling method, water is added and the mixture is poured into moulds. For pressed wheels, the dry or semidry mixture is moulded and squeezed in hydraulic presses. In the tamping process, the squeezing pressure is less than in pressed wheels. Then the wheels are baked and dried. The puddled wheels must be trimmed to size.

Grinding wheels are vitrified by being fired for several days at high temperatures, like pottery. When hard, the wheels are trued, their arbor holes are bushed with babbit metal or lead and large wheels are balanced.

(*ii*) **Shape.** The shape and size of grinding wheels depend on the design of the machine whereon it will be used, the power of the machine, the operation to be performed, the shape and size of the workpiece and the grinding conditions. According to their shape, the wheels are classified as :

1. **Straight wheels.** (Fig. 3.78 (*a*)), for cylindrical, internal, centreless and surface grinding.
2. **Bevelled-face straight.** (Fig. 3.78 (*b, c*)), for grinding thread, gear teeth etc.
3. **Straight recessed.** (Fig. 3.78 (*d*)), for cylindrical grinding and facing.

4. Abrasive-cutting types. (Fig. 3.78 (*e*)), are from 0.5 to 5 mm thick are used for abrasive-cutting off and slotting operations.

5. Cylinder, Straight Cup and Flaring Cup. (Fig. 3.78 (*f, g, h*)), for surface grinding with the end of the wheel.

(*iii*) **Size.** The principal dimensions of a grinding wheel are the outside diameter, *D*, bore *d* and the width *T*. Straight wheels are available with an outside diameter from 3 to 1100 mm and with width from 6 to 200 mm.

(*iv*) **Properties of Grinding Wheels.** The properties of a grinding wheel that determine how it acts are : kind and size of abrasive, how closely the grains are held and the kind and amount of bonding material.

(*a*) **Grade.** The grade or hardness of a grinding wheel is a measure of the resistance offered by the combined strength of the bond and the abrasive grains to the external forces striving to tear the grains from the wheel. The bonding material surrounds the individual grains and links them together by connectors called "Posts" (Fig. 3.79). The sizes and strengths of the posts depend upon the kind and amount of bonding material in a wheel. If these posts are large in cross-section, the force required to break a grain from the wheel is greater than when the posts are small. If a high dislodging force is required, the bond is said to be hard and vice versa.

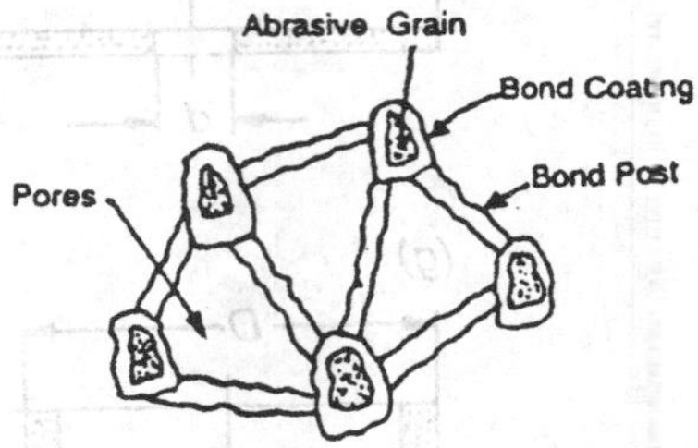

Fig. 3.79. Arrangement of Grains and Bond.

The easier a grain is torn out of the bond, the softer the wheel, and vice versa. A wheel that is too soft will crumble and rapidly loose its shape and size. If the wheel is too hard, it will load quickly with chips and there will be burns on the workpiece. Frequent dressings will be required because the dulled grains will be held in the hard bond, not allowing new sharp grains to be exposed for cutting action, that is, the wheel will not be self-dressing. Hence, soft wheels are used in grinding hard steels because the dulled grains are easily tornout, exposing sharp-edged grains underneath (self-dressing). On the opposite, in grinding soft steel, a hard wheel is used because it will require less frequent dressing. Soft coarse-grained wheels are used for grinding copper and brass because fine-grained wheels become loaded quickly.

The grades of abrasive wheels is designated by letters of the English alphabet, 'A' denoting the softest, 'Z' the hardest and 'P' the medium grade. As a rule, hard grades are used to grind soft materials, and vice versa.

1. Soft grade : A to H
2. Medium grade : I to P
3. Hard grade : Q to Z

(*b*) **Structure.** Structure of a grinding wheel is characterised by the quantitative relation and arrangement of the grains, bond and voids (pores) in a wheel (or, more simply, the spacing of the grains). The voids or spaces provide room for chips to escape during a cut and for cutting fluid to be carried into a cut. The denser the structure, the more grains fall within the surface unit and the lower the number of pores. In the open structure, the number of pores in the tool is higher and the abrasive grains are separated more widely. The denser the structure of a grinding wheel, the harder it is and more durable. However, the wheel face becomes "clogged" or "loaded" with cut chips and the wheel becomes more prone to glazing and forming of burns on the surface being ground. The open structure prevents the cut chips from clogging or loading the wheel face and hence increases the efficiency of cutting. The proportion of bond to total wheel volume can vary from 10% to 30% giving either a very close abrasive spacing or open abrasive spacing respectively.

The grinding wheels are made with a specified structure. The structures are denoted by numbers from 0 to 12 inversely proportional to the content of abrasive grains in the wheel.

1. Dense structure : 0, 1, 2, 3
2. Medium structure : 4, 5, 6, 7, 8
3. Open structure : 9, 10, 11, 12

The dense structures containing much grain and small-size pores are mainly used for lapping, finish grinding and polishing.

Medium structures 4 and 5 are applied in grinding hard and brittle materials, when a high quality of surface finish is specified. Structure 6 is used for cylindrical grinding and structures 7 and 8 for surface grinding of soft materials. The open structures allow for more productive operations but the wheel durability is lower. They are used for high-velocity grinding.

Standard marking of Grinding Wheel. All grinding wheels are marked. Markings in the form of conventional designations are applied with an indelible paint to the wheel flat surface. The marking indicates the manufacturer, abrasive material, its grain size, grade, structure, bond type etc. A typical wheel marking is shown below :

51 - A - 36 - L - 5 - V - 23

The explanation of the marking is :

51 → Manufacturer's symbol indicating exact kind of abrasive, (use optional).

A → Abrasive type : A for Al_2O_3, C for Si C and D for diamond.

36 → grain size

L → grade

5 → structure

V → Bond type

V : Vitrified

S : Silicate

R : Rubber

B : Resinoid

E : Shellac

O : Oxychloride

BF : Reinforced resinoid

RF : Rubber reinforced

23 → Private marking to identify wheel (use optional).

3.6.4. Selection of a Grinding Wheel. The selection of a grinding wheel is based on the following factors :

1. Size and shape of wheel
2. Kind of abrasive
3. Grain size of abrasive particles

4. Grade of bond

5. Structure

6. Kind of bond material

7. Function of grinding wheel

8. Other factors : wheel speed, work speed, materials to be ground and general condition of the machine.

The factors 1 to 6 have already been discussed in the previous articles. The surface speed of the grinding wheels is given in Table 3.3 :

Table 3.3. Surface speeds of grinding wheels

Type of bond	*Surface Speed m/s*
1. **Vitrified or Silicate**	
Soft	23 – 28
Medium	25 – 30
Hard	28 – 33
2. **Organic**	
Soft	23 – 33
Medium	35 – 40
Hard	38 – 50

In a summary, the general guide lines for selection of grinding wheel, are given below :

1. Aluminium oxide abrasive is best for grinding steel (both hardened and unhardened), malleable iron and soft bronzes. Black silicon carbide abrasive is most efficient in grinding cast iron, bronze and aluminium castings, and cemented carbides. Green silicon carbide is used in sharpening carbide-tipped cutting tools.

2. The harder the metal being ground, the softer the wheel should be, and vice versa, since in grinding a hard metal the grains will wear more intensively and be broken out of the wheel more readily to expose new sharp grains. So, choose a hard-grade wheel for soft materials and a soft grade wheel for hard materials.

3. Coarser wheels (a large grit abrasive) are used in grinding ductile and soft metals to avoid rapid loading of the wheel ; fine-grained wheels are used for brittle and hard metals.

4. The larger the area of contact with the work, the coarser the grain of the wheel should be. That is, choose a large grit for the maximum metal removal rate and a small grit for a good finish.

5. To avoid breaking down, harder wheels should be more coarse- grained, since, other conditions being equal, the grain of a coarse-grained wheel is subject to a higher load than that of a fine-grained wheel.

6. The longer the arc of contact with the work, the less the grain depth of cut, and the greater the tendency to load the wheel and overheat the surface being ground. Therefore, the longer the arc of contact, the softer the wheel should be.

7. Regarding the structure of the wheel :

Choose wheels of dense structure for finish and form grinding, since they retain their shape well. Choose wheels of medium-dense structure for grinding hardened steel parts, for sharpening cutting tools, for cylindrical, centreless and internal grinding and for surface grinding with the periphery of the wheel. Wheels of open structure are used in grinding soft and ductile metals and in surface grinding with the face of a cup, cylinder or segmental wheel.

8. Choose a resinoid, rubber or shellac bond for good finish and a vitrified bond for a maximum metal removal rate.

Diamond Grinding Wheels. In contrast to grinding wheels of other abrasive materials, diamond wheels are not solid throughout, but consist of a diamond layer (0.5 to 3 mm thick) secured on a body of duralumin, steel or plastic. The diamond concentration of the wheel, that is, concentration of diamond grains per unit volume of the layer [consisting of diamond grains, bond and a filler (solid minerals)] may be 50, 100 or 150 per cent. Vitrified, resinoid and metal (most frequently bronze) bonds are employed.

3.6.5. Dressing and Truing of Grinding Wheels. Grinding wheels are dressed to restore their cutting capacity which is lost when the voids between grains are loaded with the waste material of the grinding process (sintered metal dust, abrasive grit and particles of binding material) and the grains are dull from wear. The former defect is called "loading" and the grains affected as "loaded" and the latter defect is known as "Glazing" and the affected wheel face as "Glazed". Truing, an allied operation done with the same tools, restores the correct geometric shape of the wheel, that is, it makes the face of the wheel concentric to its axis of rotation and its sides flat and parallel and normal to its axis of rotation. It also makes the face and sides of the wheel concentric to its axis of rotation when a new wheel is mounted on the grinder.

Dressing and truing involve the removal of a layer of worn abrasive grains from the surface of the wheel. Wheels are dressed and trued by cutting by means of industrial diamonds in the form of multiple-stone diamond tool bits or by pressing against diamond crystals fixed in some type of setting (mechanically, by hand setting or soldering) in a steel holder. Diamond substitutes are also applied : cemented carbide or special abrasive disks (called wheel dressing rollers), steel rollers and star-type dressers. The method of crush dressing a grinding wheel by pressing a high strength roller against its surface, is very fast and particularly economical for dressing forming tools. The dressing of the grinding wheel is usually done in the grinder itself so that alignment and wheel balance are not lost.

3.6.6. Grinding Ratio. Grinding ratio is defined as the volume of material removed from the work per unit volume of wheel wear. It is a very useful term to describe "grindability" of a given material. This ratio is determined experimentally by using a set of specific boundary conditions. Thus, the ratio for a given material, will vary according to the type of operation, nature of grinding wheel, type of grinding fluid, speeds and feeds and so on.

3.6.7. Abrasive Machining. The term "abrasive machining" refers to the processes in which metal is removed by a multitude of hard, angular abrasive grains or grits which may or may not be bonded to form a tool of some definite geometric form. Grinding is a form of abrasive machining in which the abrasive tool is in the shape of a wheel, called grinding wheel.

3.6.8. Grinding Machines. Grinding machines are classified as : cylindrical, internal, centreless, surface and special grinding machines.

1. Cylindrical Grinding Machines. The term "cylindrical" is applied to the large group of grinding machines which employ centres for mounting the workpiece to be ground. Most common of the external cylindrical grinding machines is the plain grinder, which is a general purpose production machine for the grinding of axles, shafts, splines etc. A plain grinding machine is illustrated in Fig. 3.80. All the units of the machine are arranged on the base which also houses the drive mechanisms and the hydraulic drive. Sliding or traversing table (lower table) is reciprocated by the hydraulic drive to obtain the longitudinal feed at a speed from 0.08 to 10 m/min or it is traversed manually by a handwheel through a system of gearing. The worktable or the swivel table (upper table) with T slots for securing the head stock and tailstock is mounted on the sliding table.

The headstock serves to mount and drive the workpiece which is held either between centres or in a chuck. The work spindle in the headstock is powered by a motor located in the upper part of the headstock. The workpiece held between centres is rotated in the same manner as is done in a centre lathe. The tailstock supports the rear end of the work mounted between

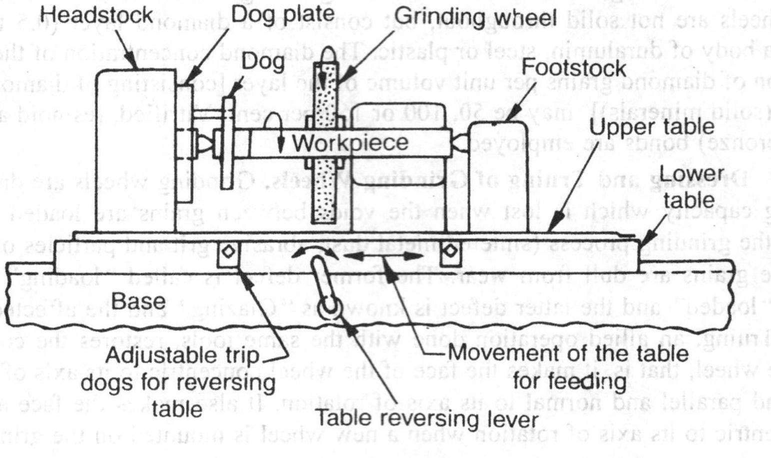

Fig. 3.80. External Cylindrical Grinding Machine.

centres. Its spindle can be advanced and retracted. The tailstock spindle with the centre is commonly advanced and clamped by a spring device. The tailstock can be adjusted along the slots of the worktable to suit the length of the workpiece and can be clamped in the required position by bolts or an eccentric clamp.

The grinding wheel is mounted and clamped on the spindle of the wheel head which travels along cross ways at the rear of the base. The wheel is driven at speeds upto 1500 rev/min from a motor in the upper part of the head.

Adjustable trip dogs clamped in longitudinal slots provided at the side of the sliding table are set up to reverse the table at the end of the stroke. Cross feed of the wheel after each stroke of the table is obtained by manually or hydraulically operated crosswise movement of the wheel head.

Chucking type cylindrical grinders are production type machines for use in rapid grinding of relatively short parts, such as ball-bearing races.

Various methods of cylindrical grinding are employed. The most extensively used are : traverse grinding, plunge-cut grinding and full-depth grinding.

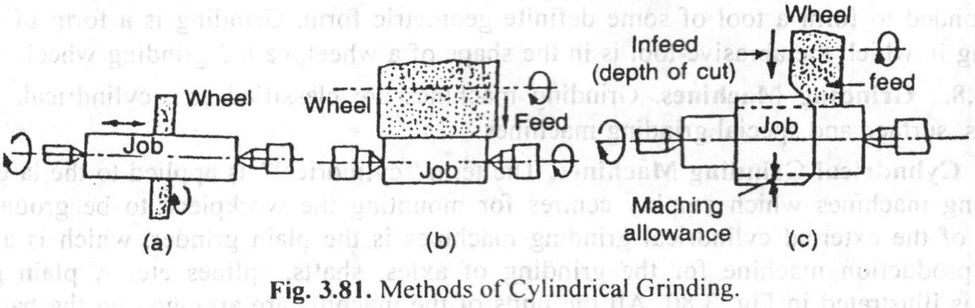

Fig. 3.81. Methods of Cylindrical Grinding.

(*i*) **Traverse Grinding.** (Fig. 3.81 (*a*)). In traverse grinding or infeed grinding, the grinding wheel is moved into the work. The desired surface is then produced by traversing the workpiece across the wheel on vice-versa. This method is used to grind workpieces of considerable length.

(*ii*) **Plunge - Cut Grinding.** (Fig. 3.81 (*b*)). The basic movement is of the grinding wheel being fed radially into the work while the later revolves on centres. It is similar to form cutting on lathe. The method is used for short workpieces where the width of the wheel overlaps the length to be ground. Short rigid workpieces can be ground by this method.

(*iii*) **Full-Depth Grinding.** (Fig. 3.81 (c)). The wheel is trued to obtain an entering taper or step, and the whole allowance is ground off in one or two lengthwise passes. The method is usually applied to relatively short surfaces of rigid shaft-type workpieces.

Tapers can be efficiently ground in centre type cylindrical grinders. Setting upto ± 10° can be made by swiveling the worktable to the required angle with respect to the axis of the grinding wheel.

The entire grinding cycle can be manual, semiautomatic or fully-automatic as desired. Plain grinders can handle parts from the smallest upto 0.9 m in diameter by 4.8 m in length, the lengths being proportional to the swing capacity of each machine.

Note. In very large grinders, the wheel is reciprocated across the workpiece held and rotated between centres, because of the massiveness of the work.

Universal Cylindrical Grinders. These grinders, in addition to the features offered by plain grinders, are provided with a swiveling headstock and a swiveling wheel head. This permits the grinding of taper of any angle, much greater than is possible in plain grinders. Universal machines are available to handle parts requiring swings upto 450 mm and centre distance of 1800 mm.

2. **Internal Cylindrical Grinding Machines.** Internal grinding is employed chiefly for finishing accurate holes in hardened parts, and also when it is impossible to apply other more productive methods of finishing accurate holes, for example, precision boring, honing etc.

There are two general methods of internal grinding : (1) with a rotating workpiece (2) with the workpiece held stationary. The first method is used in grinding holes in relatively small workpieces, mostly bodies of revolution, for example, the bores of gears and the inner surfaces of ball bearing rings. The workpiece is held in a chuck or special fixture and rotated in the same manner as in a lathe, (Fig. 3.82 (*a*)). A straight type grinding wheel is rotated and has two feeds-longitudinal feed along the wheel axis and is thus reciprocated back and forth through the length of the hole, and intermittent cross feed (radial feed) at the end of each pass, which determines the depth of cut. The grinding wheel is carried in a grinding head. In another design of the machine, the grinding wheel is rotated in a fixed position while the workpiece is slowly rotated and traversed back and forth.

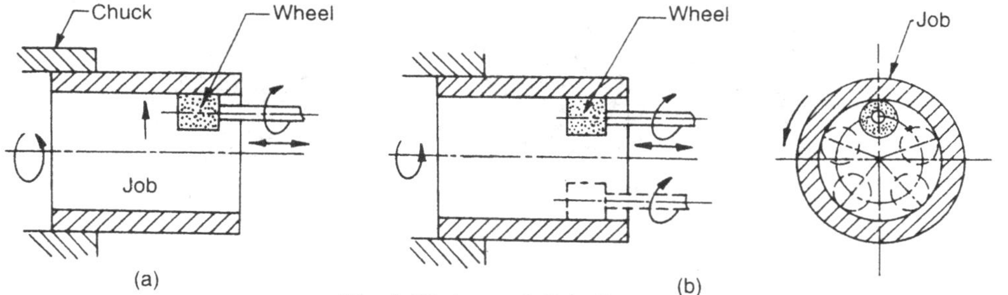

Fig. 3.82. Internal Grinding.

The peripheral speed of the wheel in grinding small holes does not exceed 10 m/s. The speed increases with the diameter of the hole to be ground, reaching 30 for diameters over 30 mm. The depth of cut (infeed per stroke) is limited by the relatively low rigidity of the wheel spindle. Infeeds range from 0.005 to 0.02 mm in rough grinding steel and cast iron, and from 0.002 to 0.01 mm per stroke in finish grinding. Lower rates of infeed are used for holes less than 40 mm in diameter and for holes with a large length-to-diameter ratio. An over-travel equal to 1/3 of the wheel width is required at each end of the stroke in internal grinding.

The second method of internal grinding is used for grinding holes in large bulky workpieces (housing-type parts) that are inconvenient or even impossible to clamp in a chuck of the grinder. They are mounted on the table of a planetary grinding machine, (Fig. 3.82 (*b*)). In addition to rotation about its axis, the wheel spindle of this type of machine also rotates with a planetary motion about the axis of the hole being ground. Axial motion of the wheel provides the longitudinal feed.

The essential difference between internal and cylindrical (external) grinding is that the former is performed with a wheel of much smaller diameter, usually from 0.7 to 0.9 of the hole diameter. Owing to their small diameters, the life of the internal grinding wheels in naturally shorter than for other kinds of grinding. Wheels of a lower grade (softer wheels) should be selected for internal grinding than for similar conditions in cylindrical (external) grinding, since the angle of contact with the surface being ground is much larger and the workpiece may be heated to a high temperature in grinding.

3. Centreless Grinding. Centreless grinding is performed on workpieces which do not or can not have centres, such as, pistons, valves, rings, tubes, balls, wrist pins, drills, bushings, drill rods, shafts and balls and rollers for bearings etc. Centreless grinding can be done both for external and internal surfaces.

(*a*) **External Centreless Grinding.** External centreless grinding can be carried out by three methods : Through feed, infeed and end feed methods.

(*i*) **Through Feed Grinding.** The principle of through feed centreless grinding is illustrated in Fig. 3.83 *a*. The workpiece rests on the workpiece rest blade and passes between two abrasive wheels : the grinding wheel and the regulating wheel. The grinding process is extended over the full length of the workpiece by its axial movement past the grinding wheel. This traverse motion is imparted to the workpiece by the regulating wheel. The regulating wheel is of rubber-bonded abrasive, having the frictional characteristics to rotate the work at its own rotational speed. The speed of this wheel, which may be controlled, varies from 0.6 to 1.0 m/s. The axial movement of the work past the grinding wheel is obtained by tilting the regulating wheel downward at the feeding end, at a slight angle (α) from horizontal. Due to this inclination, the peripheral speed of the regulating wheel is resolved into two components-work rotation speed, $V_{wk} = V_{rw} \cdot \cos \alpha$ and the rate of longitudinal feed,

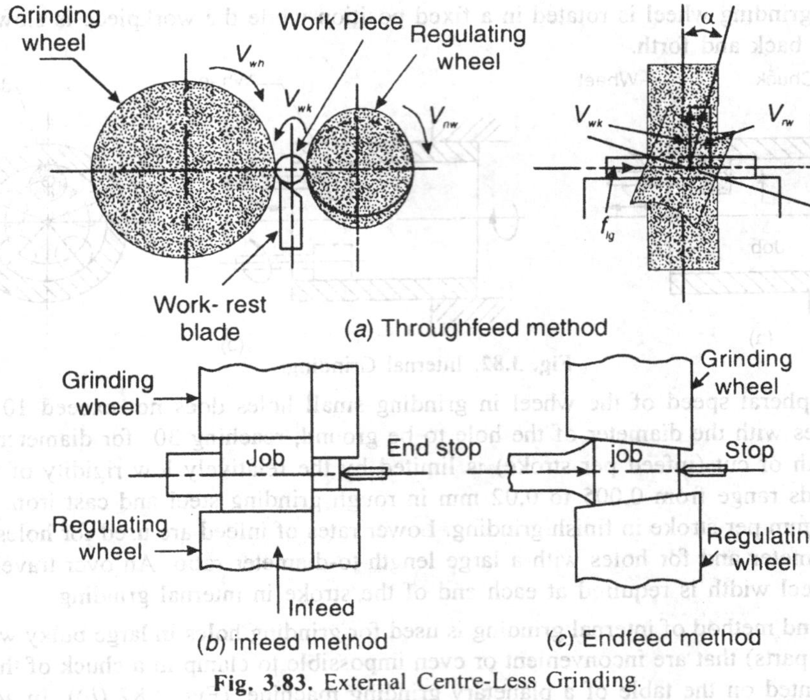

Fig. 3.83. External Centre-Less Grinding.

$$f_{tg} = V_{rw} \cdot \mu \sin \alpha$$

where μ is the co-efficient which accounts for slipping of the work with respect to the grinding wheel (μ = 0.94 to 0.98). Angle α is usually taken from 1° to 5°; the larger the angle α, the higher the rate of feed will be, and vice versa. The layer of metal removed by the grinding wheel in one pass reduces the diameter of the workpiece by from 0.02 to 0.3 mm. The periphery of the regulating wheel is not cylindrical but slightly concave to improve contact between the wheel and the work. For the same reason, the work rest blade has a angular top face to keep work against the regulating wheel face. This method is used for parallel work of any length which has no obstructions. The grinder can be automated by installing a magazine feeding device."

The centre of the workpiece is normally kept above the line connecting the centres of the grinding and regulating wheels (Fig. 3.83 (*a*)). This is to ensure that a true cylindrical surface is ground on the work. This distance is about 0.125 to 0.25 times the workpiece diameter. If it is too small, the ground surface may not be round but may resemble an equilateral triangle with arcs as sides. If it is too large, chatter and vibrations will occur. It should not be more than 10 to 12 mm.

(*ii*) **Infeed Grinding.** (Fig. 3.83 (*b*)). If the workpiece is of stepped or irregular profile and therefore can not be traversed between the wheels, infeed centreless grinding is resorted to. Here the workpiece is placed on the workrest blade after the regulating wheel is retracted. Then the regulating wheel is advanced towards the workpiece so that it is fed towards the grinding wheel. When the grinding is completed, the regulating wheel is retracted again and the finished workpiece is ejected. Since no longitudinal feed of the workpiece is required in this method, the axes of both grinding and regulating wheels are strictly horizontal, that is,

(*iii*) **End Feed Grinding.** (Fig. 3.83 *c*). In this method, both the grinding and the regulating wheels are tapered and thus produce tapered workpieces. The workpiece is fed from one side until it reaches the stop. This method results in exact taper and size."

Advantages of Centreless Grinding:

1. The main advantage is its high productivity, which is several times higher than cylindrical grinding between centres.
2. The operator need be less skilled.
3. The work is rigidly supported and there is no chatter or deflection of the work.
4. Size of the work is easily controlled.
5. As a true floating condition exists during the grinding process, less grinding stock is required.

Disadvantage:

1. Work with flats and keyways can not be ground.
2. Work having several diameters is not easily handled.
3. In hollow work, there is no assurance that the outside diameter will be concentric with the inside diameter.
4. Changeover of a centreless grinder to grind other sizes requires considerable time. Hence, it can be more expediently used in large lot production.

(*b*) **Internal Centreless Grinding.** Holes in ring type parts can be ground on internal centreless grinding machines, (Fig. 3.84). The work-piece 1 is loaded into the grinding zone where it is held between three steel rolls - support 2, pressure 3 and regulating 4 - mounted in housing 5. The lever mounted pressure roll 3 exerts a preset force on the workpiece holding it against the support and regulating rolls. The regulating roll, linked to the workdrive, rotates the workpiece at a peripheral speed of from about 0.7 to 1.0 An axial force acting on the workpiece, due to the fact that the regulating wheel axis is inclined to about holds the workpiece axially with one end face against a roller stop. The hole is ground with a wheel mounted on spindle 6. The rotation of the work on grinding is the result of the friction force between the work and the regulating roll. The workpiece is loaded and unloaded by moving the pressure roll outward. Very high accuracy can be maintained in centreless grinding. Internal tapered surfaces are ground on centreless machines by swiveling housing 5 and the workpiece to the taper angle.

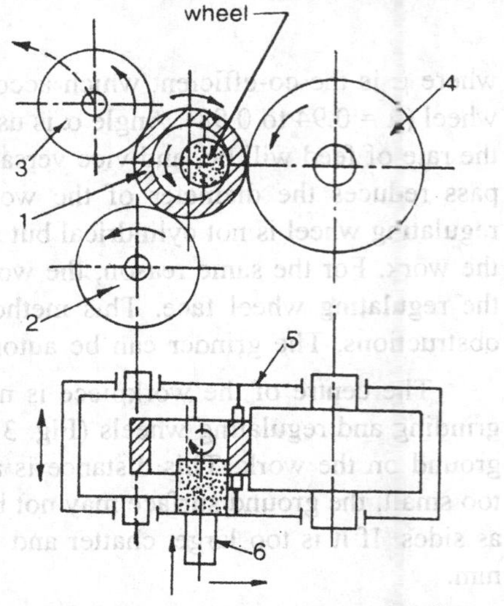

Fig. 3.84. Internal Centreless Grinding.

4. Surface Grinding. Flat surfaces are ground on surface grinding machines. Heavy workpieces are clamped on the table by means of pads, strap clamps and other devices or they are held in fixtures. Small work is usually held by a magnetic chuck. The surface grinders are available with horizontal spindle and vertical spindle and with either reciprocating table or rotary table.

In horizontal-spindle reciprocating table grinders, (Fig. 3.85 (*a*)), grinding is done with the periphery of a straight wheel. bBesides its rotary motion, the grinding wheel also has a crossfeed motion at each full stroke of the table (back and forth). The other motion is the infeed of the wheel in which it is fed to the depth of cut. Horizontal-spindle rotary-table grinders (Fig. 3.85 (*b*)) also use the periphery of the grinding wheel, but the workpieces are held, usually by a magnetic chuck, on a revolving table. Here, as before, the wheel has rotation, cross feed parallel to the surface being ground, and infeed.

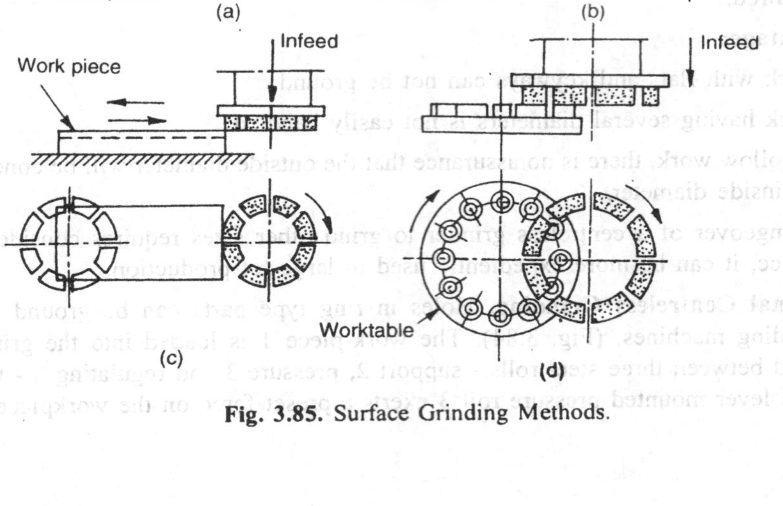

Fig. 3.85. Surface Grinding Methods.

The vertical-spindle machines use a cup-, cylinder-, or segmental wheel. They may be either of the reciprocating-table, Fig. 3.85 (*c*), or rotary-table, (Fig. 3.85 (*d*)). The movements of the work in grinding with the end face of the wheel are the same as when the wheel periphery is used. The primary rotary motion is wheel rotation about the vertical axis and the depth of cut is provided by vertical feed of the wheel along its axis.

Contoured surfaces are usually ground with specially trued form wheels.

Reciprocating grinders are primarily designed for producing long, straight surfaces of either flat or formed section. Rotary-table machines are adapted primarily to rapid production of flat surfaces on small and medium sized parts, singly or continuously. Vertical machines are capable of carrying segmented wheels upto 1050 mm diameters and with a 1500 mm magnetic rotary table chuck will handle work upto 2400 mm in diameter. Vertical-spindle rotary table machines are available also in multiple- spindle models for rough, semifinish, and finish grinding at one pass with automatic sizing and wheel advance as well as gauging.

5. Special Grinding Machines. Grinders of the cylindrical type also include form grinders, tool and cutter grinders for sharpening tools, cam grinding machines, gear grinding machines, Thread grinding machines, Grinding machines for grinding rolls for rolling mills, tool post grinder and so on.

(*a*) **Form Grinders.** Formed surfaces are finished on cylindrical or surface grinding machines depending upon the type of workpiece to be machined. This class of job also includes the grinding of bed ways of complex cross-section, Fig. 3.86.

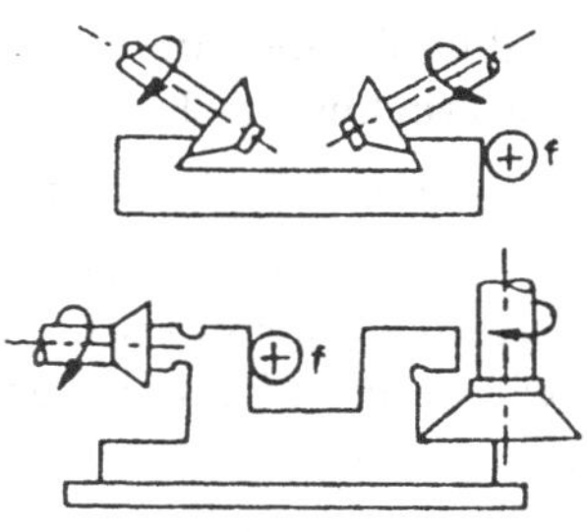

Fig. 3.86. Form Grinding.

(*b*) **Gear Teeth Grinding.** The teeth of gears are ground on gear grinding machines either by the generating process or by a forming process in which formed wheels are used.

The generating grinding method makes useof two saucer-shaped grinding wheels, Fig. 3.87 (*a*), set so that their active faces (on the spindle side) are in planes tangent to the involute curvatures of two teeth on the gear. This is accomplished by truing the wheels to an angle equal to the pressure angle of the gear being ground. In grinding, the gear has a complex rolling (generating) motion.

In formed-wheel grinding of gears, Fig. 3.87 (*b*), the contour of the wheel is trued by a special fixture so that it coincides with the profile of the tooth spaces on the gear. The adjacent flanks of two teeth are ground simultaneously.

(*c*) **Thread Grinding.** Thread is ground on thread grinding machines. These machines differ from conventional cylindrical grinding machines only in that the grinding wheel is either single-or multiple-rib wheel. The work is mounted between centres and is rotated at a definite speed. The ribbed grinding wheel is mountedon the wheel head spindle and is rotated by a separate drive. The wheel also has a longitudinal (axial) movement which amounts to one pitch of the thread per revolution of the work, (Fig. 3.87 *c*).

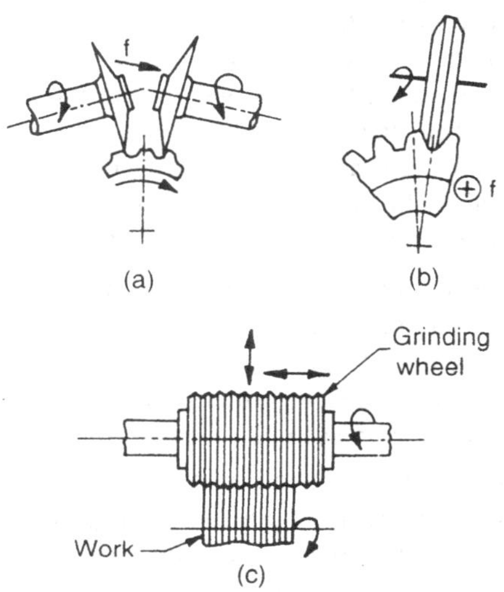

Fig. 3.87. Grinding of Gear Teeth.

(*d*) **Cam Grinders.** An arrangement for grinding the cams of a camshaft is shown in Fig. 3.88. The grinding wheel is arranged so that it can be moved radially toward or away from the workpiece in coordination with the rotation of the workpiece. A master cam imparts the desired radial movement to the workpiece relative to the grinding wheel. The grinding wheel is continuously in contact with the cam surface during operation. The camshaft has simultaneous rotary and oscillating motion, as well as an axial motion, to obtain a fine surface finish.

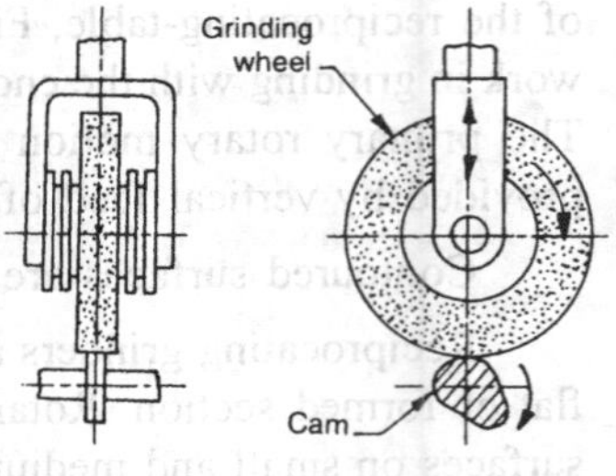

Fig. 3.88. Cam Grinding.

(*e*) **Tool Post Grinder.** Tool post grinder is used for miscellaneous and small grinding work on a lathe. The grinding wheel is held on the tool post and fed across the work, the regular longitudinal or compound rest feed being used. A common application of tool post grinder is the truing of lathe centres.

(*f*) **Tool and Cutter Grinders.** Grinding of tools and cutters is performed in special tool-grinding fixtures fitted with abrasive and diamond wheels of appropriate shapes and dimensions. Tool grinding machines are classified as :

1. ordinary 2. universal, and 3. special machines (for instance, drill sharpeners).

The ordinary machines, like emery sharpeners with a tool rest, are used for simple sharpening operations on simple tools. The universal machines are employed for sharpening a great variety of cutting tools, such as counterborers, reamers, milling cutters, special drills and special cutters.

A universal tool and cutter grinder is shown in Fig. 3.89. The machine has heavy box type base which provides it stability and rigidity. The saddle is mounted directly on the top of the base. The upper table (work table) can be swivelled on the lower table that slides longitudinally on the saddle. The saddle moves on ball bearings on hardened ways and provides cross movement. The work table has 'T' slots for mounting the various work holding devices (vice, head stock, tailstock and so on) and various other attachments. The head stock and tailstock are mounted on either side of table and hold the work (for example, arbor type cutters) in between the centres.

Fig. 3.89. Universal Tool and Cutter Grinder.

The column supporting the wheel head is mounted on the base on the back of the machine. It can be moved up and down and swivelled to either side of the column for varied set ups.

(*g*) The rolls for rolling mills are heavy, large diameter workpieces. The cylindrical grinding machines are specially designed for grinding them. Such workpieces can be rotated with more rigidity when mounted on a solid base. Instead of being supported between centres, these large workpieces may be rotated in bearings, which fit previously machined cylindrical surfaces. The cylindrical surface of the roll is ground by traversing the grinding wheel and its stand back and forth along the length of the roll.

(*h*) **Disk Grinders.** Disk grinders are the important machines among the production surface grinders. They are available in horizontal-spindle models with single or opposed spindles. Single vertical spindle machines available are used mainly for surfacing flat areas without maintaining particular dimensions. Utilizing large, solid abrasive disks or cylinder wheels, horizontal single or opposed spindle machines have oscillating, reciprocating, rotary, or through feeds and are ideally adapted for the production of parts requiring parallel, flat faces. Automatic work sizers are provided on many models, and operation can often be fully automatic. Work upto 350 mm in length can be ground in opposed wheel grinders with wheels upto 750 mm.

3.6.9. Coated Abrasives. Traditionally, the coated abrasives consist of abrasive grains attached to a flexible backing such as paper or cloth. They have been used for low-speed finishing of surfaces. However, with the development of stronger adhesives and backings, "Coated belts" have become important production tools. They operate at high speeds (upto 60 m/s) and have high metal removal rates, typically 200 cm^3/mm. cm width. Abrasive belt grinding is replacing turning, planing or milling in mass production (for example, in machining the gasket surfaces of engine blocks and cylinder heads).

Abrasive belt grinding is used for finishing metal and non-metal parts and not only for surfaces of revolution, but for flat and curved complex shaped surfaces as well, for example, turbine blades, crankshaft journals etc.

Abrasive belts are made with a cloth backing (for example, twill) by depositing grains on a layer of adhesive applied to the backing. This is called ©make coat[a]. The grains are held in place by a second layer (size coat). Bond strength is balanced to prevent stripping of new grains while allowing release of worn grains. Grains with sharp edges and of elongated shapes are electrostatically aligned to give cutting edges of low negative rake angles, spaced some 10 times farther apart than in grinding wheels. The adhesive may be hide glue etc. Aluminium oxide abrasive grains are used to grind steel, malleable iron and bronze, while silicon carbide is used for grey cast iron, brass and aluminium alloys. Multi-layer coated abrasive belts are seldom employed owing to difficulties of their manufacture and their high cost.

The thickness of the belt (together with the layer of abrasives) does not usually exceed 2 mm. The width of the belt is selected in accordance with the length of the surface to be ground. For example, in grinding surfaces of revolution (shafts) the belt width equals the length of the shaft. The length of the abrasive belt also depends upon the size of the workpiece and the requirements made to the output of the grinding operation. The longer the belt, the greater the number of grains participating in the cutting process during one revolution of the belt (for the given belt width). The coated abrasive belts used in practice have a width from 10 to 3000 mm and are from 500 to 7000 mm long.

The methods of abrasive belt grinding differ in the way the belt is pressed against the surface to be ground. Various methods are illustrated in Fig. 3.90. Depending upon its shape, the surface can be ground with the unsupported section of the belt, with a part supported by the driving contact wheel or by an intermediate contact roller whose shape is followed by the belt.

Grinding with the unsupported section of the abrasive belt has the highest production capacity as the belt has an increased area of contact, covering the largest part of the workpiece surface. This enables all kinds of surfaces, with the exception of cylindrical and curvilinear surfaces with a small radius of curvature, to be ground.

Grinding with a part of the belt supported by a contact wheel requires a much higher pressure to attain the same output. The output in this case is inversely proportional to the diameter of the contact wheel. It is not advisable, however, to use wheels less than 150 mm in diameter.

Sometimes two supports, arranged at the two sides of the workpiece, are more efficient than a contact wheel. They increase the angle of contact with the workpiece and, consequently, the area of contact. The tension of an abrasive belt is adjusted by idlers.

The contact pressure between the abrasive belt and the surface being ground should preferably be :

0.05 to 0.2 N/mm^2 for steel and cast iron

and ≯ 0.04 N/mm^2 for non-ferrous alloys and aluminium.

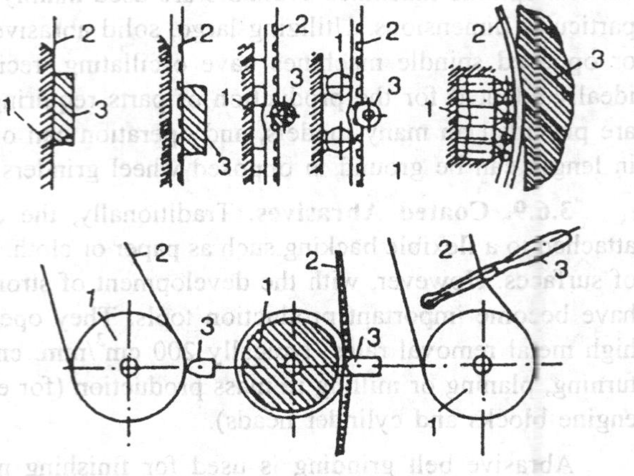

Fig. 3.90. Abrasive Belt Grinding.

Cutting speeds depend upon the workpiece material, type of operation being performed (roughing or finishing) and other factors. In rough abrasive belt grinding of external surfaces of revolution on work-pieces of steel with a tensile strength of less than or equal to 800 N/mm^2 the recommended cutting speed is 25 to 30 m/s and steels with tensile strength greater than 800 N/mm^2 and for cast iron and bronze workpieces, it is 15 to 20 m/s and for grinding aluminium, it is 45 to 50 m/s.

Mineral oils, kerosene, emulsions and fatty oil pastes are used as cutting fluids in abrasive belt grinding operations.

Advantages.

1. The working surface of the abrasive belt is much greater than that of the grinding wheel which promotes dissipation of the heat generated during grinding. This lengthens belt life and almost completely excludes distortion of the workpiece.
2. More uniform surface of the belt in comparison with a grinding wheel, as a result of uniform application of the abrasive grains by the electrostatic field method so that they are oriented vertically and embedded in the glue covered backing.
3. Increased number of abrasive grains per unit area compared to that of the grinding wheel, thereby increasing the grinding capacity and output.
4. Absence of vibration and impact loads, ensuring a higher class of surface finish.

3.6.10. Grinding Points. Grinding points, [Fig. 3.91] are employed for internal grinding and to clean up hard-to-reach spots in making moulds and dies, when grinding wheels are

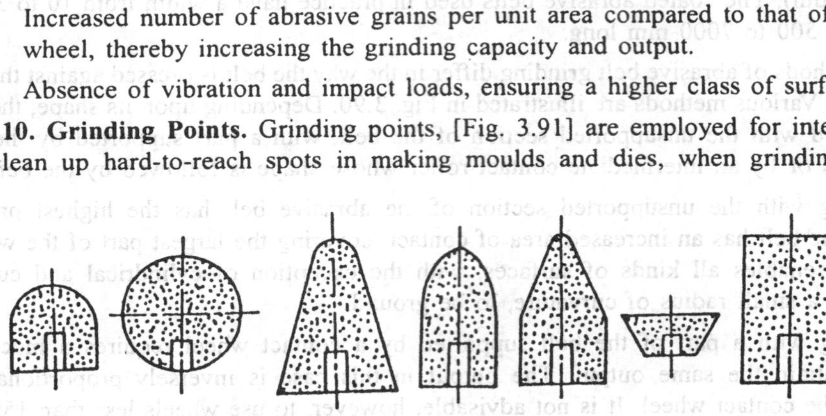

Fig. 3.91. Grinding Points.

unsuitable. These points do not have through holes and are cemented onto a special stem. Grinding points range from 3 to 40 mm in diameter.

3.6.11. Grinding Segments. A segmental grinding wheel consists of several abrasive segments, clamped in some manner in chuck. The segments have a length from 125 to 300 mm. Some of the shapes of the grinding segments are shown in Fig. 3.92. Segmental wheels are used in surface grinding. An essential advantage of a segmental wheel is that a damaged segment can be readily replaced, without discarding the whole wheel. The spaces between the segments facilitate the delivery of cutting fluid to the grinding zone and the disposal of chips and worn-out particles of the wheel. Owing to the smaller area of contact with the wheel, the workpiece is heated less in grinding with a segmental wheel.

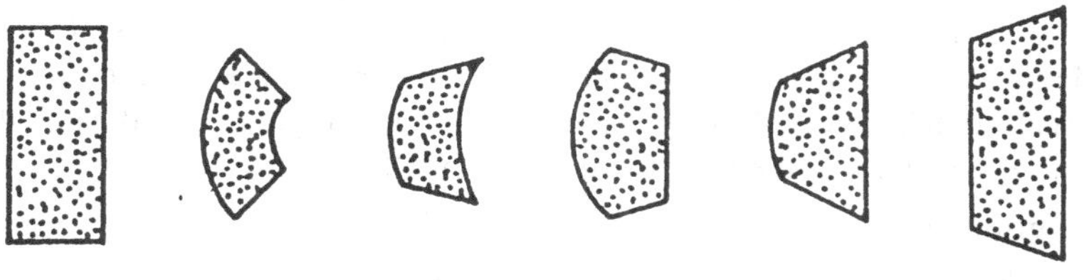

Fig. 3.92. Shapes of Grinding Segments.

3.6.12. Abrasive Sticks. Abrasive sticks are used for hand fitting operations and for honing and super-finishing operations performed in special machines for micro-finishing. The various shapes of the abrasive sticks are shown in Fig. 3.93. Honing and super-finishing are done with sticks of square (first), flat (second) and special flat (last one) honing shapes, which are mounted in special heads.

Fig. 3.93. Abrasive Sticks.

3.6.13. Grinding Process Variables

Wheel Speeds. Proper grinding-wheel speeds have been fairly well established, and most machines can be adjusted to attain the desired cutting efficiency. Major problem in the past has been with the internal grinding where proper peripheral wheel speeds were difficult to obtain. Today, however, high-cycle, direction-driven wheel heads are rapidly supplanting previous types. Efficient peripheral wheel speeds, observing manufacturers' maximum limits for specific wheels, are generally as follows :

Types of Grinding	*speed, m/s*
Cylindrical grinding	
Resinoid	27.5 – 60.0
Vitrified	27.5 – 32.5
Surface grinding	20.0 – 25.0
Internal grinding	10.0 – 30.0
Diamond wheels	25.0 – 32.5

Work Speeds. For obtaining the desired finish as well as maximum stock removal commensurate with good wheel life, proper work speeds are necessary. At the line of cutting, the work and the abrasive wheel pass in opposite directions except in internal centreless grinding. Work speed at this point should be set according to the material being ground. The low end of This range is used for roughing cuts and the high end for finishing.

Material	*speed, m/s*
Soft steel	0.15 – 0.25
Hard steel	0.35 – 0.50
Cast iron	1.00 – 2.00
Aluminum	0.50

Stock Removal. Most efficient stock removal is obtained with coarse-grit wheels, and while surface finish is distinctly rougher, commercially acceptable dimensional accuracy can be maintained. Maximum economy in manufacture thus depends to a large extent upon judicious selection of surface finish. The following stock removal per single pass indicates the comparative rates, though these will vary somewhat for different materials :

Rough grinding	0.05 – 0.50 mm
Semifinish grinding	0.025 – 0.25 mm
Finish grinding	0.0125 – 0.125 mm

3.6.14. Balancing of Grinding Wheels: One way of mounting of a grinding wheel is shown in Fig. 3.94. It is clamped by flanges (one each on its either side) directly on the spindle. Thick compressible washers (0.5 to 3 mm thick) made of cardboard, rubber or leather etc. should be inserted in between the flanges and the wheel.

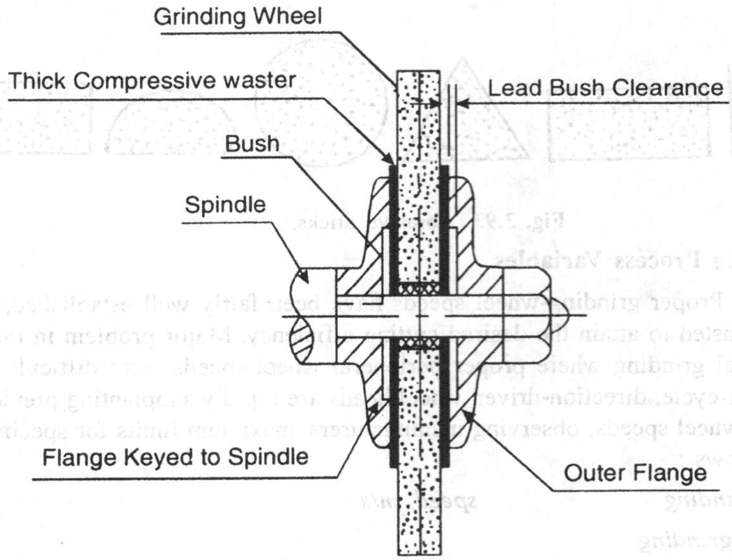

Fig. 3.94. Mounting of Grinding Wheel.

The grinding wheels rotate at very high speeds. Therefore, any out of true and out of balance *wheel will* produce large unbalanced forces which will lead to excessive vibrations. This will result in poor surface finish of workpiece and reduced wheel life. Thus, before mounting on the grinding wheel spindle, each wheel with its sleeve should be balanced. The test should be carried out after truing up the wheel face. The wheel is mounted at the centre of a perfectly straight and round arbor. The assembly is then rested on straight knife edged ways or roller rests of the balanced machine, Fig. 3.95. Out of balanced wheel will not remain stationary. It will come to rest with its heavy side underneath. The wheel is balanced by cutting some of the lead from the heavy side of the bush.

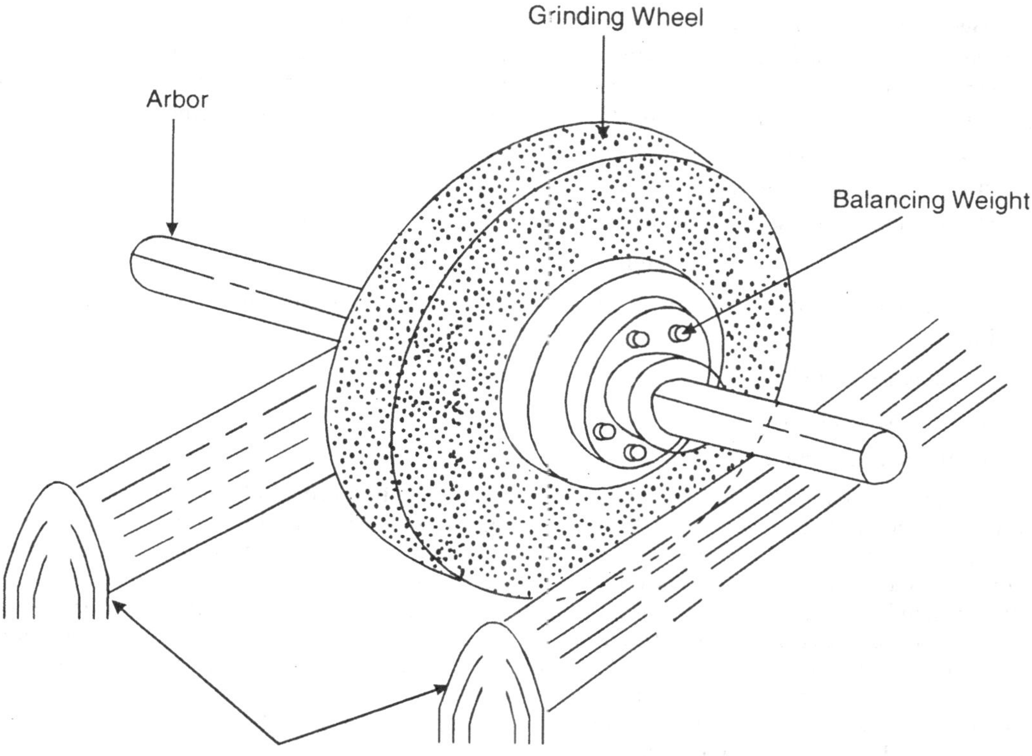

Fig. 3.95. Static Balancing Device.

Another method of balancing the wheel is shifting three balanced weights in an annular groove of the wheel sleeve (or mounting flange). These adjustable segments can be fixed around a circle at any position by tightening the screws.

3.6.15. High Speed Grinding : Conventional grinding is basically a finishing process. It is employed to obtain high dimensional and geometric accuracies so as to obtain a good surface finish. Due to this, the metal removal rate (MRR) in conventional grinding process is quite low as compared to turning and milling operations. In high speed grinding, it is possible to achieve increased MRR comparable to turning and milling operations and at the same time obtain accuracy and surface finish equivalent to that for conventional grinding.

High speed grinding (High-efficiency grinding) differs from conventional grinding in that the grinding wheel speeds are quite high (usually 60 m/s and above upto 80 m/s) as compared to about 30 m/s for conventional grinding. The work speeds and indeed rates are also comparably higher since the force on the work is reduced (from the power, force, velocity relationship).

High speed grinding is not a simple matter of increasing the wheel speed on conventional grinding machines. The process variables (wheel speed, work speed and infeed rate) must be compatible with the operating variables, such as machine condition, wheel structure, coolant application, surface quality required and the hardness of the work-material.

The grinding machines used for high speed grinding must be rigid (that is, with vibration free structures) and must have higher machine and spindle stiffness, powerful wheel and work drives and proper and strong safety guards.

The abrasive used for fabricating high speed grinding wheel is Al_2O_3 and the bond used is either a special high strength vitrified bond or polyurathene bond.

Because of the increased metal removal rate, the wheel structure has to be quite open. The problem of wheel loading is less with open structure. The feeding of the coolant is also more efficient with open structure. To ensure proper cooling of the interface, a high pressure coolant systems with special nozzle is used. Also, an oil coolant in place of an emulsion will greatly increase the metal removal rate.

To increase the metal-removal rate, the infeed rate must be increased proportionately with the wheel speed to present a full layer of material for each revolution.“

From above, it is clear that high speed grinding results in:“

1. Reduced thickness of chip removed by individual abrasive grain leading to reduced grinding forces.

2. Increased MRR.

3. Increased grinding ratio.

4. Improved surface finish.

3.6.16. Abrasive Machining : The conventional grinding process being a finishing process, its metal removal rate is very less. When the metal removal rate is the main consideration and surface finish and accuracy are secondary, such grinding operations are included under the term “Abrasive machining”. It includes all the operations where cost is a major factor in deciding between grinding and other machining methods. Increased metal Removal Rate can be obtained by increasing process parameters (feed and/or depth of cut). The process compares favourably with other machining processes such as milling, turning and broaching etc. The geometric tolerances and surface quality obtained are of the same order as those obtained by these machining operations. However, the power consumption per unit MRR is about 3 times in Abrasive machining as compared to milling. But it is much less as compared to conventional grinding process.

However, Abrasive machining has become competitive with other machining operations due to the following factors :

1. Any material, such as metals, ceramics, refractory alloys and super-alloys can be machined by this process.

2. Work surface conditions do not have the adverse effects on the process.

3. Rough and finishing cuts can be done in one operation.

4. Handling cost is greatly reduced, because the machine set up is not changed for rough and finish cuts.

5. Work-holding devices/fixtures can be eliminated or made simpler as compared to other machining processes such as milling.

6. Less machining allowance needs to be allowed for abrasive machining than for other machining processes. There is thus initial saving (in material).

The machines used for abrasive machining are grinding machines of high power and rigidity.

Product Applications : Valve manifolds, cylinder heads, steam turbine casings, Aluminium piston skirts, Steering knuckles.

3.6.17. Surface Finishing Processes :

1. Lapping: Lapping is a surface finishing process used on flat or cylindrical surfaces (mainly external). Lapping is the abrading of a surface by means of a lap (which is made of a material softer than the material to be lapped), which has been charged with the fine abrasive particles. When the lap and the work surface are rubbed together with the fine abrasive particles between them, these particles become embedded in the softer lap. It then becomes a holder for the hard abrasive. As a charged lap is rubbed against a hard surface, the hard particles in the surface of the lap remove small amounts of materials from the harder surface. Thus it is the abrasive which does the cutting and the soft lap is not worn away, because the abrasive particles become embedded in its surface, instead of moving across it.

In lapping, the abrasive is usually carried between the lap and the work in some sort of a vehicle. The vehicle or lubricant controls to some extent the cutting action and prevents scoring the work and caking of the abrasive. Some of the vehicles used include: kerosene plus a small amount of machine oil, greases, fine sperm oil for fine job, olive oil, lard oil, spindle oil, and soapy water. Naphtha is used to clean the laps.

Laps may be made of almost any material soft enough to receive and retain the abrasive grains. They may be made of soft cast iron, wood, leather, brass, copper, lead or soft steel. The most common lap is fine grain cast iron. Copper is used rather often and is the common materials for lapping diamonds. For lapping hardened metals for metrolographic examination, cloth laps are used.

For steel surfaces, artificial corrundum is used as an abrasive for preparatory lapping and again used in a final state for finishing. Silicon carbide gives good results on cast iron and alumina for finest lapping. For lapping small components, diamond dust or boron carbide in the finest grain size give good results. The other abrasives used are: rough (Ferric oxide Fe_2O_3), green rough (chromium oxide Cr_2O_3) and crocus powder. The abrasive particles are from 120 grit up to the finest powdered sizes. In nearly all cases, a paraffin lubricant is used. The exception being for soft materials when a soluble oil or water lubricant is used.

In lapping, the material removal is usually less than 0.025 mm, although rough lapping may remove as much as 0.075 mm and finish lapping as little as 0.0025 mm. Commercial lapping operations can produce parts to limits of 0.000625 mm. Since it is such a slow metal removal process, it is used only to remove scratch marks left by grinding or honing or to obtain very flat or smooth surfaces such as required on gauge blocks or liquid tight seals where high pressures are involved. Materials of almost any hardness may be lapped. However, it is difficult to lap soft materials since the abrasives tend to become embedded.

Thus, lapping is done:

(*i*) to produce geometrically true surface.

(*ii*) to correct minor imperfections in shape.

(*iii*) to obtain fine dimensional accuracy to provide a very close fit between the contact surfaces.

(*iv*) to secure a fine surface finish.

Lapping Methods:– Lapping may be done by hand or mechanically with the help of special lapping machines.

(*a*) *Hand lapping for flat work* : Here, the lap is a flat similar to a surface plate. Grooves are usually cut across the surface of a lap to collect the excessive abrasive and chips. For finishing of the work surface, either the lap or the workpiece is held by one hand and the irregular rotary motion of the other by the second hand, enables the abrading of the two surfaces in contact. The work is turned frequently to obtain uniform cutting action. The method is used for lapping: press work dies, dies and metallic moulds for castings etc., surface plate, engine valve and valve seat, gauge blocks and piston rings etc. Piston rings are customarily made parallel and to high precision by hand lapping. Manual lapping is used to bring gauge blocks to their final stage of dimensional accuracy and parallelism, a finish of 0.025 to 0.050 m and a tolerance as small as ± 0.000025 mm.

(*b*) *Hand lapping for external Cylindrical work* (*Ring Lapping*) : An external lap for external workpieces (round) is shown in Fig. 3.96. It is split by a saw cut and can be closed in by tightening one or more screws. The diameter of the hole is made the same as that of the piece to be lapped, and the hole is, of course bored before the saw cut is made. Internal laps are made to expand.

The ring lap is reciprocated over the workpiece surface. The method is usually used for stepped plug gauges or gauges made in small quantities.

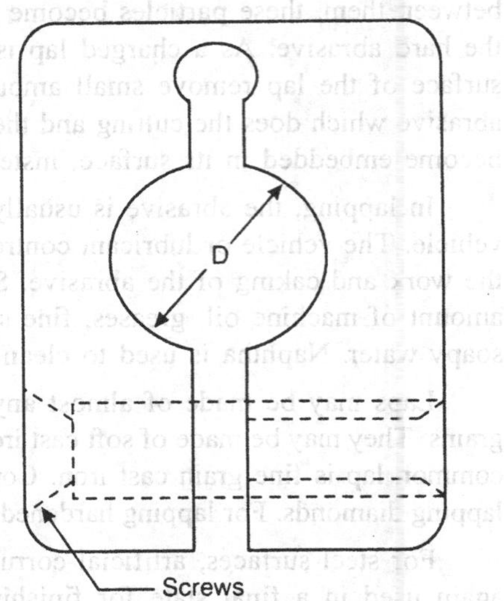

Fig. 3.96. External Lap.

(*c*) *Machine Lapping* : Mechanical lapping is a high production process, for example gudgeon pins 25 mm diameter and 75 mm long are lapped at the rate of 500 pieces per hour, removing 0.05 to 0.075 mm of material with a limit of accuracy of roundness, straightness and size within 0.025 mm. Mechanical lapping machines are of vertical construction with the work holder mounted on the lower table which is given an oscillating motion. The upper lap is stationary and floating, while the lower one revolves at about 60 rev/min. Several types of lapping machines are available for lapping round surfaces. A special type of centreless lapping machine is made for lapping small parts such as piston pins, ball bearing races etc.

A general purpose machine for lapping both cylindrical and flat surfaces is shown in Fig. 3.97. A number of workpieces are placed between the upper and the lower lap, whose surfaces have previously been lapped flat. The workpieces are placed in slots in a work holder so that their axes X-X are not quite radial. The shape of the slots will depend upon that of the workpieces. The two laps are rotated and the work holder is given as oscillation of about 25 mm amplitude. A stream of vehicle in which fine abrasive flour is suspended is fed to the centre of the laps and flows outwards and the workpieces are thus gradually lapped to size.

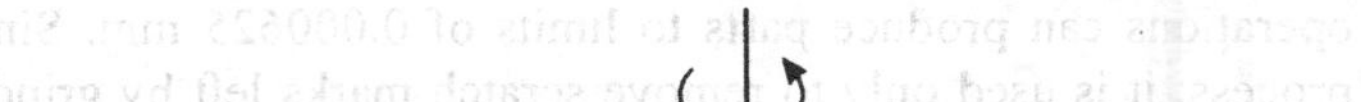

Fig. 3.97. Machine Lapping.

During machine lapping a pressure of 0.007 to 0.02 N/mm^2 for soft materials and upto 0.07 N/mm^2 for hard materials is satisfactory. Mechanical lapping machines can be used for lapping: (*i*) External cylindrical surfaces, and (*ii*) flat surfaces.

The following are examples of work done by lapping: aircraft piston pins, automotive wrist pins, diesel engine injector-pump parts and spray nozzles, plug gauges, certain dies and moulds, gauge blocks, refrigerator-compressor parts, oil-burner parts, micro-meter spindles, roller bearings, taper rollers, worm and worm gears, crankshafts, camshafts, ball bearing race-ways etc.

Lapping has become a common production process with the demand for hardened surfaces having only a few micrometers of surface finish. However, because, it is such a slow method of metal removal, it is obviously relatively expensive and is not economically justified unless operating requirements make such surface finishes absolutely necessary.

Lapping and polishing differ in the following manner—Polishing is meant to produce a shiny surface whereas a lapped surface does not usually have a bright shiny appearance. Lapping definitely removes metal from lapped surface, whereas, polishing as a rule does not remove any appreciable amount of metal. Lapping improves the geometrical shape of the body, whereas, polishing does not. Lapping is essentially a cutting process, while polishing consists of producing a kind of plastic flow of the surface crystals so that the high spots are made to fill the low spots.

As written above, lapping is normally adopted for external surfaces. However, it can be used for internal surfaces also.

2. Honing. Honing is a grinding or abrading process. In it, a very little material is removed. This process is used primarily to remove the grinding or tool marks left on the surface by previous operations. The cutting action is obtained from abrasive sticks (aluminium oxide or silicon carbide) mounted in a mandrel or fixture. A floating action between the work and the tool prevails so that any pressure exerted on the tool is exerted and transmitted equally on all sides.

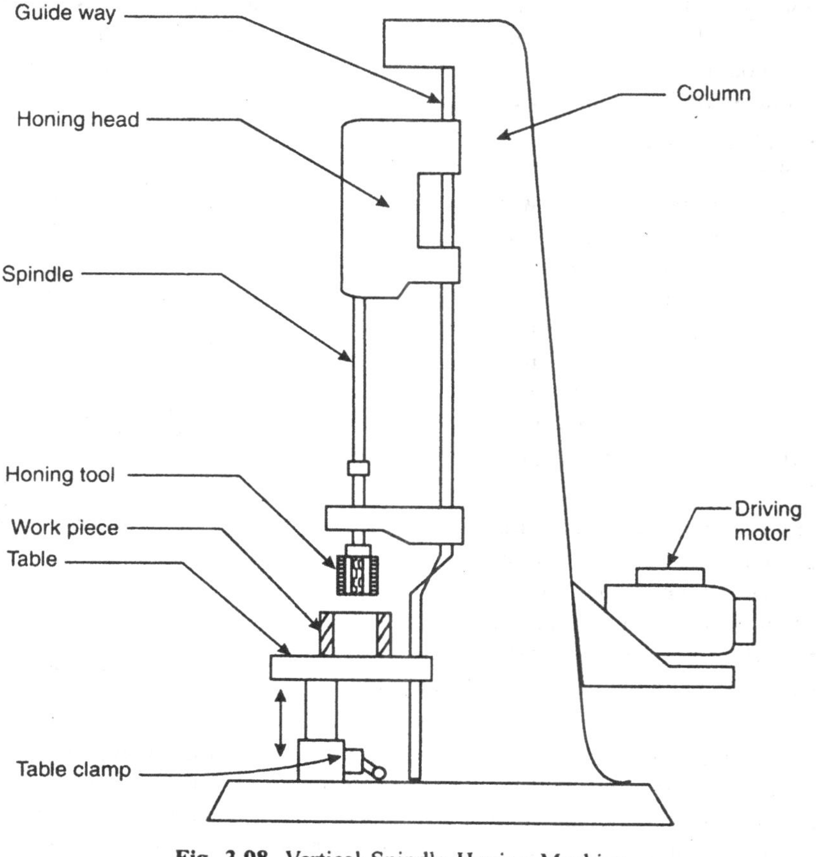

Fig. 3.98. Vertical Spindle Honing Machine.

The honing tool is given a slow reciprocating motion as it rotates, having resultant honing speeds from 15 to 60 mpm. This action results in rapid removal of stock and at the same time, the generation of a straight and round surface. Defects such as slight eccentricity, a wavy surface, or a slight taper caused by previous operations can be corrected by this process. Parts honed for finish remove only 0.025 mm or less. However, when certain inaccuracies must be corrected, amounts upto 0.50 mm represent usually practice. Collants are essential to the operation of this process to flush way small chips and keep temperatures uniform. Sulphurised mineral base or lard oil mixed with kerosene is generally used. Paraffin is also used.

Most honing is done on internal surfaces, or holes, such as automobile cylinders. There are a few applications of honing to external surfaces. Parts can be of any shape, but the surface must be cylindrical. Practically any material can be honed. Soft materials which cannot be lapped, can be honed because of the use of bonded abrasive. Hard and soft cast iron, steel, carbides, bronze, aluminium, brass and silver, as well as glass, ceramics and some plastic can be honed.

Honing machines are similar in general construction to vertical drilling machines Fig. 3.98 but the spindle reciprocation is usually by hydraulic means. The rotary motion may be from a hydraulic motor or by gearing. The speed ratio of two motions affects the work finish and may be varied throughout the operation or for different materials. For cast iron, the speed ranges from 60 to 150 mpm for rotation; with 15 to 21 mpm for reciprocation. The corresponding speeds for steel are : 45 to 60 mpm for rotation and 12 mpm for reciprocation. The reciprocation motion distributes the wear over the whole length of the sticks and keeps the bore cylindrical. Semi-automatic honing machines used in the finishing of automobile cylinder bores are of vertical type. Both single and multi spindle machines are used for this operation. The abrasive stones are mounted on a honing tool. In order to expand the abrasive stones outward to fit the hole to be honed, the conical wedge is moved relative to the honing tool.

A general arrangement of a vertical honing machine is shown in Fig. 3.99. The abrasive sticks (upto 8 in number) are expanded while honing takes place, if required, by micrometercontrolled, mechanical or hydraulic, means. The honing tool will follow the axis of the original hole, therefore, the honing tool or fixture must be free to float. This is done by using universal joints as shown in the figure. Due to this, the honing tool becomes self centring and it is not necessary to line up the hole and hone axes precisely. Vertical machines have been designed for work upto 500 mm diameter.

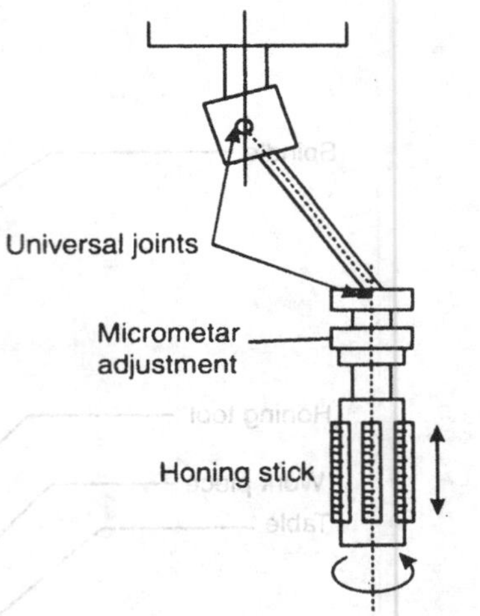

Fig. 3.99. Vertical Honing Machine.

Horizontal honing machines are used only for honing large, long gun barrels and similar work. The workpiece is held on the left and the tool is rotated and reciprocated by the head on the right end of the bed of the machine. Machines of this type are made with strokes of upto 22.5 m and hone holes as large as about 1 m in diameter.

All honing gives a smooth finish with a characteristic cross hatch appearance. The depth of these hone marks can be controlled by variations in pressures, speed and type of abrasive used. Accurate dimensions can be maintained by the use of automatic size controlled devices in connection with honing. Typical applications of honing are finishing of automobile engine cylinders, bearings, gun barrels, ring gauges, piston pins. Shafts and flange faces.

The grit size of abrasive material used in abrasive sticks is 80 to 180 for primary honing and 300 to 500 for secondary honing. Surface finish of the order of 0.05 m R can be obtained by honing.

3. Super-finishing. Super-finishing is a micro finishing process that produces a controlled surface condition on parts which is not obtainable by any other method. The operation which is also called 'microstoning' consists of scrubbing a stone against a surface to produce a fine quality metal finish. The process consists of removing chatter marks and fragmented or smear metal from the surface of a dimensionally finished parts. As much as 0.03 to 0.05 mm of stock can be efficiently removed with some production applications, the process becomes most economical if the metal removal is limited to 0.005 mm.

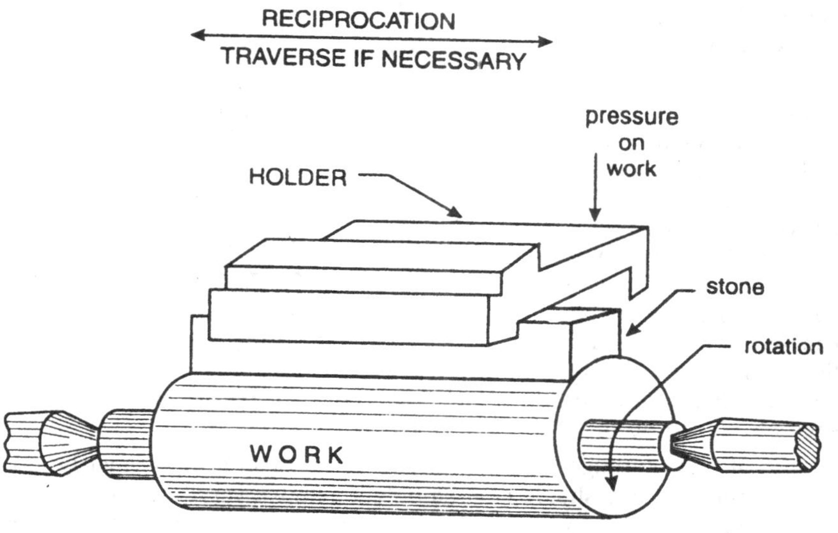

Fig. 3.100. Super Finishing

The method is performed by rapidly reciprocating a fine grit stone with a soft bond and pressing it against a revolving round work-piece, Fig. 3.100. The stone quickly wears to conform to the contour of the work-piece. The work-piece and tool are flooded with a cutting fluid to carry away heat and particles of metal and abrasive. The time needed for super finishing is quite small. Parts may be super-finished to a smoothness of 0.075 mm as rapidly as 15 to 50 seconds. However, to obtain a better finish 2 or 3 minutes may be required. Product applications of super-finishing are: computer memory drums, sewing machine parts, automotive cylinders, brake drums, bearings, pistons, piston rods and pins, axles, shafts, clutch plates, tappet bodies, guide pins etc.

4. Polishing. Polishing is a surface finishing process to obtain a smooth and lustrous Surface. The process is carried out with the help of resilient wheels or belts made of cloth, leather or felt and coated with fine abrasive particles of Al_2O_3 or diamond. The fine lustrous (shining) surface is obtained due to the cutting action of fine abrasive particles and the softening and smearing of surface layers by frictional heating during the process.

There are three variations of the polishing process : roughing, dry fining and fine finishing. Roughing and dry fining are carried out with dry wheels, whereas in fine finishing operation, oil, tallow or beeswax is used. The grit size of the abrasive is :

20 to 80 for roughing

90 to 120 for dry fining

and 150 for fine finishing

The limitation of the process is that parts with irregular shapes, sharp corners, deep recesses and sharp projections are difficult to polish.

5. Buffing. Buffng is a polishing operation in which the workpiece is brought in contact with a revolving cloth buffing wheel, that usually has been charged with a very fine absrasive Fig. 3.101. The polishing action in buffing is very closely related to lapping in that when a polishing medium such as 'rouge' is used, the cloth buffing wheel becomes a carrying vehicle for the fine abrasives. In (this acton the abravsive removes amounts of metal from the workpiece, thus eliminating the scratch marks and producing a very smooth surface. When softer metals are buffed, particularly without the use of an abrasive, there is some indication that a small amount of metal flow may occur which helps to reduce the high spots and produce a high polish.

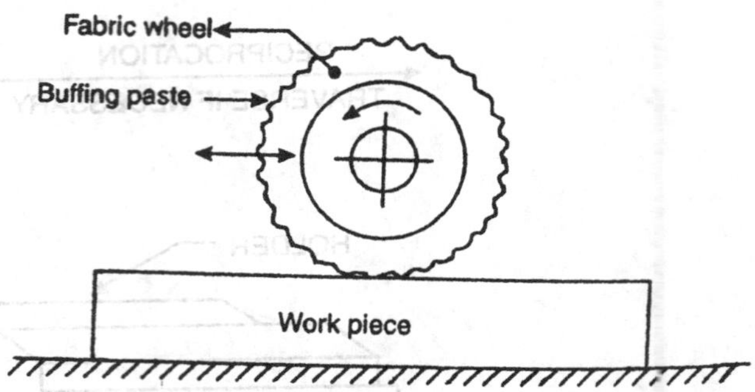

Fig. 3.101. Buffing.

Buffing wheels are made of discs of linen, cotton, broad cloth and canvas. They are made more or less firm by the amount of stitching used to fasten the layers of the cloth together. Buffing wheels for very soft polishing or which can be used to polish into interior corners may have no stitching, the cloth layers being kept in position by the centrifugal force resulting from the rotation of the wheel. Buffing wheel speeds are in the range of 32.5 to 40 m/s.

Various types of buffing rouges are available. Most of them being primarily ferric oxide in some soft type of binder. Buffing should be used only to remove very fine scratches or to remove oxide or similar coatings which may be on the work surface. It ordinarily, is done manually, the work being held against the rotating wheel. This procedure is apt to be relative expensive because of the labour cost. There are semi-automatic buffing machines available consisting of a series of individually driven buffing wheel which can be adjusted to the desired position so as to buff different portions of the workpiece. The workpieces are held in fixtures on a rotating circular worktable so as to move past the buffing wheels. If the workpieces are not too complex in shape, very satisfactory results can be achieved with such equipment and the buffing cost will be low.

Product applications of buffing process which produces mirror-like finish are : objects used on mobile homes, automobiles, motor-cycles, boats, bicycles, sporting items, tools, store fixtures, commercial and residential hardware and household utensils and appliances.

3.6.18. Abrasive Jet Machining (AJM). In abrasive jet machining or grinding method, the material is removed from the surface of a workpiece, by impinging a focussed jet of fine abrasive particles carried by a compressed gas which imparts kinetic energy to the stream of fine abrasives. The stream leaves through a nozzle at a velocity of the order of 300 m/s and strikes the surface of the workpiece, producing impact loading on it. Severe plastic deformation or micro-cracks occur in the vicinity of the impact. Due to repeated impacts, small chips of material get loosened and a fresh surface gets exposed to the jet.

The carrier gas can be air, nitrogen or carbon dioxide, but never oxygen. The air must be filtered to remove water, oil and other contaminants. Nozzle pressure can range between 0.20 N/mm^2 and 0.85 N/mm^2 but is usually about 0.5 N/mm^2. Higher pressure results in rapid nozzle wear, low pressure gives slow metal removal rates.

To resist abrasion and wear of nozzles, these are made of hard materials such as tungsten carbide and synthetic sapphire. The useful life of sapphire nozzles is about 10 times that of tungsten carbide nozzles. The sizes of the nozzles are usually :

Tungsten Carbide

Rough	: 0.0125 to 0.80 mm diametre
Rectangular	: 0.075 mm × 0.50 mm to
	: 0.175 mm × 3.75 mm
Square	: 0.65 mm square
Sapphire nozzles are made	
only round	: 0.20 to 0.65 mm square

The various abrasive materials used in AJM include : aluminium oxide, silicon carbide, glass powder or brads, dolomite (calcium magnesium carbonate) and specially prepared sodium bicarbonate. When sodium bicarbonate is used, it must be kept dry. The cutting performance will depend upon : hardness, strength, particle size and particle shape of the abrasive. Al_2O_3 is a general purpose abrasive and is used in sizes of 10, 25 and 50 microns. SiC is used for faster cutting on extremely hard materials. It is used in 25 and 50 microns sizes. Dolomite is used in 200 mesh size for light cleaning and etching. Sodium bicarbonate is used for extra find cleaning and glass beads (0.635 to 1.27 mm diameter) are used for light polishing and fine deburring.

In general, larger sizes are used for rapid removal rate while smaller sizes are used for good surface finish and precision.

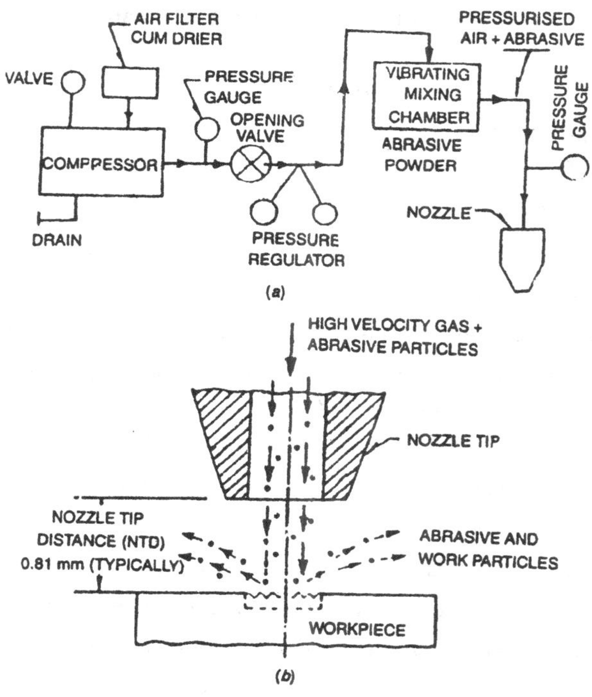

Fig. 3.102. Schematic Diagram of AJM.

The metal removal rate will depend upon : the diameter of nozzle, composition of abrasive gas mixture, hardness of abrasive particles and that of the work-material, particle size, velocity of jet and distance of workpiece from the jet. The typical metal removal rates vary from 0.016 to 0.02 cm^3/min.

Masks defining the machining area are sometimes used to prevent stray cutting. Copper is a good all purpose masking material. Glass gives excellent definition but has short life. Rubber has long life but gives poor definition.

Application of AJM include : abrasive, cutting, or cleaning (for example electrical contacts) and for machining semi-conductors such as silicon, gallium or germanium, for making holes and slots in glass, quartz, sapphire, mica and ceramics. Other applications are : cleaning hard surfaces, deburring, scribing, grooving, polishing and radiusing. A dimensional tolerance of ± 0.05 mm can be obtained. Surface finish obtained is in the range of 0.5 to 1.2 μ m CLA. Typical power input is 0.25 kW. The method is used chiefly to cut materials that are sensitive to heat damage and thin section of hard materials that chip easily and to cut intricate holes that would be more difficult to produce by other methods. The other advantage is its low capital cost.

The disadvantages of the method are : low metal removal rate, possibility of stray cutting, embedding of abrasive particles in soft workpiece and dust control requirements.

A schematic diagram of AJM is shown in Fig. 3.102. Dry and filtered gas (in this case air) is raised to a high pressure in the compressor. The pressurised air flows to the vibrating mixing chamber containing abrasive powder. The mixture of pressurised air and abrasive powder than flows to the nozzle. It then impinges on the workpiece after shooting out of the nozzle exit. The pressure regulator regulates the gas flow and its pressure. The feed rate of abrasive powder is controlled by amplitude or vibrations of the mixing chamber. The movement of the nozzle towards the workpiece or *vice versa* is controlled by a cam mechanism or a pantograph mechanism.

PROBLEMS

1. Define grinding operation.
2. What is the difference between 'rough' grinding and 'precision' grinding?
3. Why is grinding so important in modern production ?
4. Why the natural abrasives are not suitable for making grinding wheels?
5. How is aluminium oxide abrasive produced ? Write its field of application.
6. How is silicon carbide abrasive produced ? Write its field of application.
7. Under what conditions are diamond, boron carbide and cubic boron nitride used as abrasive materials for making grinding wheels ?
8. What is meant by 'grain size' of an abrasive material ?
9. Discuss the various types of bonding materials used for making grinding wheels ?
10. Define a grinding wheel.
11. Discuss the various methods of making grinding wheels.
12. Sketch the various shapes of grinding wheels and write their fields of application.
13. What is meant by "grade" and "structure" of a grinding wheel ?
14. What is meant by standard marking of grinding wheels ?
15. How the grinding wheel is selected for a particular job ?
16. What is meant by dressing and truing of grinding wheels ?

17. Define "grinding ratio".
18. Sketch and explain the working of an external cylindrical grinding machine.
19. Sketch and explain the three methods of cylindrical grinding.
20. What is the difference between plain and universal cylindrical grinders?
21. Describe in detail how an internal grinder operates.
22. What are the advantages and disadvantages of centreless grinding ?
23. Sketch and explain the three methods of external cylindrical centreless grinding.
24. Sketch and explain internal centre less grinding process.
25. Sketch and explain the various methods of surface grinding.
26. How the jobs are held during surface grinding operation ?
27. Sketch and explain the following grinding processes : Form grinding, Gear tooth grinding, Thread grinding and Cam grinding.
28. What is a tool post grinder ?
29. What kind of grinding can be done on the lathe ?
30. Sketch and discuss a tool and cutter grinder.
31. What are disk grinders ?
32. What are coated abrasives ?
33. Explain abrasive belt grinding.
34. Write the drawbacks of abrasive belt grinding.
35. Discuss the various grinding process variables.
36. Write a short notes on : High speed grinding, Abrasive Machining.
37. Write a short notes on :- Lapping, Honing, Super-finishing.
38. Write a short note on "Buffing" process.
39. Write a short note on "Polishing" process.
40. Write a short note on Abrasive Jet Machinig or Grinding process.

3.7. BROACHING MACHINES

Broaching is a process of machining a surface with a special multipoint tool called a "broach", whose teeth remove the whole machining allowance in a single stroke. The process differs from other machining processes in that the only motion is the primary cutting motion of the tool. The feed is obtained by placing the teeth progressively deeper within the tool, thus, each tool edge takes off a successive layer of the material. The primary cutting motion is translational along the broach axis.

Broaching is used for machining through holes of any cross sectional shape, straight and helical slots, external surfaces of various shape, external and internal toothed gears, splines, keyways and rifling. Broaching of inside surfaces is called "internal or hole broaching" and that of the outside surface as "Surface broaching". Typical internal broaching operations are : the sizing of holes, cutting of serrations, straight or helical splines, keyways and gun rifling.

Although the cutting speeds used in broaching are relatively low (2 to 15 m per min), the production capacity is very high since the total length of the cutting edges that are simultaneously in operation is very great. The output in broaching can be raised still higher if broaching machines with a continuous working motion are used, in conjunction with automatic workpiece loading and unloading.

Owing to the high output and machining accuracy, as well as fine surface finish produced, broaching is finding wider and wider application in the engineering and metal-working industries. Broaches are expensive tools, however, and their use is justified, in the main, only in large lot and mass production.

3.7.1. Details of an Internal Broach (Hole Broach). A typical internal broach is shown in Fig. 3.103. To machine an internal hole, the broach is gripped by a puller at the shank end. The front pilot centres the broach in the hole before the teeth begin to cut. The front taper (5 to 20 mm) facilitates the insertion of the front pilot in the hole. The first set of teeth behind the front pilot, removes most of the material and are called "roughing teeth". These are followed by a few teeth called "semi-finishing teeth" where the depth of cut of individual tooth is quite small. Finally, there are finishing or sizing teeth which are all of the same size and have the shape of the finished hole. Sometimes, a few burnishing teeth may be provided after the finishing teeth. These have no cutting edges but are button shaped and from 0.025 to 0.075 mm larger than the size of the hole. The resulting rubbing action smooths and sizes the hole. They are used primarily on cast iron and non-ferrous metals. The "rear pilot" supports the broach after the last tooth leaves the hole.

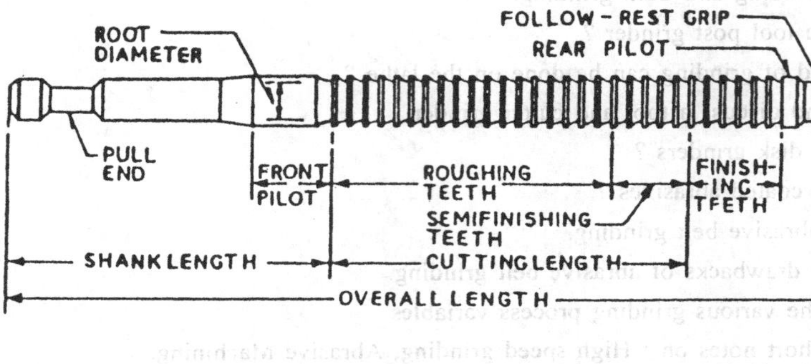

Fig. 3.103. A Typical Internal Broach.

(*a*) **Tooth Elements.** The tooth shape of a typical broach is shown in Fig. 3.104. The front rake angle (face angle or hook angle) refers to rake angle of a single point cutting tool and the back -off angle (relief angle) is provided to prevent rubbing of tool with workpiece.

Hook angle: 15° to 20° for steel

6° to 8° for C.I.

Back-off angle: 1° to 3°

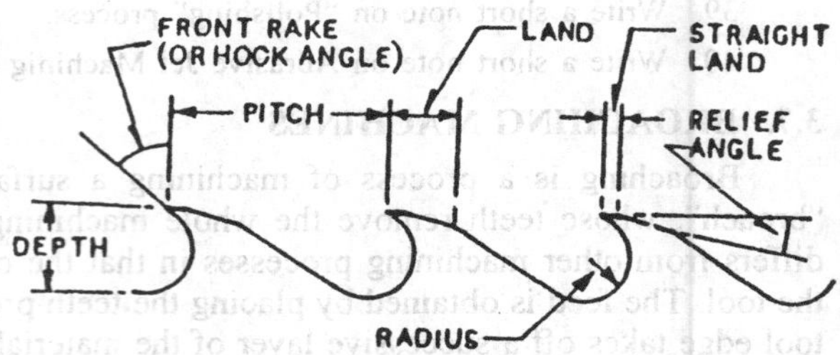

Fig. 3.104. Tooth Shape of a Broach.

(*b*) **Material.** High Speed Steel (H.S.S.) is by far the most widely used material for the broaches. Brazed carbides or disposable inserts are sometimes used for the cutting edges when machining cast iron parts which require close tolerances and production rates. Carbide tools are also used to an advantage on steel casting to offset the damaging effect of local hard spots.

(*c*) **Construction.** A broach may be either solid or assembled or built up from shells, replaceable sections or inserted teeth. Replaceable sections, teeth or shells make a broach easier to repair.

3.7.2. Types of Broaching Machines. A broaching machine consists of a work-holding device (fixture), a broaching tool, a drive mechanism and a suitable frame. The broach is usually secured in the main slide of the broaching machine and travels with the slide.

A broaching machine may be a horizontal or a vertical one depending upon whether the slide travels in the horizontal plane or in the vertical plane. A horizontal machine has the following characteristics :

(*i*) Any part of it, especially, the work station can be reached readily from the floor.

(*ii*) A long slide can be supported at many points and levelled to keep it straight.

A vertical machine has the following characteristics :

(*a*) It occupies less floor area.

(*b*) For a machine with a long stroke, a pit should be sunk or there should be a platform for the operator to reach the workstation.

1. Internal Broaching Machines. We have pull broaching machines and push broaching machines and these machines may be horizontal or vertical. In pull broaching machines, the force is applied to the shank and the body of the broach is in tension, (Fig. 3.105 (*a*, *b*)). If the force is applied to the rear end of the broach, it is a "push broach" and is in compression, (Fig. 3.105 (*c*)) To avoid buckling, a push broach should be shorter than a pull broach and its length usually does not increase 15 diameters. A vertical push broaching machine is mostly used for internal broaching such as hole sizing, and key way cutting. In horizontal pull broaching machine, Fig. 3.105 (*a*), the work is held tightly against the platen in the broaching operation by the cutting force.

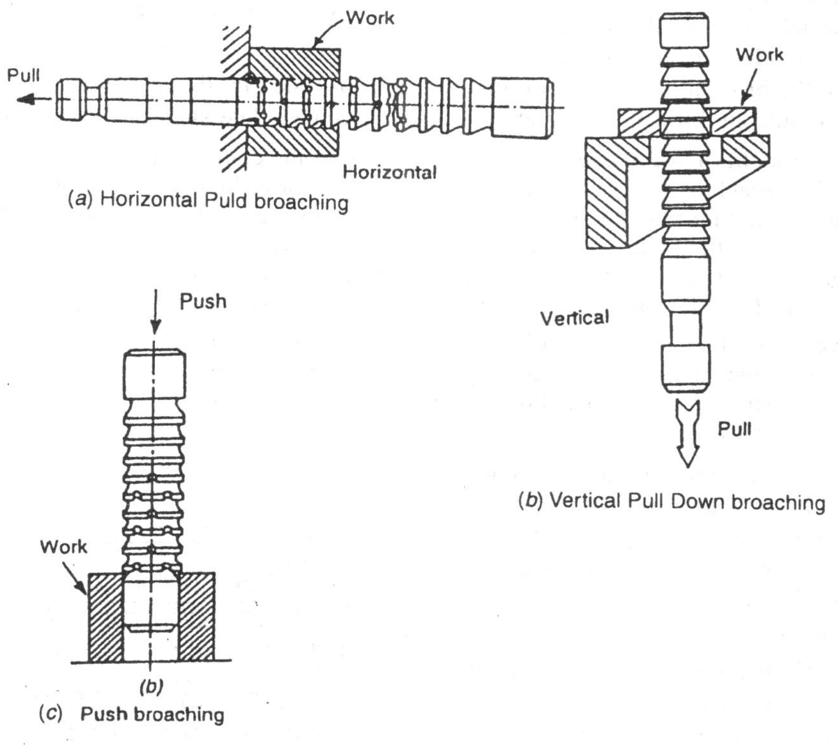

Fig. 3.105. Internal Broaching.

Vertical pull broaching machines may be : vertical pull-down machines or vertical pull up broaching machines.

(*a*) **Vertical Pull Down Broaching Machines.** In these machines, the broach, instead of being pushed, is pulled through the job (Fig. 3.105 *b*). The pulling mechanism is in the base of the machine. The broach is suspended above the work-table by an upper carriage. To start the broaching process, the broach is lowered through the workpiece held in a fixture on the worktable. The broach is automatically engaged by the pulling mechanism and is pulled down through the job. After the operation is completed, the broach returns to its starting position.

(*b*) **Vertical Pull Up Broaching Machine.** Here, the pulling mechanism is above the worktable and the broach is in the base of the machine. The broach enters the job held against the underside of the table and is pulled upward. At the end of the operation, the work is free and falls down into a container.

2. Surface Broaching Machines. External surface broaches are classified in respect to the surface they machine into : flat (slab), peripheral and contour broaches. The common surface broaching machines are discussed below :

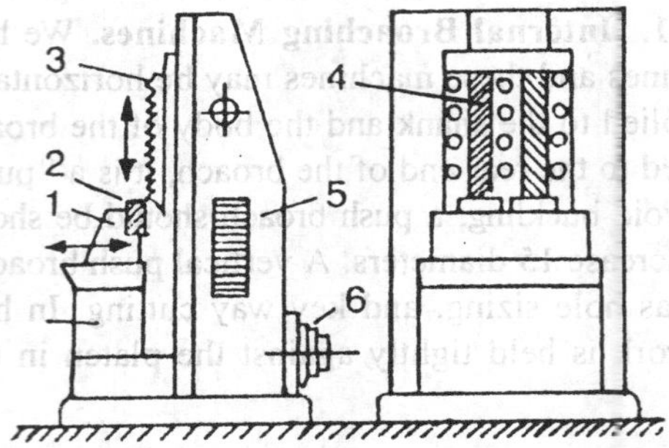

Fig. 3.106. Vertical Surface Broaching Machine.

(*a*) **Vertical Broaching Machine.** A vertical broaching machine, (Fig. 3.106) has a box shape column 5 which houses electric drive 6 with the units of hydraulic drive. Broaching tools 4 are mounted on slide 3 which is hydraulically driven and accurately guided on the column ways. Slide with the broaches travels at various speeds control-led by the hydraulic drive. Its stroke is adjusted to suit the broaching operation to be per-formed. The slide has a rapid return stroke.

The worktable 1 is mounted on the base in front of the column. It can be retracted to unload and load the broaching fixture and advanced to a preset broaching position. Workpiece 2 is held in a broaching fixture mounted on the table. After advancing the table to the broaching position, it is clamped and the slide with the broach travels downwards, machining the workpiece. The table then recedes to load a new workpiece and the slide returns to its upper position. The cycle is then repeated in the same order. Such a machine is called as "Single slide machine".

In vertical "double slide" surface broaching machine, there are two slides which operate opposite each other. The work is held on shuttle worktables which move out during the unloading and loading operations while the slide returns to its starting position. While this is going on, the other slide is at work.

(*b*) **Horizontal Surface Broaching Machine.** Here, the broach is pulled over the top surface of the workpiece held in the fixture on the worktable. The cutting speed range from 3 to 12 mpm with a return speed upto 30 mpm.

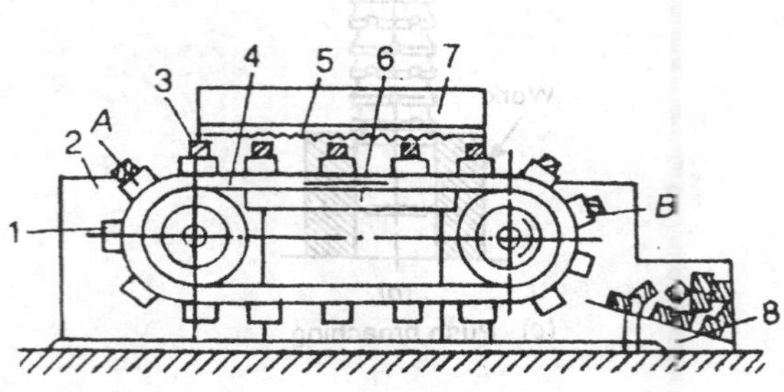

Fig. 3.107. Continuous Surface Broaching Machine.

(*c*) **Continuous Surface Broaching Machine.** A continuous chain type surface broaching machine is shown in Fig. 3.107. A continuous chain 4 travels in a horizontal plane over sprockets mounted in box-shaped base 2. Fixtures 1 for locating and holding workpieces 3 are mounted at intervals on the chain. Broach 5 is fixed horizontally above the chain, under bracket 7 mounted on the base. A Rigid guiding member 6 is arranged under the chain in the zone where the workpieces pass under the broach and provides for horizontal motion of the chain so that a definite amount of stock is removed from the workpieces.

The workpieces are loaded into the fixtures at station A where they are clamped either manually or automatically. The workpieces, located and clamped on the travailing chain, passes between the broach and the guiding plate and then are automatically unclamped and ejected at station B where they drop into hopper 8 of the machine. Continuous surface broaching machine increases the productivity.

3.7.3. Broaching Drives. The drives used to reciprocate the broaching tools are : Hydraulic and Electro-mechanical. Hydraulic drives cost less, give smoother action and are more popular. An electro- mechanical drive employs a screw engaged with a recirculating ball nut. This type of drive is preferred for longer strokes and high speed machines.

Most H.S.S. broaches cut at less than 0.15 m/s and the carbide broaches around 0.50 to 1.50 m/s.

3.7.4. Size of Broaching Machines. The size of a broaching machine is given by two numbers : $a \times b$ where

a = Force the machine can apply in tonnes

b = stroke of the slide, mm.

3.7.5. Advantages of broaching.

1. Rapid operation since the whole machining allowance is removed in a single stroke of the broach.
2. Very high production capacity.
3. High degree of surface quality.
4. The process can be used both for internal as well as external machining.

Limitations

1. Cost of tool is very high.
2. A separate broach has to be made for each shape and size, therefore, broaching is primarily a method for mass production.
3. The workpiece must be rigidly held and the broach firmly guided.
4. Large amounts of metals can not be removed.

PROBLEMS

1. Define a broaching operation.
2. Sketch and discuss a typical Internal Broach.
3. Sketch a typical tooth shape of a Broach.
4. Write briefly about the materials of broaching tools.
5. Write briefly about the construction of broaching tools.
6. With the help of neat sketches, explain the horizontal pull broaching operation and the vertical push broaching operation.
7. With the help of a neat sketch, discuss the working of a surface broaching machine.
8. With the help of a neat sketch, discuss the working of a continuous surface broaching machine.
9. Write the advantages and limitations of broaching process.

3.8. SAWING MACHINES

The various construction materials (stock) available in shop stores are of standard forms and sizes. The stock is cut to correct length to make a product. The operation of cutting stock to correct length is called "sawing". The process can be accomplished either by hand (hand hack-saw) or by power operated machines. Sawing by hand is very slow and laborious. These troubles are overcome by using power operated sawing machines.

All the sawing machines use saw blades to cut the stock. Although different in overall form, they all contain a series of cutting teeth that operate in the same basic way. When the teeth are laid out into a straight line, one obtains a hacksaw, or, if the sawblade is flexible and made into an endless loop, a band saw. All these tools are form tools and the cut progresses by positive infeed or by the pressure exerted on the tool.

3.8.1. Saw blade. The important features of a saw blade are : Material, Tooth form, Tooth set, Tooth spacing and Size.

1. Material. Blades are made of carbon steel, Tungsten alloy steel, High Speed tungsten steel, Molybdenum steel, High speed molybdenum steel and Cemented carbide. Most saws are solid, some have teeth of a hard material backed by flexible steel, and many have tipped or inserted teeth.

2. Tooth Form. Two commonly used tooth forms for saw blades are shown in Fig. 3.108. The straight tooth form, (Fig. 3.108 *a*) is suitable for finer pitches whereas the tooth form shown in (Fig. 3.108 *b*) is used with coarser pitches. This tooth form is theoretically better because the cutting edges are backed up by more metal. However, it is more difficult to obtain this tooth form on smaller teeth.

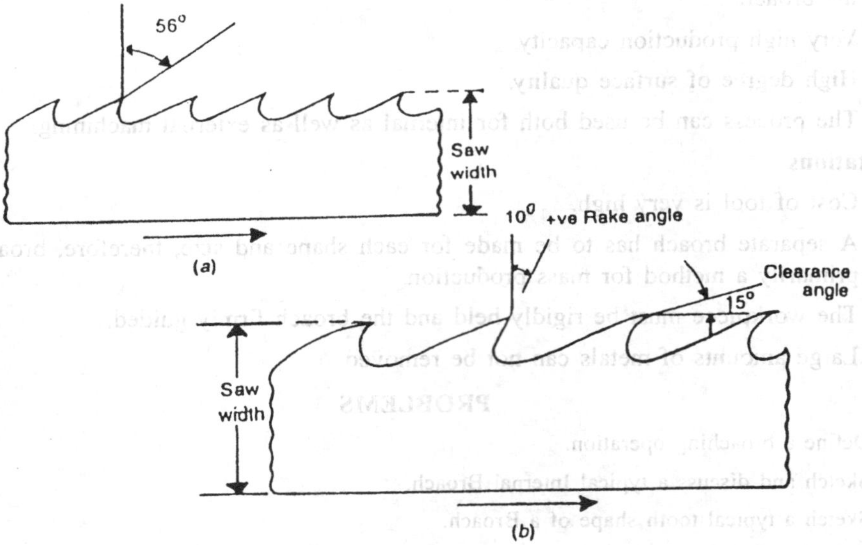

Fig. 3.108. Tooth Forms.

3. Tooth Set. Tooth set refers to how the teeth are bent or offset to one side or the other, and the amount of the offset. The set makes the cut or "kerf" wider than the blade thickness so that the blade will not stick or bind. We have three types of tooth sets, Fig. 3.109.

(*a*) **Straight Set.** When the teeth are offset alternately to the right and left, (Fig. 3.109 *a*), it is called straight set or alternate set. This set is most suitable for non-ferrous metals and non-metals.

(*b*) **Wavy Set.** When several teeth are offset in one direction, and then several are offset in the opposite direction, it is called a wavy set, (Fig. 3.109 (*c*)). This set is employed with small teeth for sawing thing sheets and sections.

(*c*) **Raker Set.** One tooth is offset to the left, the next tooth is offset to the right and the third tooth is unbent (Fig. 3.109 (*b*)). This form is repeated and we get a raker set. This set is used for cutting ferrous metals unless the metal thickness is too thin.

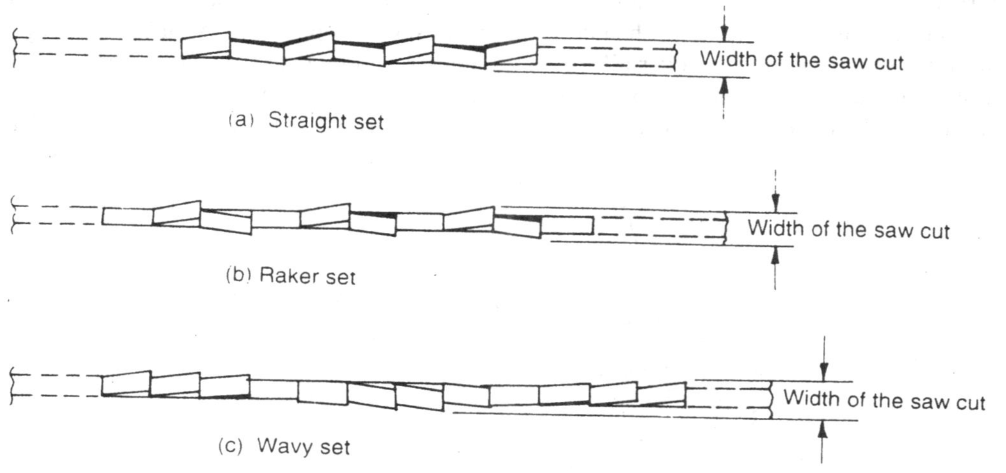

Fig. 3.109. Tooth Sets.

4. **Tooth Spacing.** Tooth spacing or pitch has an important influence upon saw performance. Selection of pitch for a sawing job will depend upon the length of saw in contact with the job and the work material. Longer saw cuts require coarser pitched teeth because these cuts will produce longer chips needing more space for chips between the teeth. Softer materials will also produce longer chips and so will need coarser pitched saws. Finer pitched teeth are used for sawing harder materials. For cutting thin workpiece or thin sections, at least 2 or 3 teeth should always be in contact with the workpiece to avoid snagging of the teeth.

The commonly used hand hacksaw blades are furnished with 14, 18, 24 and 32 teeth per inch. Power hacksaw blades are coarser (from 4 to 14 teeth per inch). Band saws are also with coarser pitches and may have as few as 2 teeth per inch. The pitch may run from 0.20 inch on small circular saws to 2 inches on large diameter saws for soft materials."

5. **Size.** The size of hand hacksaw blades is :

thickness = 0.64 mm, width = 12.7 mm

Length = 254 and 304.8 mm.

The size of power hacksaw blades is

thickness = 1.27 to 2.54 mm, width = 25.4 to 50.8 mm

Length = 304.8 to 609.6 mm.

The heavier the cut, the longer should be the blade. Most band saws are welded in continuous loops and commonly are from 0.020 to 0.060 inch thick and $\frac{1}{16}$ to 1 inch (a few upto 2 inch) wide. The narrower blades are required to cut around small radii. Circular cold saws are tooth disks from 1/32 to 1/12 inch thick and 8 to 80 inch in diameter.

6. **Cutting Speed.** Cutting speed for tungsten H.S.S. blades :

Mild steel	= 0.75 m/s
Cast iron	= 0.50 m/s
Brass and	= 1.5 m/s Aluminium
Bronze	= 1.25 m/s
Thin sections,	= 1.5 m/s. pipes and tubes

3.8.2. Types of Sawing Machines. The common sawing machines are discussed below :

1. Power Hacksaw (Reciprocating Sawing Machine). The power hack saw is a power driven tool for cutting off stock, (Fig. 3.110). The machine consists of a base with a worktable and vise on it. The vise is adjustable for cutting at right angles and at several other angles. A saw frame holds the blade. In operation, the stock is held in a vise. The sawing takes place only on the forward, or draw, stroke. The frame lifts up slightly on the return stroke. There are switches to turn on the power and a clutch handle to start the sawing operation. A sufficient supply of cutting fluid should be flooded over all metals except Cast iron, at the cut. At the completion of a cut, an automatic switch may stop the sawing motion of the machine. The size of the saw is determined by the largest diameter (or square) that can be cut. A common size is 3 inch by 3 inch.

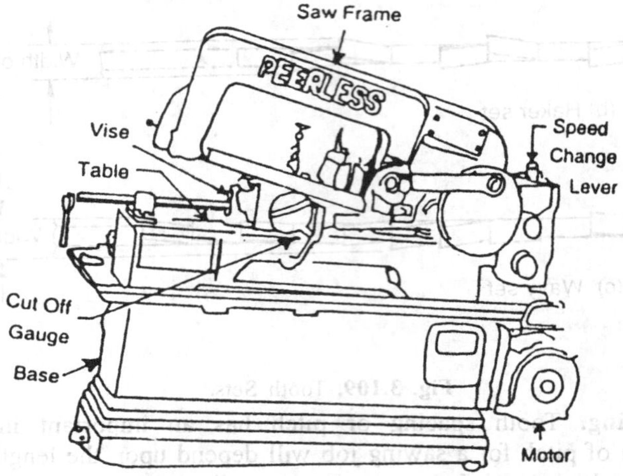

Fig. 3.110. Power Hacksaw.

2. Circular Sawing (Cold Sawing Machine). A saw for cold sawing of a metal is a profile sharpened cutter for milling a narrow slit. Saws of small diameter may be made in the form of

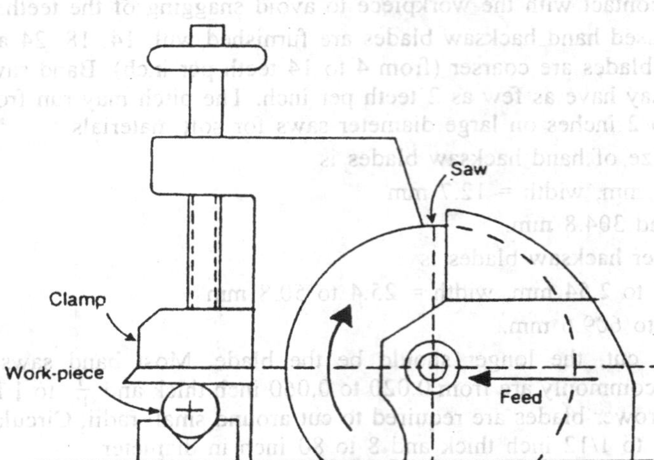

Fig. 3.111. Circular Sawing Machine.

a solid disk with teeth located on the periphery. At present, circular saws upto 250 mm in diameter are usually made solid with integral teeth. Circular saws over 250 mm in diameter are of either segmental or inserted-blade construction. The working of a circular sawing machine is illustrated in Fig. 3.111.

3. Band Saws. Band saws with an endless saw blade are being used to a greater and greater extent for cut off and straight sawing operations for metals since they enable the width of the kerf to be substantially reduced. They are used in special power bandsawing machines. A continuous saw blade runs over the rims of two wheels on the machine.

(*a*) **Horizontal Band Saw.** The horizontal band saw is a small metal cutting band saw that can be used with the blade in a horizontal position, (Fig. 3.112). The saw consists of a base with a worktable and vise. The saw frame is hinged at one end. On the frame are two wheels over which the continuous blade operates. Most of the blade is covered. Machine size is determined by the wheel diameter. The saw carried on a frame is fed downward through the workpiece held in the vise. Except for the continuous cutting action of the endless band saw blade, horizontal band saws incorporate most of the features of the power hacksaw.

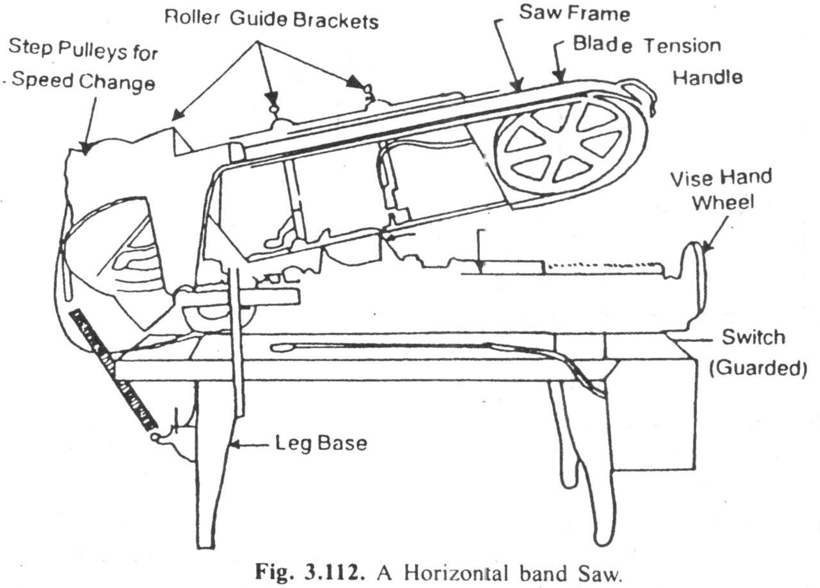

Fig. 3.112. A Horizontal band Saw.

(*b*) **Vertical Band Saw.** In this machine, (Fig. 3.113), the endless band saw blade runs over two wheels in a vertical position. Saw guides help support the sides of the saw band near the sawing. Cutting speeds may range from 0.25 to 7.5 depending upon the metal to be cut. The blade tooth form, (Fig. 3.108 *a*), is commonly used for band saws. However, for sawing softer metals such as aluminium and lead alloys, the tooth form shown in Fig. 3.114 (buttress or skip-tooth form) is employed. This provides more space for chips and, therefore, the saw will not become overloaded. In the Fig. 3.113, the power is given to the power wheel from a variable speed motor.

3.8.3. Field of Application of Sawing Machines. Hacksaw machines are relatively slow and find usage in low-output production and toolroom and maintenance work.

Cold Saw. Where production is in the medium or high ranges, cutoff and similar sawing operations are normally done on the cold saw. Manually operated cold saws are used for low production work and automatic-cycle and special units are used in the higher production ranges. Much larger sizes and cuts are possible than with hack saws. Standard machines commonly

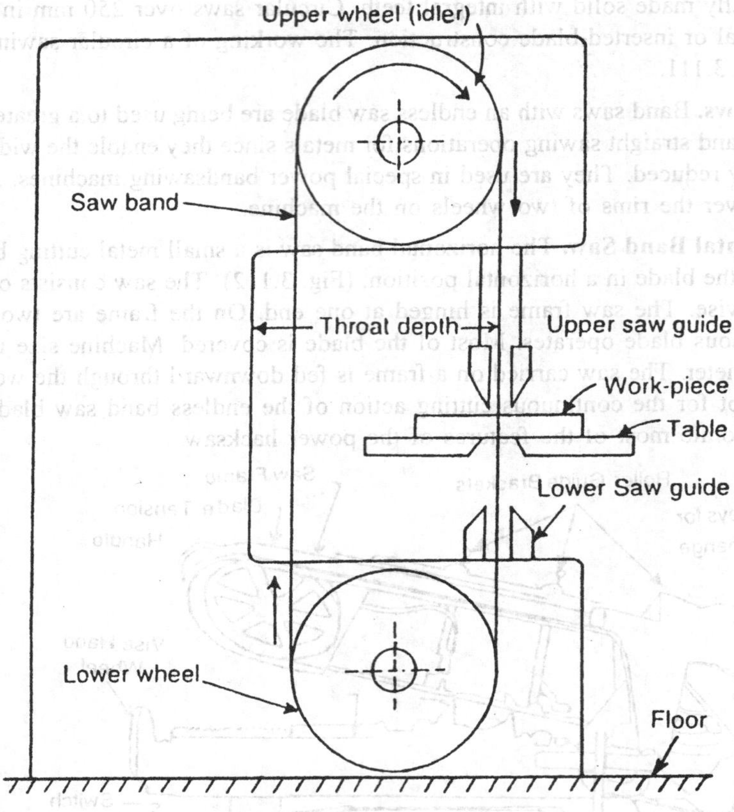

Fig. 3.113. Vertical band Saw Machine.

available employ blades with diameters ranging from 1016 to 1143 mm. Some machines use blades as much as 3.05 m in diameter. A typical machine with a 1143 mm blade will cut a full 406.4 mm square or a rectangle 330 × 644 mm.

Band Saw. Capable of cuts other than straight, the band saw is useful for production operations other than cut off. With conventional plain contour band sawing, it is practicable to produce blank parts in quantities upto about 500 pieces, eliminated the usual die costs for such parts. Sheet material is compressed and welded together to simplify the sawing, and intricate or three-dimensional shapes are practical.

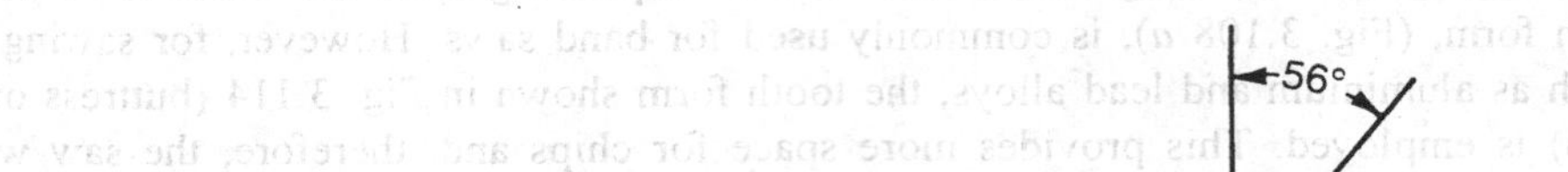

Fig. 3.114. Skip-Tooth Form.

PROBLEMS

1. Define "sawing operation".
2. Discuss the materials for saw blades.
3. Sketch and discuss the various tooth-forms used for saw blades.
4. What is "tooth-set" ? Sketch and discuss the various tooth-sets used for saw blades.
5. How the pitch for a saw blade is selected ?
6. Sketch and explain the working of a Power Hacksaw.
7. Sketch and explain the working of a Circular Sawing Machine.
8. Sketch and explain the working of a Horizontal Band Saw.
9. Sketch and explain the working of a Vertical Band Saw.
10. Discuss the field of applications of the various sawing machines.
11. Explain why broaching is a commonly used process ? Give some typical applications.
12. Explain why hacksaws are not as productive as horizontal band saws ?
13. What is the function of burnishing teeth on broaches?
14. Write briefly on Broaching drives.
15. How is the size of a broaching machine specified?
16. Write briefly on Abrasive sticks.
17. What are grinding points? Sketch the various grinding points?
18. Write briefly on Grinding segments.
19. What is skip-tooth form? For which materials it is preferred?

Specifications of Grinding Machines :–

Grinding Machine	*Main Specifications*
External cylindrical grinding	Swing ϕ , centre distance
Internal cylindrical grinding	Maximum grinding bore ϕ .
Centreless grinding external	Job ϕ
Centreless grinding, internal	Bore ϕ
Surface grinder, Reciprocating Table	Table size (Length × Width)
Surface grinder Rotary Table	Table ϕ
Specifications of Sawing Machines	
Power Hacksaw Machines	Job ϕ
Circular Sawing Machine	Blade ϕ
Band Saw (Hor. Vert.)	Job ϕ (Throat Depth)

4

MACHINING VARIABLES AND RELATED RELATIONS

4.1. GENERAL

As already discussed under Art. 1.6, the three machining elements of any machining process are : feed, f (mm/rev.), depth of cut, d (mm), and cutting speed, V (m/min.).

1. Lathes. Refer to Art. 1.6 and Fig. 1.13,

$$V = \frac{\pi DN}{1000}, \quad \text{m/min.} \qquad \text{... (1)}$$

Area of uncut chip : The cross-sectional area, A_C of the layer of the work material being removed is

$$A_C = \text{width of chip} \times \text{thickness of uncut chip.}$$
$$= b \times t$$

Now if C_s = side cutting edge angle, then

$$b = \frac{d}{\text{Cos } C_s} \qquad \text{and } t = f \text{ Cos } C_s$$

$$\therefore \quad b \times t = f.d \text{ mm}^2$$

∴ Metal removal rate, (MRR) is given as

$$\text{MRR} = \pi.D.A_c.N \text{ mm}^3/\text{min}$$

$$\therefore \quad \text{MRR} = 1000\, A_c.V \text{ mm}^3/\text{min} \qquad \text{....(From Equation .1)}$$
$$= 1000\, fdV, \text{ mm}^3/\text{min}$$

Machining time : The machining time is given as,

$$T_m = \frac{L}{fN} \text{ min, per pass (cut)}$$

where L = Total length of travel of the tool or of the workpiece
= (Length of surface to be machined) + (Tool approach) + (Tool overtravel)

Tool approach = Distance a tool is fed from the time it touches the workpiece, until it is cutting to the full depth.

Tool overtravel = Distance a tool is fed, while it is not cutting.

It is the distance over which the tool idles, before it enters and after it leaves the cut. Approach of most of the single point cutting tools is negligible.

Refer to Fig. 4.1

$$L = l + x + y$$

where x = Tool approach = $d \tan C_s$ or $d \text{ Cot } \lambda$ mm

y = Tool overtravel = 1 to 2 mm

λ = Tool approach angle = $90 - C_s$

For facing operation,

$$L = D/2 + x + y, \text{ mm},$$

Now, Total machining time $= T_m \times i$

where i = number of passes or cuts $= \dfrac{\text{Total machining allowance}}{\text{Material removed per cut}}$

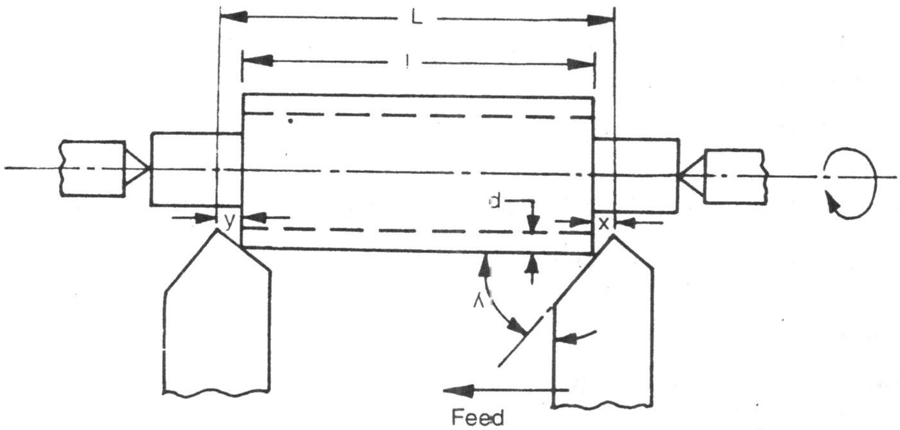

Fig. 4.1. Turning Operating.

Now, total machining allowance $= \dfrac{D_i - D_f}{2}$, for turning, and $= \dfrac{D_f - D_i}{2}$, for boring

where D_i = initial diameter of the workpiece

D_f = finished diameter of the workpiece

Material removed per cut = Depth of cut

Depth of cut is half the difference between the work diameter at the start of the cut and the diameter of machined surface obtained after the cut.

$$\therefore \quad d = \frac{\text{Total machining allowance}}{i}$$

In practice, the depth of cut per pass is not constant. For roughing operations, it is much greater than for the finishing operations.

2. Shapers, Planers, Slotters : Refer to art. 3.2.4.

$$\text{MRR} = f.d.L.N, \text{ mm}^3/\text{min}.$$

Here, L = Length of the workpiece, mm

N = Number of the complete strokes /min.

f = feed, mm/stroke

Machining Time, $T_m = \dfrac{B + x + y}{fN}$, min.

B = Width of the machined surface

x = Side approach of the tool $= d \cot \lambda$

y = Side over travel of the tool = 2 to 3 mm.

Another method of calculating the Machining Time

$$T_m = \frac{B}{f}\left(\frac{L}{V_w \times 1000} + \frac{L}{V_r \times 1000}\right), \text{ min}$$

where L= Length of stroke (for Shaper) or length of Table travel (for Planer), mm
= Length of workpiece + Approach + Overtravel
V_w = working stroke speed or cutting speed, m/min
V_r= Return stroke speed, m/min; See Fig. 3.39
B = Width of job, mm
f = Cross-feed per full stroke of table (For Shaper) and of tool head (for Planer), mm.

V_w can be determined as,

$$V_w = \text{Cutting speed} = \frac{L \times N \times (1 + K)}{1000}, \text{ m/min.}$$

V_w is 'V' as under Art. 3.2.4.

where N = Number of full (double) strokes/min, that is, working plus return strokes

K = Ratio of return time to cutting time, or $= \frac{V_w}{V_r}$

On an average, K $= \frac{2}{3}$ to $\frac{3}{4}$

Now, if V_m = Mean or Average speed, then

$$T_m = \frac{B}{f} \times \frac{2L}{V_m \times 1000}, \text{ min.}$$

V_m is given as, $V_m = \frac{2V_w \times V_r}{V_w + V_r}$, m/min.

3. Drilling :

Cutting speed, $V = \frac{\pi DN}{1000}$, m/min.

Here D = Drill diameter, mm
N = rotational speed of the drill, / min

$$\text{MRR} = \frac{\pi D^2}{4} f.N, \text{ mm}^3/\text{min}$$

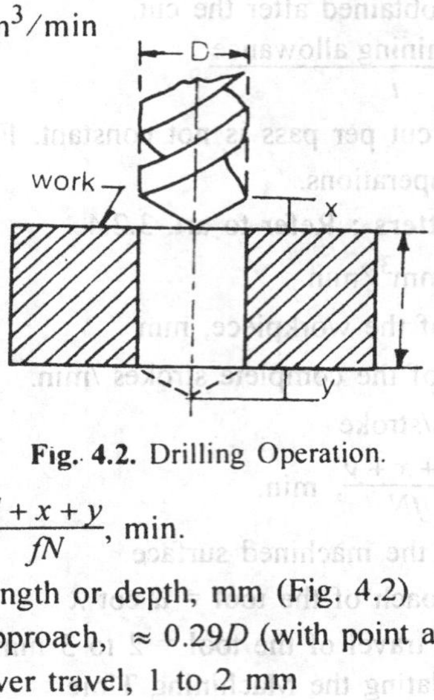

Fig. 4.2. Drilling Operation.

Machining time, $T_m = \frac{L}{fN} = \frac{l + x + y}{fN}$, min.

where l = hole length or depth, mm (Fig. 4.2)
x = Tool approach, $\approx 0.29D$ (with point angle of 118°)
y = Tool over travel, 1 to 2 mm

4. Milling : See Art. 3.5.6.

(*a*) **Peripheral Milling** : See Fig. 3.65 (*a*).

$$MRR = b.d.F \text{ mm}^3/\text{min.}$$

where b = cutting edge engagement = width of job

d = depth of cut

F = table feed, mm/min.

Now $F = f$ (mm/rev of cutter) × N $= f_t \times n \times N$, mm/min.

f_t = feed per tooth, per rev., mm.

n = number of teeth in the cutter

Machining time, $T_m = \dfrac{l + x + y}{F}$, min.

where, l = length of job

x = cutter approach

y = over travel = 1 to 5 mm

x is given as :

$$x = \sqrt{d\,(D - d)} \quad \text{, [See Fig. 4.3 (a)].}$$

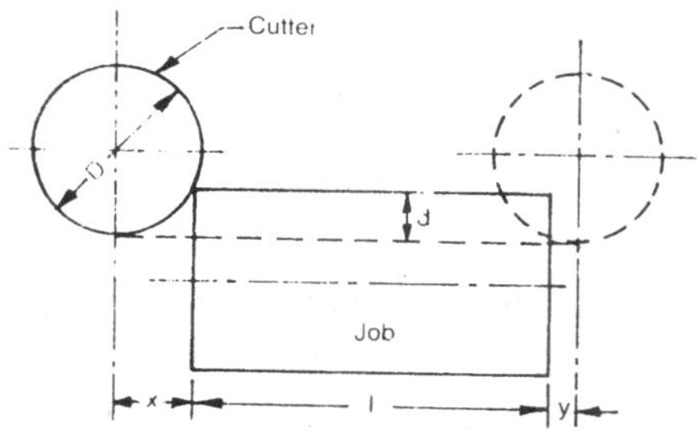

(a) Slab Milling.

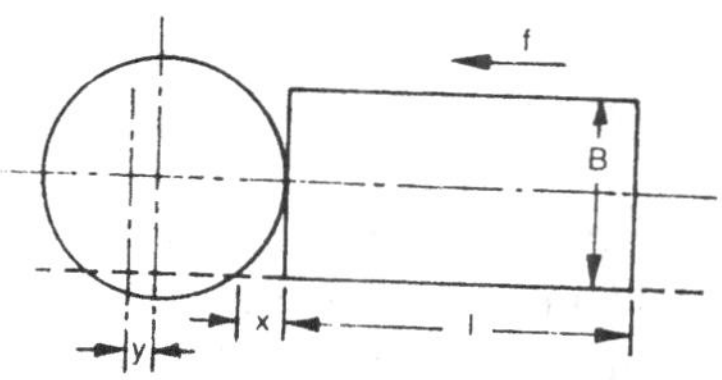

(b) face milling

Fig. 4.3. Milling Operation.

(*b*) **Face Milling :** See Fig. 3.65 (*b*)

For face milling, the tool approach is given as, [Fig. 4.3 (*b*)]

$$x = 0.5\left(D - \sqrt{D^2 - B^2}\right)$$

B is the width of the job.

5. Broaching: The cross - sectional area, Ac, of uncut chip per tooth is

$A_c = f_z.b$ mm^2 for a keyway broach

$= f_z.b.n.$ mm^2, for a multiple-Spline" $= f_z.\pi.D$, mm^2 for a round broach

where f_z = cut per tooth mm

= Difference in height between two successive cutting teeth

b = width of broach

$n =$ number of splines

$D =$ diameter of round broach, mm

MRR/pass = MRR/tooth × z

∴ MRR/tooth = 1000.A_c.V. mm^3/min

Where z = number of teeth simultaneously in operation.

$$\text{Machining time, } T_m = \frac{l_w + l_b}{V}$$

where l_w = Length of workpiece

l_b = Length of broach

However, machining time per piece

$$T = \frac{L}{1000\ V} + \frac{L}{1000\ V_r}$$

where L = Length of stroke, mm

V_r = Return velocity of broach.

or
$$T = \frac{L}{1000\ V} k.$$

where k = Co-efficient, taking into account the ratio of working and return stroke speeds = 1.4 to 1.5

6. Grinding :

(*a*) **Cylindrical grinding :** In Centre-type cylindrical grinding, we have two types of grinding operations; traverse grinding and plunge-cut grinding.

(*i*) **Traverse grinding :** In traverse grinding, the depth of cut is the layer of metal removed by the grinding wheel in one traverse stroke parallel to the axis of the job. It is also called "in feed". So, depth of cut is,

$$d = \frac{\text{Work diameter} - \text{final diameter}}{2} = \frac{D_w - D_f}{2}$$

where d = 0.01 to 0.025 mm for rough grinding

= 0.005 to 0.015 mm for finish grinding

The traverse feed is the longitudinal feed, parallel to the axis of the job, mm/rev for rough grinding,

f = (0.3 to 0.5) × width of wheel, for $D_w < 20$ mm

= (0.7 to 0.85 × width of wheel, for $D_w \geq 20$ mm

For finish grinding, f = (0.2 to 0.4) × width of wheel. Table traverse will be,

$f_t = f.\ N_w$, mm/min

where N_w = work speed, rev/min.

Peripheral speed of work,

$$V_w = \frac{\pi\ D_w\ N_w}{1000}, \text{ m/min.}$$

Peripheral speed of the wheel (Cutting speed),

$$V_g = \frac{\pi\ D_g\ N_g}{1000}$$

where $D_g =$ grinding wheel diameter, mm

$N_g =$ grinding wheel speed, rev/min.

When the work and the wheel rotate in opposite direction to each other, the cutting speed can be taken as,

$$V = V_g + V_w$$

Metal removal rate : It can be taken approximately as

$$\text{MRR} = \pi\, d\, D_w.f.N_w, \ \text{mm}^3/\text{min}$$

Machining Time :

$$T_m = \frac{L}{fN_w} \times i \times k, \ \text{min}$$

$i =$ number of passes

$$= \frac{\text{machining allowance on each side}}{\text{depth of cut (infeed)}}$$

$k =$ sparking out factor

This factor takes into account the removal of metal even in passes in which there is no in-feed. This occurs due to: the springing of the work, non-uniform wear of the wheel, breaking out of grains and the absence of continuous cutting edge. All this leads to an increase in machining time.

k is about 1.2 for rough grinding, and = 1.4 for finish grinding

(*ii*) **Plunge-cut grinding :**

$$\text{MRR} = \pi\, D_w.B_g.f_p \ \text{mm}^3/\text{min.}$$

where $B_g =$ wheel width

$f_p =$ rate of radial feed, mm/min.

$$\text{Machining time, } T_m = \frac{\text{Machining allowance}}{f_p}$$

$$f_p = f \times N_w, \ f \text{ is in mm / rev.}$$

(*b*) **Surface grinding :**

$$\text{Cutting speed, } V = \frac{\pi\, D_g\, N_g}{1000}, \ \text{m/min.}$$

$$\text{MRR} = f_p\, .d f_t$$

where $f_p =$ cross-feed per stroke, mm/stroke

$f_t =$ table traverse rate, mm/min

$d =$ depth of cut per pass.

$$\text{Machining time } T_m = \frac{L.B}{V.1000.f_p}\, .\, i.k$$

Where L= Length of workpiece to be ground

B = Width of workpiece to be ground

V = Traverse Velocity of table, m/min.

$f_p =$ Crossfeed of grinding wheel, mm/stroke

SOLVED EXAMPLES

Example 4.1 : *Determine the cutting speed for machining a workpiece 120 mm in diameter at a rotational speed of 500 rev./min.*

Solution : The cutting speed is given by the relation,

$$V = \frac{\pi DN}{1000 \times 60}, \text{ m/s}$$

$$= \frac{\pi \times 120 \times 500}{1000 \times 60} = \boldsymbol{\pi} \textbf{ m/s}$$

Example 4.2 : *Determine the rotational speed of the lathe spindle in machining a workpiece 80 mm in diameter at the cutting speed of 3.6 m/s.*

Solution : From the relation,

$$V = \frac{\pi DN}{1000 \times 60}$$

$$N = \frac{V \times 1000 \times 60}{\pi \times D}$$

$$= \frac{3.6 \times 1000 \times 60}{\pi \times 80} = \textbf{859.33 rev./min.}$$

Example 4.3 : *Determine the feed per min. in turning a workpiece on a lathe at a spindle rotational speed of 1000 rev./min. The tool feed per spindle revolution is 0.26 min./rev.*

Solution : Now,

$$\text{Feed per min.} = \text{Feed per revolution} \times \text{No. of revolutions/min.}$$

$$= 0.26 \times 1000$$

$$= 260 \text{ mm/min.}$$

Example 4.4 : *Determine the depth of cut in boring a hole 55 mm in diameter to 60 mm in diameter by one pass on a lathe.*

Solution : Now material removed per cut (on one side) = depth of cut, d

$$\therefore \quad d = \frac{\text{Total machining allowance}}{i} = \frac{D_f - D_i}{2 \times i}$$

In the example, $D_i = 55$ mm, $D_f = 60$ mm, $i = 1$

$$\therefore \quad d = \frac{60 - 55}{2 \times 1} = \textbf{2.5 mm}$$

Example 4.5 : *Determine the depth of cut in turning a workpiece 150 mm in diameter to 140 mm diameter in two passes. Roughing pass is from 150 to 142 mm in diameter.*

Solution :

(*a*) **Roughing pass :**

$$d = \frac{150 - 142}{2 \times 1} = \textbf{4 mm}$$

(*b*) **Finishing pass :**

$$d = \frac{142 - 140}{2 \times 1} = \textbf{1 mm}$$

Example 4.6 : *Determine the machining time in facing a workpiece/cutting off a ring from a tube like workpiece on a lathe. The outside diameter of the workpiece* is 100 mm, its inside diameter 84 mm, spindle speed 250 rev./min., tool feed is 0.14 mm/rev.

Solution : We know that,

$$T_m = \frac{L}{f \times N}$$

$$L = \text{Tool travel} = \frac{D_i - D_f}{2} + (1 \text{ to } 2 \text{ mm})$$

$$= \frac{100 - 84}{2} + 2 = 10 \text{ mm}$$

$$\therefore \quad T_m = \frac{10}{0.14 \times 250} = \mathbf{0.286 \text{ min.}}$$

Example 4.7 : *A through hole 37 mm is diameter is to be enlarged to diameter 40 mm on a vertical driling machine depth of hole = 50 mm. Determine the depth of cut and the machining time. Take cutting speed as 0.30 m/s and feed as 0.8 mm/revolution.*

Solution : (*a*) Depth of cut, $d = \frac{40 - 37}{2} = \mathbf{1.5 \text{ mm}}$

(*b*) $$T_m = \frac{L}{f \times N}$$

$L = l + x + y$

x = tool approach = $d \cos 60^\circ = 0.9$ mm

y = tool overtravel = 1 to 3 mm, say 2 mm

$\therefore L = 50 + 0.9 + 2 = 52.9$ mm

$$N = \frac{V \times 1000 \times 60}{\pi D} = \frac{0.30 \times 1000 \times 60}{\pi \times 40} = 134 \text{ rev./min.}$$

$$\therefore \quad T_m = \frac{52.9}{0.8 \times 134} = \mathbf{0.49 \text{ min.}}$$

Example 4.8 : *A workpiece of 76 mm diameter is to be machined on a lathe to 68 mm diameter. The total length of the workpiece is 250 mm. The recommended values of the machining elements are : feed = 0.25 mm/rev, a cutting speed of 60 mpm, and a maximum depth of cut of 2.5 mm. How long will it take to finish, the part? Also, determine the metal removal rate.*

Solution : The machining time is given as,

$$T_m = \frac{L}{f \times N} \times i, \text{ min.}$$

where, $L = l + x + y$

x = Tool approach = $d \cot \lambda$

Now $d = 2.5$ mm, Let $\lambda = 75^\circ$ ($C_s = 15$ to 30°)

$\therefore \quad x = 2.5 \cot 75^\circ = 0.67$ mm

y = over travel = 2mm (say)

$f = 0.25$ mm/rev.

$$N = \frac{1000\,V}{\pi D} = \frac{1000 \times 60}{\pi \times 76} = 251 \text{ rev./min.}$$

$$i = \frac{76 - 68}{2 \times 2.5} = 2$$

$$\therefore \quad T_m = \frac{250 + 0.67 + 2}{0.25 \times 251} \times 2 = \mathbf{8.05 \text{ min.}}$$

$$\text{MRR} = 1000 f d V$$

$$= 1000 \times 0.25 \times 2.5 \times 60 = 37.5 \times 10^3 \text{ mm}^3/\text{min.}$$

Example 4.9 : *A part of 25cm in diameter and 50 cm length is to be turned down to 23.5 cm for the entire length. The suggested feed is 1 mm per rev. and the cutting speed is*

135 mpm. The maximum allowable depth of cut is 5.0 mm. What are the feed speed, spindle rpm, MRR, and cutting time ? Assume the over travel is 12.5mm.

Solution : Spindle rpm $= \dfrac{1000\,V}{\pi\,D}$

$$= \frac{1000 \times 135}{\pi\,D} = 172 \text{ rev./min.}$$

Feed speed $= f \times N = 1 \times 172 = 172$ mm/min.

MRR $= 1000\,.\,f\,.\,d\,.\,V = 1000 \times 1 \times 5 \times 135 = 67.5 \times 10^4\ \text{mm}^3/\text{min}$

Cutting time $T_m = \dfrac{L}{fN} \times i$

Here, $i = \dfrac{250 - 235}{2 \times 5} = 2$

and $L = 500 + 12.5 = 512.5$ mm

$$\therefore \quad T_m = \frac{512.5}{172} \times 2 = \mathbf{5.96\ min}$$

Example 4.10 : *A hole of 25 mm diameter and 62.5 mm depth is to be drilled. The suggested feed is 1.25 mm per rev. and the cutting speed is 60 mpm. What are the feed speed, spindle rpm, and cutting tim ? Assume the clearance height is 5 mm. Also find the MRR.*

Solution : Spindle rpm, $N = \dfrac{1000\,V}{\pi\,D}$

$$= \frac{1000 \times 60}{\pi \times 25} = \mathbf{764\ rev./min.}$$

Feed speed $= fN = 1.25 \times 764 =$ **955mm/min.**

Cutting time $= \dfrac{L}{fN}$

Now $L = 62.5 + 5 = 67.5$ mm.

$$\therefore \quad T_m = \frac{67.5}{955} = \mathbf{0.0707\ min.}$$

MRR $= \dfrac{\pi\,D^2}{4} f \times N = \dfrac{\pi \times 625}{4} \times 955 = \mathbf{46.8974 \times 10^4\ mm^3/min.}$

Example 4.11 : *A 50 cm × 15 cm surface of cast iron block 50 cm long × 15 cm wide × 10 cm thick, is to be machined on a shaper. Ram speed is 20 mpm, length of stroke is 510 mm, depth of cut is 4 mm, feed is 1.5 mm/stroke, stock to be removed is 6 mm, side cutting edge angle of the tool is 45°, ratio of time taken in return stroke to the time taken in cutting stroke is 0.5. Determine the metal removal rate and the total cutting time.*

Solution: Now refer to Art. 3.2.4,

$$V = \frac{LN(1 + K)}{1000}$$

$$\therefore \quad N = \frac{1000 \times 20}{510 \times 1.5} \cong 26. \text{ full strokes/min.}$$

MRR $= f\,.\,d\,.\,L\,.\,N = 1.5 \times 4 \times 510 \times 26$

$= \mathbf{79.56 \times 10^3\ mm^3/min.}$

Cutting Time, $T_m = \dfrac{B+x+y}{fN}$

$B = 150$ mm

$x = d \cot \lambda = 4$ mm

$y = 3$ mm (say)

Now number of passes $= \dfrac{\text{Stock to be removed}}{\text{depth of cut}} = \dfrac{6}{4} = 2$ (say)

$\therefore$ Total cutting time, $T_m = \dfrac{150+4+3}{1.5 \times 26} \times 2 = 8.05$ min.

Example 4.12 : *Evaluate the cutting parameters for the slab milling operation for the following data: Diameter of milling cutter = 100 mm, cutter speed = 500 rpm, width of cutter = 100 mm, depth of cut = 5 mm, table feed = 100 mm/min , length of workpiece = 50 cm, width of workpiece = 80 mm, number of teeth in the cutter = 8.*

Solution : Cutter diameter = 100 mm. Cutter speed = 500 rpm.

Cutting speed, $V = \dfrac{\pi D N}{1000} = \dfrac{\pi \times 100 \times 500}{1000} = 157$ m/min

MRR $= b \,.\, d \,.\, F$

where b = width of job = 80 mm

d = depth of cut = 5 mm

F = Table feed = 100 mm/min.

Machining time, $T_m = \dfrac{l+x+y}{F}$

$l = 500$ mm.

Here $y = 4$ mm (say)

$x = \sqrt{d(D-d)} = \sqrt{5 \times 95} = 21.8$ mm.

$\therefore \quad T_m = \dfrac{500+21.8+4}{100} = \mathbf{5.258}$ **min.**

Example 4.13 : *A premachined cylindrical box, 55 mm in diameter and length 62 mm is to be broached on a horizontal broaching machine. Determine the machining time. Take cutting speed as 8 m/min., and return speed as 20 m/min. Overall length of the broach is 570 mm and the shank length is 265 mm.*

Solution : The broaching operation is similar to shaping and planing operations, in that the return stroke (idle stroke) is faster than the cutting stroke. Total time is given as,

$$T_m = \frac{L}{\text{Cutting speed}} + \frac{L}{\text{Return speed}}$$

or This relation can be written as,

$$T_m = \frac{L}{1000\, V q} \,.\, K \,.\, i$$

where L = Length of working stroke of the broach

$= l_b + l +$ Overtravel (30 to 50 mm).

l_b = Length of the working portion of the broach

= Over-all length of broach − shank length of the broach,

l = Length of workpiece

q = No. of workpieces broached in one pass,

i = No. of passes (runs)
V = Cutting velocity, m/min.
K = Co-efficient accounting for the rapid return stroke
$= 1 + \frac{V}{V_r}$; V_r is the return stroke velocity.

In the problem,

$$l_b = 570 - 265 = 305 \text{ mm}$$

$$\therefore \quad L = 305 + 62 + 50 \text{ (say)} = 417 \text{ mm}$$

$$q = 1, \ i = 1, \ V = 8 \text{ m/min}, \ V_r = 20 \text{ m/min}.$$

$$\therefore \quad K = 1 + \frac{8}{20} = 1.4$$

$$\therefore \quad T_m = \frac{417}{1000 \times 8 \times 1} \times 1.4 \times 1$$

$$= \mathbf{0.073 \text{ min.}}$$

Example 4.14 : *Determine the machining parameters for the broaching operation from the following date; cutting speed = 12 m/min., return speed of broach = 25 m/min., length of workpiece = 200 mm, length of broach = 300 mm, length of stroke = 600 mm, width of workpiece = 10 mm, rise per tooth = 0.10 mm.*

Solution : MRR $= 1000 f_z \cdot b \cdot V$.

f_z = cut per tooth = 0.10 mm

b = width of contact = 10 mm

V = 12 m/min

$$\therefore \quad \text{MRR/tooth} = 0.10 \times 10 \times 1000 \times 12 = 12000 \text{ mm}^3\text{/min.}$$

$$\text{Machining time, } T_m = \frac{l_w + l_b}{V}$$

$$l_w = 200 \text{ mm}, \ l_b = 300 \text{ mm}$$

$$\therefore \quad T_m = \frac{200 + 300}{12 \times 1000} = 0.042 \text{ min.}$$

$$\text{Machining time per piece} = \frac{L}{1000\, V} + \frac{L}{1000\, V_r}$$

Now L = 600 mm, V_r = 25 m/min.

$$\therefore \quad \text{Time} = \frac{600}{1000 \times 12} + \frac{600}{1000 \times 25} = \mathbf{0.074 \text{ min.}}$$

Example 4.15 : *Determine the machining parameters for the rough grinding of medium steel shaft from the following date : Work diameter = 38 mm, length of part = 200 mm, Total stock = 0.25 mm, grinding wheel = 50 mm face, depth of cut = 0.025 mm, cutting speed = 15 m/min.*

Solution: Now $D_w = 38$ mm

$\therefore f$ = feed/rev. = (0.7 tp 0.85) × width of wheel

Width of wheel = 50 mm

$\therefore f = 0.8 \times 50 = 40$ mm.

Number of cuts $\quad i = \dfrac{\text{Total stock}}{\text{Depth of cut per pass}} = \dfrac{0.25}{0.025} = 10$

Wheel speed $\quad = \dfrac{1000\,V}{\pi D_w} = \dfrac{1000 \times 15}{\pi \times 38} = 125$ rev./min.

$$\therefore \quad T_m = \frac{L}{f\,N_w} \times i \times k = \frac{200}{40 \times 125} \times 10 \times 1.1 = \mathbf{0.44\ min.}$$

Example 4.16. *A workpiece 200 mm × 300 mm is to be machined on a shaper. Calculate the machining time. Take,*

V_w = 10 m/min, V_r = 20 m/min, f = 5 mm/full stroke, min clearance of each end = 50 mm.

Solution: $B = 200$ mm

$L = 300 + 2 \times 50 = 400$ mm

$$\therefore \quad T_m = \frac{B}{f}\left(\frac{L}{V_w \times 1000} + \frac{L}{V_r \times 1000}\right)$$

$$= \frac{200}{5} \times \left(\frac{400}{10 \times 1000} + \frac{400}{20 \times 1000}\right)$$

$= \mathbf{2.4\ min.}$

Now Average speed,

$$V_m = \frac{2 \times V_w \times V_r}{V_w + V_r} = \frac{2 \times 10 \times 20}{10 + 20} = \frac{40}{3} \text{ m/min}$$

$$\therefore \quad T_m = \frac{B}{f} \times \frac{2L}{V_m \times 1000}$$

$$= \frac{200}{5} \times \frac{2 \times 400 \times 3}{40 \times 1000}$$

$= \mathbf{2.4\ min.}$

Example 4.17. *A plain surface, width = 100 mm and length = 320 mm is to be face-milled on a vertical milling machine. The milling allowance is 4 mm. The cutter has 16 teeth and the feed per tooth is 0.25 mm. The spindle speed is 125 rev./min. Calculate the machining time. Diameter of cutter = 160 mm.*

Solution. $\quad T_m = \dfrac{l + x + y}{F}$, min.

Where $\quad l$ = length of job = 320 mm

x = cutter approach or Added table travel

$= 0.5\ (D - \sqrt{D^2 - B^2})$

$= 0.5(160 - \sqrt{160^2 - 100^2}) = 17.5$ mm

Let $\quad y$ = over travel be = 3 mm.

F = Table feed per min, mm/min.

= Feed per tooth per rev. (mm)

× number of teeth in cutter × spindle revolution min.

$$= 0.25 \times 16 \times 125 = 500 \text{ mm/min.}$$

$$\therefore \quad T_m = \frac{320 + 17.5 + 3}{500} = \mathbf{0.68 \text{ min.}}$$

Example 4.18. *The top surface of a slab, 520 mm wide and 4000 mm long is to be planed on a planer. Take cutting speed as 18.8 m/min. and return speed as 75 m/min. Take machining allowance as 10 mm. The tool approach angle is 45°. Calculate the machining time. Cross feed of tool = 3 mm/full stroke.*

Solution. Refer Art. 3.2.4, the cutting speed or working speed is given as,

$$V = \frac{LN(1+K)}{1000}, \text{ m/min}$$

Where L = Length of table stroke (for planer)

= length of job + table overtravel on both sides

= 400 + 325 (say) = 4325 mm.

N = Number of full strokes/min

$$K = \frac{\text{Return time}}{\text{Cutting time}} = \frac{\text{Cutting speed}}{\text{Return speed}}$$

$$= \frac{18.8}{75} \cong 0.25$$

$$\therefore \quad N = \frac{1000 \times 18.8}{4325\,(1.25)} = 3.5$$

Now, Cutting time,

$$T_m = \frac{B + x + y}{fN}$$

B = Width of job = 520 mm

x = Side approach of tool = $d \cot \lambda$

= 10. cot 45° = 10 mm

y = Side overtravel of tool = 2 to 3 mm, Take 3 mm

$$\therefore \quad T_m = \frac{520 + 10 + 3}{3 \times 3.5} = \mathbf{51.7 \text{ min.}}$$

Example 4.19. *A premachined cylindrical bore 55 mm in diameter and length 62 mm is to be broached on a horizontal broaching machine. Overall length of broach is 570 mm and shank length of broach is 265 mm. Cutting speed is 8 m/min. and return speed is 20 m/min. Take tool overtravel as 50 mm. Determine the broaching time.*

Solution.

$$T_m = \frac{L}{1000V} . k$$

$$k = 1 + \frac{V}{V_r} = 1 + \frac{8}{20} = 1.4$$

L = Length of stroke

= Length of working part of broach + length of job + tool overtravel

= (570 – 265) + 62 + 50 = 417 mm

$$\therefore \quad T_m = \frac{417}{1000 \times 8} \times 1.4 = \mathbf{0.073 \text{ min}}$$

Example 4.20. *A shaft of length, 210 mm and 40 mm in diameter is to be longitudinally ground in one pass on a cylindrical grinding machine. The allowance per side is 0.2 mm. Wheel diameter is 600 mm and its width is 63 mm. Cutting speed is 35 m/min. Determine the grinding time.*

Solution.

$$T_m = \frac{L}{f.N_w}.i.k$$

L = Length of table stroke

= Length of workpiece + wheel overtravel on each side

$= 210 + 2 \times 0.5\ B_g$

Now, $B_g = 63$ mm (width of grinding wheel)

$\therefore$ $L = 273$ mm

$$N_w = \text{Work speed} = \frac{1000\,V_w}{\pi\,D_w} = \frac{1000 \times 35}{\pi \times 40} = 280 \text{ rpm}$$

f = traverse feed, mm/rev.

$= (0.2 \text{ to } 0.4) \times B_g$ for finish grinding

$= 0.3 \times 63 = 18.9$ mm/rev.

i = number of passes

$$= \frac{\text{machining allowance on each side}}{\text{depth of cut (infeed)}}$$

For finish grinding, depth of cut = 0.005 to 0.015 mm

= 0.005 mm (say)

$$\therefore \quad i = \frac{0.2}{0.005} = 40$$

Let $k = 1.4$

$$\therefore \quad T_m = \frac{273}{18.9 \times 280} \times 40 \times 1.4$$

$= \mathbf{2.89}$ **min.**

Example 4.21. *A flat surface of a workpiece is to be ground, workpiece width is 110 mm and length is 280 mm. Machining allowance per side = 0.35 mm. Traverse velocity of table is 16 m/min. Downward feed of the grinding wheel (depth of cut) per pass is 0.015 mm per reverse of the grinding wheel. Cross feed of grinding wheel = 32 mm/table stroke. Determine the total machining time.*

Solution.

$$T_m = \frac{L.B}{1000.V.f_p}.i.k$$

$L = 280$ mm

$B = 110$ mm

$V = 16$ m/min

$f_p = 32$ mm/table stroke

i = number of pass

$$= \frac{0.35}{0.015}$$

$k = 1.4$ (say)

$$\therefore \quad T_m = \frac{280 \times 110}{1000 \times 16 \times 32} \times \frac{0.35}{0.015} \times 1.4$$

= **1.965 min.**

Example 4.22 : *A 150 mm long 12.7 mm diameter stainless steel rod is being turned to 12.19 mm diameter on a centre lathe. Spindle speed = 400 rev./min. Axial speed = 203.20 mm/min. Determine : Cutting speed, MRR and Machining time.*

Solution :

(*i*) Cutting speed, $V = \dfrac{\pi DN}{1000 \times 60}$

$= \dfrac{\pi \times 12.7 \times 400}{1000 \times 60} = 0.26$ m/s

Depth of cut, $d = \dfrac{12.7 - 12.19}{2} = 0.255$ mm

Feed, $f = \dfrac{203.20}{400} = 0.508$ mm/rev.

(*ii*) Metal removal rate, MRR = Area of cut × V

Now area of cut = $d.f.$

∴ MRR = $d.f.\ \pi$ Dave . N

Dave = 12.445 mm

∴ MRR = 0.255 × 0.508 × π × 12.445 × 400

= **2024 mm³/min.**

(*iii*) Machining time, $T = \dfrac{L}{fN}$, L is tool travel

$= \dfrac{150}{0.508 \times 400} = \mathbf{0.72}$ **min.**

Example 4.23 : *The following data is given for slab milling of a 300 mm long, 100 mm wide mild steel block :*

Feed = 0.25 mm/tooth; depth of cut = 3.2 mm

Cutter diameter = 50 mm

Number of cutter teeth = 20

Cutter speed = 100 rev./min.

Cutter is wider than the job. Calculate

(i) MRR *(ii) Machining Time*

Solution :

(*i*) MRR = Width of cut × depth of cut × Table feed

Now width of cut = 3.2 mm

Table feed = feed rate, mm/min.

= Feed per tooth × number of teeth in cutter × cutter speed

= 0.25 × 20 × 100

= 500 mm/min.

∴ MRR = 100 × 3.2 × 500

= 16000 mm³/min.

= **2667.7 mm³/s**

(*ii*) Machining time $= \dfrac{\text{Total table travel}}{\text{Table feed, mm/min.}}$

Now Total Table travel $= \text{Length of job} + \text{Added Table Travel}$

Added Table Travel $= \sqrt{Dd - d^2}$

$= \sqrt{50 \times 3.2 - 3.2 \times 3.2}$

$= 12.24$ mm

$\therefore$ Machining time $= \dfrac{300 + 12.24}{500} = \mathbf{37.5\,s}$

PROBLEMS

1. A 100 mm diameter brass bar of 250 mm length is to be turned to a new diameter of 99.75 mm.
   (*a*) Determine the rpm setting for the lathe.
   (*b*) Determine the cutting time assuming an allowance of 6.25 mm.
   (*c*) Determine the metal removal rate.
   Take cutting speed = 150 mpm and feed = 0.25 mm/rev.
2. A 19 mm diameter hole is to be drilled in a piece of C-20 steel of thickness 25 mm. Taking cutting speed as 400 rev/min. and feed as 0.50 mm/rev.
   (*a*) Calculate the cutting speed in mpm.
   (*b*) Determine the machining time assuming an allowance of 12.5 mm.
   (*c*) Determine the metal removal rate.
3. A 100 mm diameter aluminium bar of 300 mm length is to be turned to a new diameter with a speed of 157.5 mpm, feed of 0.50 mm/rev. and 0.75 mm depth of cut.
   (*a*) Determine the rpm to be set on the machine tool.
   (*b*) Determine the machining time assuming an allowance of 6.25 mm.
   (*c*) Determine the metal removal rate.
4. A slot 6.25 mm wide and 6.25 mm deep is to be milled in one pass in a component. The component is 450 mm long and milled with a 150 mm diameter, 10-tooth cutter. Take cutting speed = 45 mpm and feed = 0.125 mm/tooth
   (*a*) Determine the rpm setting.
   (*b*) Determine the feed rate in mm/min.
   (*c*) Calculate the machining time by taking allowance equal to one cutter diameter.
   (*d*) Calculate the metal removal rate.
5. A 63.5 mm diameter plain milling cutter having 6 teeth is used to facemill a block of aluminium 18 cm long and 3 cm wide. The spindle speed is 1500 rpm and the feed is 0.125 mm/tooth/rev.
   (*a*) Determine the table feed in mm/min.
   (*b*) Total table travel.
   (*c*) Cutting time.
6. Calculate the time required on the shaper to complete one cut on a plate 600 mm × 900 mm, if the cutting speed is 6 m/min. The return time to cutting time ratio is 1 : 4, and the feed is 2 mm/stroke. The clearance at each end is 75 mm.
7. Determine the machining time for one pass longitudinal turning of a shaft from 70 mm to 64 mm over a length of 200 mm. The speed of lathe spindle is 600 rev./min. Tool feed is 0.4 mm/rev. and plan approach angle of tool is 45°. [**Ans.** 0.85 min.]
8. Determine the machining time in facing a workpiece of 165 mm diameter on a lathe in one pass. The speed of the spindle is 480 rev/min. and tool feed is 0.3 mm/rev. Machining allowance is 3.5 mm and plan approach angle of tool is 45°. [**Ans.** 0.61 min]
9. A surface 90 mm wide and 200 mm long is machined on a Shaper. The machining allowance is 2 mm. Take: Cutting speed = 23.8 m/min. Ratio of cutting speed to return speed = 0.8, cross-feed of table = 0.6 mm/full stroke, tool overtravel on each side as 25 mm, plan approach angle of the tool = 45°. [**Ans.** 2.95 min.]
10. A through hole of diameter 20 mm is to be drilled to a depth of 80 mm on a vertical drilling machine. Take cutting speed as 22.3 m/min. and drill feed as 0.4 mm/rev. [**Ans.** 0.62 min.]

11. A plain surface 70 mm wide and 600 mm long is to be face milled on a vertical-spindle milling machine. The machining allowance is 3.7 mm, to be removed in one pass. Cutter diameter is 110 mm and the cutting speed is 172.7 m/min. Number of teeth on the cutter is 4 and the feed is 0.2 mm/tooth/rev. Find the machining time. **[Ans. 1.54 min.]**
12. A plain surface 100 mm wide and 320 mm long is to be face milled on a vertical milling machine. Cutter diameter is 160 mm and number of cutter teeth is 16. Cutting speed is 63 m/min. Feed per tooth is 0.25 mm/tooth. Determine the machining time. **[Ans. 0.68 min.]**
13. The peripheral milling of a plain surface, 75 mm wide and 300 mm long is carried out on the horizontal milling machine. Cutter diameter is 90 mm, cutting speed is 28.3 m/min, number of cutter teeth is 8 and feed per tooth is 0.2 mm. The machining allowance is 3 mm to be removed in a single pass. Determine the machining time. **[Ans. 1.98 min]**
14. A plain surface 65 mm wide and 225 mm long is to be machined on a horizontal milling machine. The machining allowance is 1.5 mm. Cutter diameter is 63 mm, cutting speed is 49.5 m/min, number of cutter teeth is 14 and feed per tooth is 0.11 mm. Determine the machining time. **[Ans. 0.59 min.]**
15. A slot 32 mm wide, 15 mm deep and 250 mm is to be rough machined in a single pass by a disk-type milling cutter on a horizontal milling machine. Cutter diameter is 150 mm, cutting speed is 37.6 m/min, Number of teeth on cutter is 16 and the feed per tooth is 0.078 mm. Determine the machining time. **[Ans. 3 min.]**
16. A through slot 32 mm wide, 15 mm deep and 300 mm long is to be milled by an end mill cutter on a vertical milling machine. The cutter diameter is 32 mm and it has 6 teeth. Cutting speed is 20 m/min. and the feed per tooth is 0.083 mm/tooth. Determine the machining time. **[Ans. 3.19 min. Hint: Take tool approach = $\frac{D}{2}$]**
17. The following data is given for slab milling :

    Cutter diameter = 150 mm
    No. of teeth = 10
    Width of cut = 60 mm
    Length of job = 500 mm
    Depth of cut = 3 mm
    Table speed = 0.6 m/min.
    Cutter speed = 100 rev./min.

    Find : Feed per tooth and MRR. **[Ans. 0.6 mm/tooth, 1800 mm^3/s]**
18. A steel bar 100 mm outside diameter is being turned on a centre lathe. The spindle speed is 100 rev./min. Determine :
    (*a*) Cutting velocity at outside diameter.
    (*b*) Cutting velocity when the diameter is 50 mm. **[Ans. 5.24 m/s; 2.62 m/s]**
19. A 10 mm diameter hole is to be drilled in a block of Magnesium alloy. The feed is 0.2 mm/rev. and the spindle speed is 800 rev./min. Determine : MRR. **[Ans. 209.47 mm^3/s]**

5

TOOL LAYOUT FOR CAPSTANS AND TURRETS

5.1. GENERAL

Both Capstan and Turret lathes are sometimes called simply as "Turret lathes". A simple block diagram of a turret lathe is shown in Fig. 5.1. It differs from an engine lathe in that, the tail stock is replaced by a hexagonal turret which can be swivelled about a vertical axis and

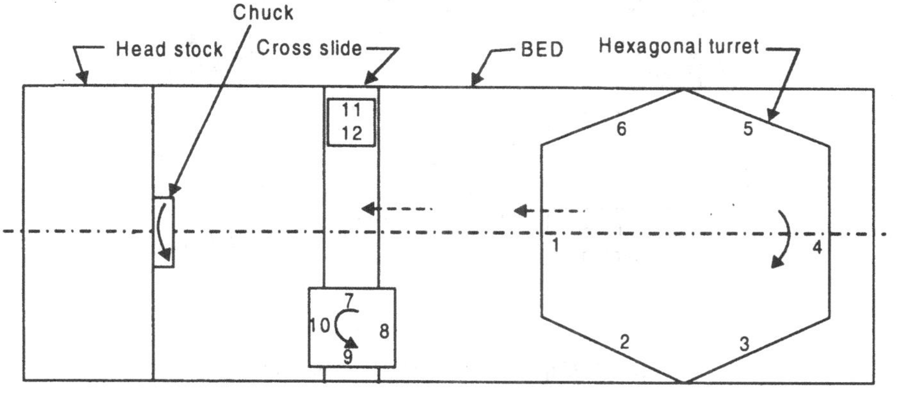

Fig. 5.1. Block Diagram of a Turret Lathe.

moved longitudinally along the bed to bring the tools into cutting positions, the regular tool post is replaced by a square cross-slide turret and the cross-slide may support a square turret on the rear. In this way a minimum of 14 tools can be mounted at one time. As will be described later, this number can be increased by using multiple tooling at each position. The advantages of incorporating multiple tool holders is the chief distinguishing feature of a turret lathe. This enables the setting up of all tools for a job. Except for sharpening, they don't need further handling. The various elements of the Turret lathe are shown in Fig. 5.2.

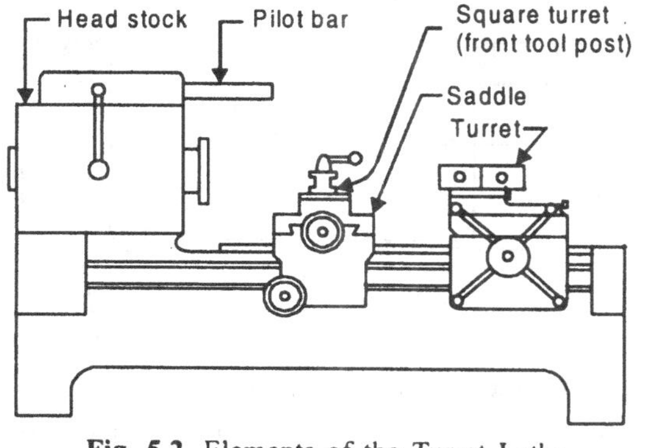

Fig. 5.2. Elements of the Turret Lathe.

Turrets take their place between centre lathes and automatics with regard to the volume production. They are essentially meant for medium batch production.

5.2. TYPES OF TURRET LATHES

The two main types of turret lathe are : (*a*) Ram type turret lathe, and (*b*) Saddle type turret lathe.

Ram type turret lathe. In this type of turret lathe, the revolving turret is mounted on a ram or a slide carried in a base which can be clamped in any position along the bed of the machine, as shown in Fig. 5.3. This type of lathe is somewhat lighter in construction and therefore can be quickly and easily operated. This lathe is also called as 'Capstan Lathe'.

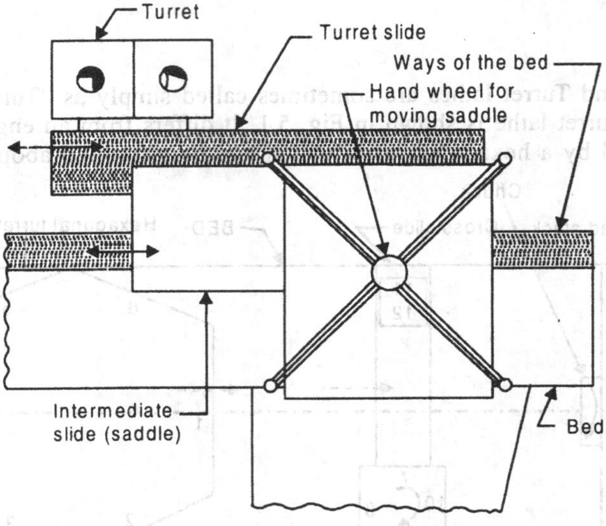

Fig. 5.3. Capstan Lathe Slide Arrangement.

Saddle type turret lathe. On this lathe, the revolving turret is mounted on a saddle which moves back and forward directly on the machine, Fig. 5.4. This design in also known as simply turret lathe. This type is heavier, has a longer stroke, and is more rigid.

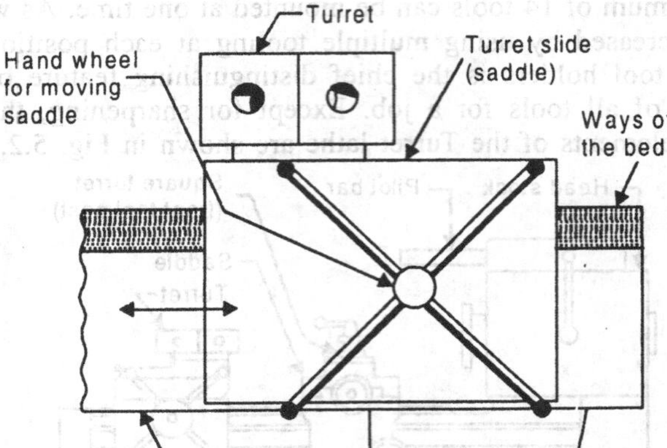

Fig. 5.4. Turret Lathe Slide Arrangement.

5.2.1. Turret Lathe Size. The size of turret lathe is designated by a number that indicates the diameter of the workpiece that can be swung and the diameter of the bar that can be passed through the hole spindle. For example, if the capacity of a turret lathe is designated as : 40-80-

150, then it means that the bar diameter is 40 mm, swing over cross-slide is 80 mm and swing over bed is 150 mm.

5.2.2. Applications of Capstan Lathes. The ram or slide is lighter and can be moved more quickly than a saddle but lacks some rigidity. Because of convenience and speed and limited travel of the turret, the ram type construction is ideally suited for the production of parts relatively short in length which are turned from the bar, and where the ram overhang can be kept short. Forgings, castings and stampings can also be machined by using jaw chucks.

The ram type turret lathe (capstan lathe) is made in the following size ranges :

Bar diameters = 12 mm to 60 mm

Swing over bed = 200 mm to 300 mm

Swing over cross-slide = 100 mm to 200 mm

Maximum turret movement = 75 mm to 350 mm

5.2.3. Applications of Turret Lathes. The saddle type construction, *i.e.*, the turret lathe is heavier, has a longer stroke and is more rigid. It provides good support for the tools and the means to move the tools a long way when necessary. It is more suitable for longer and heavier chucking work which requires longer and heavier cuts, *i.e.*, for the machining of forgings and castings.

The saddle type lathe (turret lathe) is made usually in the following size ranges :

Bar diameter = Small lathes upto 50 mm, and
large lathes upto 100 mm

Swing over bed = 350 mm to 900 mm and over

Swing over cross-slide = 250 mm to 650 mm

Maximum turret movement = 750 mm to 2000 mm.

5.3. MAIN PARTS

The main parts of a turret lathe are as given below :

Head-Stock. The head-stock of a turret lathe is similar in construction to that of the engine lathe. However, it is larger and heavier in construction and with a wider choice of speeds and is fitted with a more powerful motor. Speed may range from 30 to 2000 r.p.m. There are two principal types of headstocks : (*a*) an electric head in which a variable speed motor is mounted directly in the head-stock (*b*) the all geared head which gives a wider range of speed and allows heavier cuts to be taken. For this, the electric motor is sometimes located in or on the motor leg below the head-stock and connected to the geared-head pulley by V-bets or a flange mounted motor may be directly connected to the drive shaft in the head.

The spindle is hollow and bar stock can be fed through it into a collet chuck. Direction of spindle rotation is reversible to enable the withdrawal of a solid tap from a tapped hole or the removal of a button die after the threading operation. The various spindle speeds, as well as forward and reverse movements of the spindle are controlled by levers extending from the head.

Turret. On all but very small capstan lathes, the turret is a hexagonal block mounted on vertical spindle. On each face, there is a hole to receive the shank, pilot or register of a flange mounting tool holder. To fasten the tool holder in position, four set pins are passed through the drilled holes in the flange into matching drilled and tapped holes in the turret face. On small capstan lathes, the turret is circular with six holes equally spaced around the circumference. The tool holders are provided with parallel circular shanks which fit into these holes. A clamp bolt is used to lock it in position.

A large star wheel at the front of the machine is used for moving the turret forward for each tool position. The forward movement is controlled by a preset stop and then the turret is returned to its starting position. As it returns, the turret is automatically indexed one-sixth of a revolution, bringing the next tool in the cutting position. There are six adjustable stops, one for each of the six turret faces and each stop is adjusted during the set up. These six preset stops are indexed automatically whenever the turret is indexed, so that the stop corresponding to a cutting tool is in use. The turret is moved to the workpiece by hand, and feeding may be done by hand or by power feed. For power feeding when the desired turret travel is reached, an automatic trip lever stops further movement of the tool by disengaging the drive clutch.

Cross-Slide. There are two types of cross-slides which can be used for turret lathes : (a) side hung type, and (*b*) reach over type. In side hung type, the carriage does not extend across the entire top of the bed, but is supported entirely on the front way and another way located on the lower front portion of the bed. This design gives greater swing capacity across the cross-slide, but the rear tool post on the cross-slide is lost. This type of cross-slide is best suited for saddle type lathes which are used for large diameter jobs. However, this design has the disadvantage of lower rigidity, which reduces the number of tools that can be used simultaneously.

The reach over or bridge type of cross-slide is supported on the two bed ways in the conventional manner. In addition it may also be supported by a lower rail. This design is preferable for small capstan lathes (ram type machines) engaged in bar work where a large swing across cross-slide is not necessary, since this design restricts the swing across cross-slide. This type has advantage that a rear tool post may be provided which is frequently used for cut off operations. Also, this design is more rigid than the side hung type.

The tool post mounted on the cross-slide at the front of the machine, *i.e.*, the front tool post is a square turret with four faces for the mounting of the cutting tools. This turret can move in two directions; perpendicular and parallel to the rotational axis of the spindle. Stops are there for each motion of each cutting tool. These stops are indexed automatically when the square turret is indexed, through 90. This brings the next turret face with its tool or tools into position for the next machining operation. Feeding or movement in either direction may be by hand or by power feed. The rear tool post (when provided) usually carries the cut off or parting tool. Since the surface of workpiece at the rear end is moving upward, the cut off tool is fitted upside down in the rear tool post.

5.3.1. Indexing of Turret and Cross-Slides. As already noted, the saddle, in the case of Capstan Lathes, is mounted on the bed and the turret slide or ram moves back and forth in the guides of the saddle. In Turret Lathes, the turret slide (saddle) travels directly along the bed ways. The turret slide gets its movement from a rack and pinion drive. The shaft on which the large star-wheel is mounted (the starwheel is for providing handfeed to the turret slide) carries a pinion on its inner end. This pinion engages with a rack on the slide to provide longitudinal motion in either direction. For power feed, a shaft passes through the apron and is driven from the feed gearbox at the base of the headstock.

For controlling the linear dimensions of length or diameter, the forward movement of the turret slide is limited by adjustable stops (which are six in number) mounted on a rotating circular stop plate or roll. The stop roll is linked to the turret through bevel gearing and rotates automatically as the turret indexes, thus bringing a separate stop screw into working position for each turret face. These stop screws act as automatic feed trips, and dead stops for hand feed. The movement of the turret slide (travel of the tool) is fixed by the adjustment of the stop screw which is locked in position by the locking

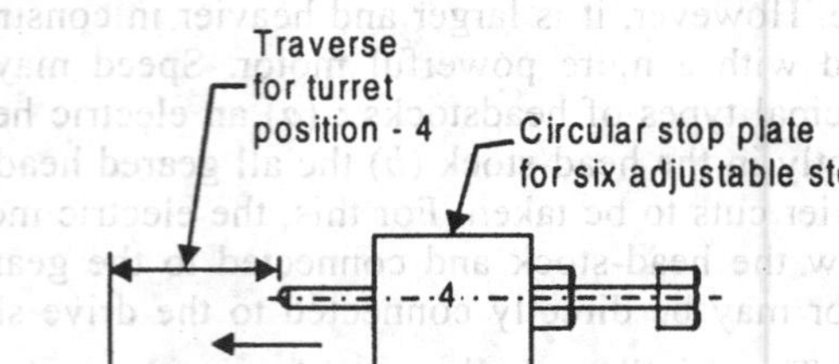

Fig. 5.5. Turret Stops.

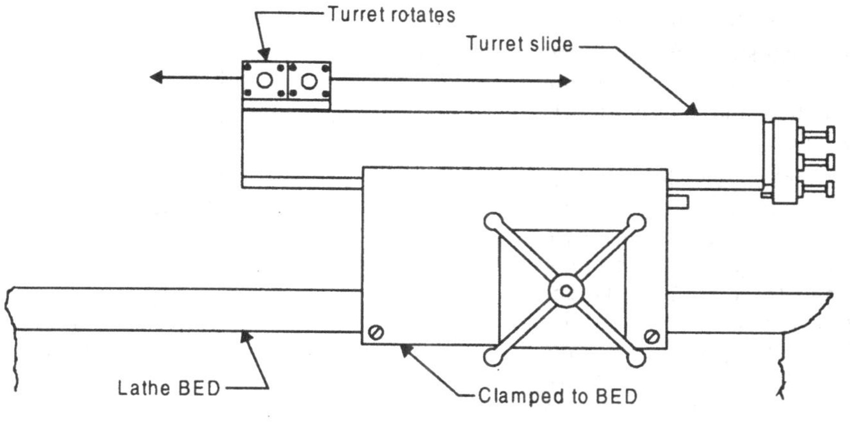

Fig. 5.6

nut, Fig. 5.5. When the turret slide reaches the end of its travel for a certain position of the turret, the corresponding stop screw runs against the fixed stop pin mounted in the saddle, Fig. 5.6. By means of a shaft, this stop pin actuates the mechanism located in the apron of the turret saddle for disengaging the working feed. This mechanism disconnects the feed gear train by means of a claw clutch and the slide (ram) stops in the preset position. When high precision work is required on high quality turret lathes, dial indicator type stops are used in place of stop screw, Fig. 5.7. This arrangement affords a visual indication to the operator, about exact end of the turret slide movement. At that time, the pointer of the dial indicator will read zero. This also avoids the variation of pressure against the fixed stop.

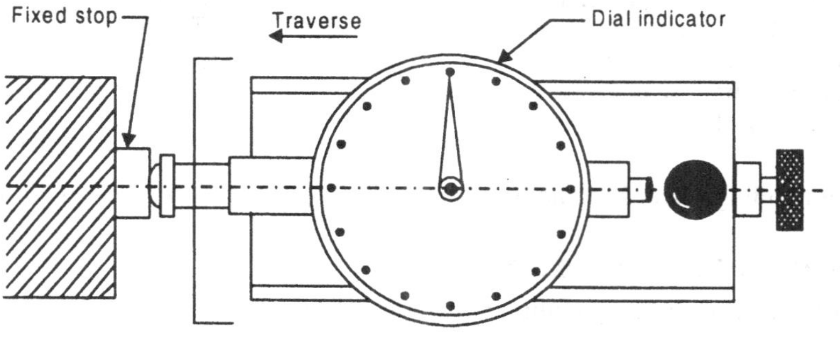

Fig. 5.7. Indicator Type Stop.

A tapered locking pin ensures that the turret holes are in strict alignment with the spindle axis after each indexing of the turret. The pin enters one of the six bushings fitted in the base of the turret. A clamping ring serves to clamp the turret rigidly after each indexing. It also relieves the locking pin of shearing stress due to the cutting force and the consequent torque developed in turret.

The turret of the capstan lathe being lighter, is indexed automatically, whereas the turrets of turret lathes do not index automatically, the saddles carrying them being heavy. The automatic indexing mechanism may vary from manufacturer to manufacturer. Here, we shall discuss a simple indexing mechanism. When the slide or ram is retracted at the end of its cutting travel, a withdraw finger encounters a cam on the turret rest and the lockpin is withdrawn from its slot. At the same time, a dog engages a cam and releases the turret clamp. Further withdrawal of the capstan slide causes the turret to rotate one-sixth of a revolution by a rotating finger which engages a six-toothed ratchet. Once the turret has indexed one sixth of a turn the locking pin enters the next bush in the base of the turret and relocks the turret in its new position.

Another mechanism known as 'Geneva mechanism' which is extensively used for automatic indexing on turret lathes and automatics is explained below :

The number of slots in the Geneva plate, Fig. 5.8, depends upon the number of indexing stations (here 6). The indexing plate or the crank disc, during its rotation, gets engaged with the Geneva plate by means of the crankpin or roll. This pin engages a slot of the Geneva plate and turns it through one-sixth of its revolution per revolution of the disc.

As mentioned earlier, the turrets of the turret lathes do not index automatically. The saddle carrying the turrets are heavy and they have long travel. To bring the saddle back to an automatic indexing position, will mean a considerable wasted movement. Due to this, indexing is not done automatically, but is done by hand. The turret clamping mechanism is released by a lever and the turret is indexed by hand to the desired position, after which the locking pin is relocked and the turret clamped in place.

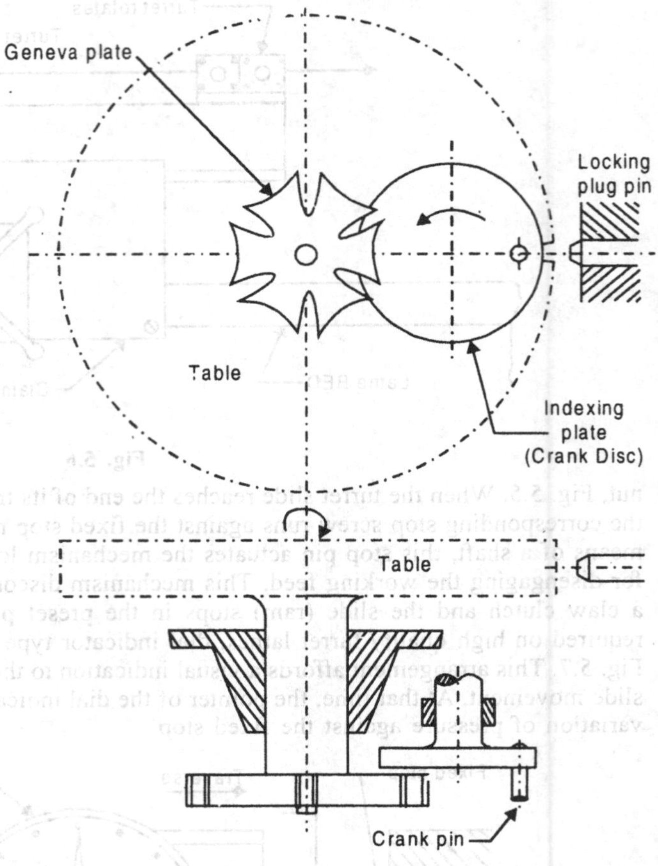

Fig. 5.8. Geneva Indexing Mechanism.

The cross-sliding turret had indexing and clamping mechanism similar to those explained above.

5.4. WORK HOLDING EQUIPMENT

The variation work holding equipments used in turret lathes are : Collets, Jaw Chucks and Fixtures.

Collets. Collets or collet chucks are used mainly to hold bar stock especially in the smaller sizes. A collet is a circular steel shell having three or four equally spaced slits extending the greater part of its length, Fig. 5.9 (*a*). These slits impart springing action to the collet. That is why, collets are also known as "Spring Collets". The collet nose is made thicker to form the jaws. The inside of the collet is made according to the shape of the work to be held. The various types of collect chucks are shown in Fig. 5.9 (*b-d*).

1. **Drawback Collet.** In drawback collet, Fig. 5.9 (*b*), the spring collet is pulled by the thrust tube or collet tube, to the left into the taper bore of the spindle nose. This action puts pressure on the tapered sections of the collet, forcing them inward and tightly clamping the bar stock. The *thrust tube* is placed behind the collect in the spindle. It can be moved axially a short distance by either a handwheel or a lever. The bar to be machined is passed through the tube and the collet jaws.

2. **Push-out Collet.** In push-out collet, Fig. 5.9 (*c*), the thrust tube pushes the spring collet to the right into the tapered seat in the spindle rose. This clamps the bar stock in the same manner as before.

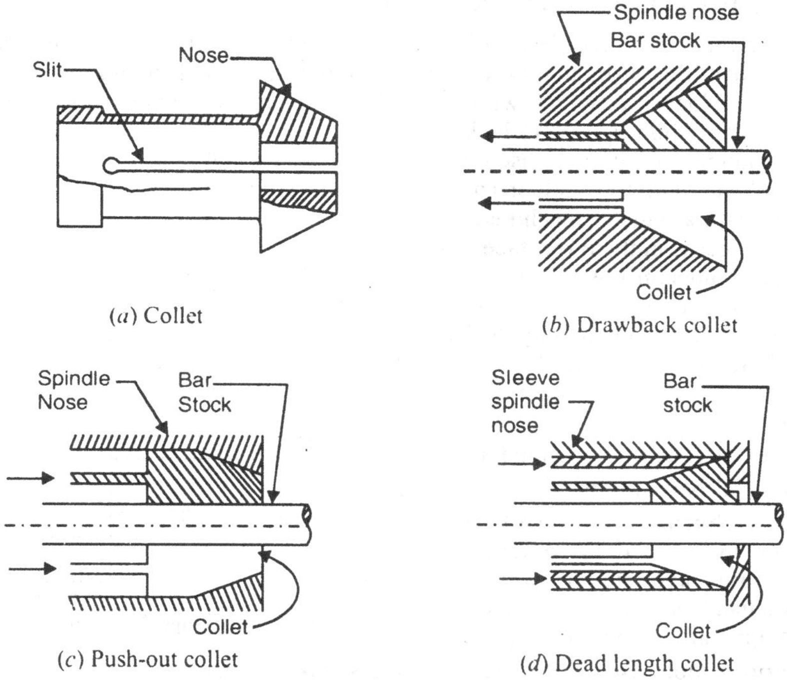

Fig. 5.9. Collet.

3. **Dead length Collet.** In this design, Fig. 5.9. (*d*), the spring collet has no axial movement during operation. This chuck is closed when the thrust tube pushes a sleeve with an internal taper onto the taper of the collect forcing it inward to clamp the work.

Collets are designed to take only one size of work. Standard sizes are available for bars upto about 50 mm diameter.

Jaw Chucks. The types of jaw chucks commonly used on capstan lathes and turret lathes are:

(*a*) Two-Jaw Chucks (*b*) Three-Jaw Chucks (*c*) Four-Jaw Chucks.

(*a*) **Two-Jaw Chucks.** These chucks are of the self-centring type and are used for bar work. Many of these chucks may have blank jaws to which may be attached specially shaped holding devices or jaws for quickly locating and clamping irregular or odd-shaped pieces, for example, small castings and forgings. Two-jaws hold the irregular work more readily since the clamping is at two points which are diametrically opposite. Two-jaw chucks are available in size from about 125 mm to 250 mm outside diameter to hold bar stock diameter of about 20 mm to 45 mm diameter.

(*b*) **Three-Jaw Chucks.** Three-jaw universal chucks are used for holding round or hexagonal bar stock or other symmetrical work. They are of self-centring type and are used mostly for machining of forgings and castings. The size of these jaw chucks ranges from 100 mm to 750 mm diameter and they can hold work upto about 650 mm diameter.

(*c*) **Four Jaw Chucks.** These chucks are used for holding rough castings and square or octagonal work. The jaws move independent of one another. These chucks are available in sizes upto about 1000 mm diameter.

Two-jaw chucks and three-jaw chucks are generally power operated. They may be either pneumatically or hydraulically operated. Four-jaw chucks are manually operated.

Fixtures. A special fixture may also be designed and used for holding the work, if the quantity of parts to be made is sufficient. The fixture is mounted on the machine spindle in place of a chuck or a collet.

5.5. STANDARD EQUIPMENT AND TOOLS

The manufacturers have designed standardised set of tools for use on turret lathes. Tool holders are either 'Shank mounting' or 'flange mounting'. Shank mounting is used for smaller sizes. The tool holder is provided with a parallel circular shank which fits, into the bore in the turret face. A clamp bolt is used to lock it in position. Flange mounting is used for larger tool holders. The tool holder has a square flange with a register machined on it to locate in a recess in the turret face. To fasten the tool holder in position four set pins are passed through drilled holes in the flange (Fig. 5.10) into matching drilled and tapped holes in the turret. There are two sets of permanent tool set ups for turret lathes :

1. Universal chucking equipment
2. Universal bar equipment

Universal chucking equipment. This permanent set up is used for processing forgings, castings and similar work. A typical permanent set up of universal chucking equipment is shown in Fig. 5.11.

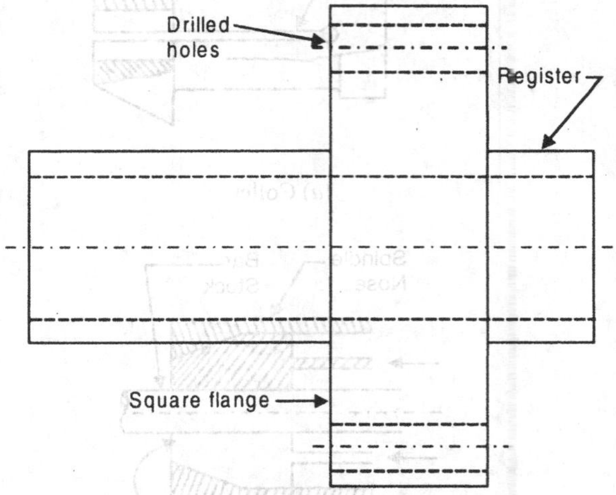

Fig. 5.10. Flange Mounting Tool Holders.

The various tools used in this set up are discussed below.

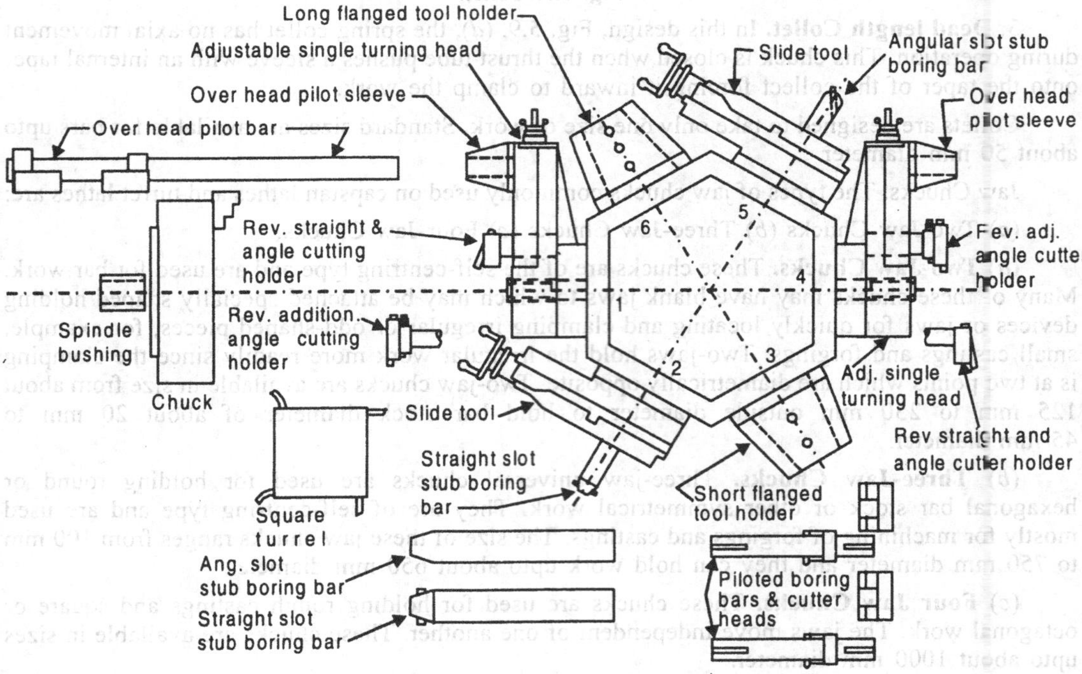

Fig. 5.11. A Typical Permanent Set up of Universal Chucking Equipment.

1. **Adjustable Single and Multiple Turning Heads.** These are shown at stations 1 and 4 (Fig. 5.11). They are used for heavy roughing. A pilot bar is used in combination with this over-head turning head. It is generally mounted over the headstock, Fig. 5.12, but in some machines it may be mounted over the turret at each tool position requiring it. The pilot bar engages a sleeve as it travels (or the tool does) and provides support and rigidity as well as piloting the tool to

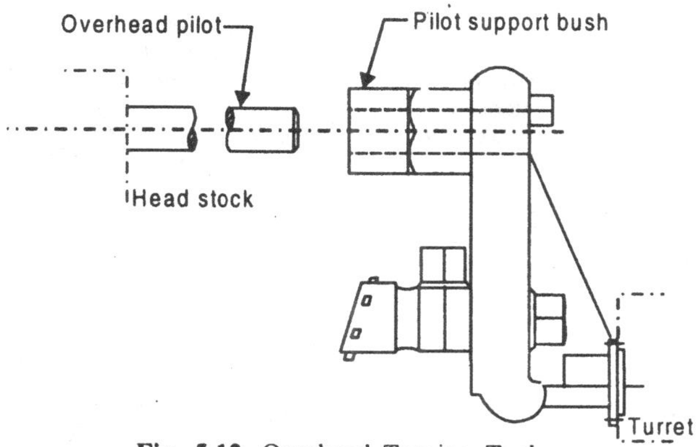

Fig. 5.12. Overhead Turning Tools.

increase accuracy. This is particularly important for heavy cuts which must also be accurate. The cutters for these overhead cuts are held in shank-type holders. It is also possible to use boring bars, drills or core drills in the centre holes of these heads working with the overhead turning cutters.

2. **Ajustable Slide Tool.** This is a tool which has a limited amount of travel in a direction usually perpendicular to the rotational axis of the spindle. The movement of this slide can be used for adjusting depth of cut or for feeding when facing, cutting grooves, or cutting recesses. The amount of movement can be accurately adjusted with the micrometer dial on the screw which moves the slide. Recessing with a slide tool is shown in Fig. 5.13.

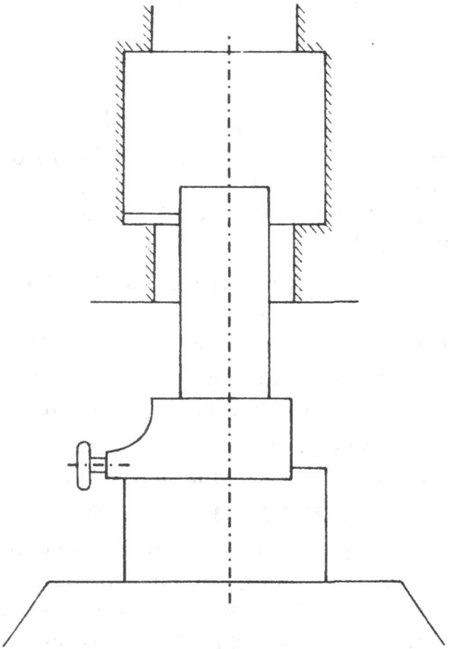

Fig. 5.13. Recessing with a Slide Tool.

The flanged tool holders (stations 3 and 6) provide positions for shank-type tools, for example, drills, taps, reamers etc. that can quickly be inserted during set up.

Universal Bar Equipment. A typical permanent set up of universal bar equipment is shown in Fig. 5.14. The various tool holders are discussed below :

1. **Box Tools.** Box tools are the predominant type of turning tools on turret lathes because of the lack of end support. The box tool supports the work with rollers or a crotch while the tool is cutting, thus enabling heavy cuts to be taken with accuracy and without springing. This tool has the limitation in that the length of cut taken is limited. The single cutter turner in position 6, is the simplest example of a box tool. The roll support the work at the point of cut. If the rollers are set behind the cutter, they will burnish the work to a fine finish and help in holding accurate sizes. The rolls are also used ahead of the cut if this surface has already been machined. Rolling ahead will give a turned diameter concentric to the rolling diameter but will also result in holding a tolerance only as good as that on the rolling diameter. When rolling ahead of the cut, lighter feeds must be used for finishing, as there is no burninshing. Box tool at station 1 is a multiple cutter turner.

2. **Die head.** On turret lathes, most external threads are cut using a self-releasing threading die head. As it cuts, it screws itself onto the threaded workpiece. The turret carrying the die head stops moving just prior to the full length of thread being reached, but a portion of the die head continues to be pulled ahead by the thread being cut. At a predetermined location, the cutters automatically open outward, clearing the thread, so that the die head and turret can be moved

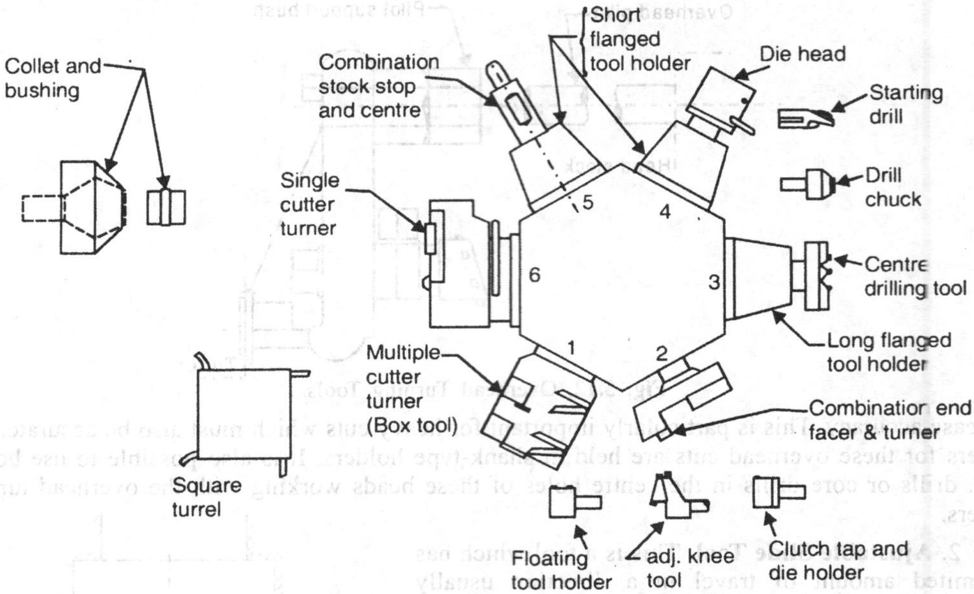

Fig. 5.14. A Typical Permanent Set up of Universal Bar Equipment.

back to the starting position without stopping and reversing the rotation of the workpiece. A lever is provided which sets the chasers (thread cutters) by operating the scroll to move the chasers either inwards or outwards. Thread diameters from about 6 mm to about 100 mm can be cut by this method.

3. **Adjustable Knee Tool Holder.** Adjustable knee tool holder is used for turning combined with drilling, boring or centring on short pieces. The turning tool is carried in an overhead arm, whereas the drill or boring bar is secured in a central bore that is concentric with the turret bore.

4. **Combination Stock Stop and Starting Drill.** This tool which is shown at position 5 (Fig. 5.14) is a two purpose tool that saves one turret face and one index.

The other tools which are used for bar work on turret lathes include : Drill chuck. Floating tool holder for reaming and knurls etc.

5.6. MACHINE OPERATIONS

A turret lathe can perform internal or external work such as turning, boring, drilling, facing, recessing, reaming, tapping, threading, forming, knurling, chamfering and parting off etc. The operations which are usually performed from the hexagonal turret include : turning, boring, drilling, reaming, tapping and threading. The operations which are done from cross-slides include : facing, forming, knurling, chamfering, recessing and parting off. Internal cuts are almost always made by tools in hexagonal turret.

The cuts which can be taken on a turret lathe are of three types :

1. **Combined cut.** A combined cut is one where tools in both the hexagonal turret and square-slide turret are made to cut at the same time.

2. **Multiple cut.** A multiple cut is one where two or more tools are applied at the same time from the turret station.

3. **Successive cut.** A successive cut is one where cuts are made from successive faces of the turret consecutively.

Most threads are cut on a turret lathe from the hexagonal turret. Self-opening dies and collapsible taps allow rapid retraction of the tool after the thread is cut. The length of cut, however, is limited. Threading is done mostly on the end of the workpiece. For longer threads, a single point threading tool in the front cross-slide must be used. This method is much slower when a good, clean and accurate thread is desired.

In the production of holes, provision must be made for the proper order of internal cuts. For example, a roughing operation using a drill is usually performed first to remove metal quickly. Then another roughing operation is carried out with a boring bar, after which a reamer is used to size the hole quickly and give a good finish.

There are three methods to obtain taper on a turret lathe—(1) with a form cutter : This method is suitable only for short, steep tapers, (2) with a taper turning attachment as on an engine lathe, and (3) with a roller-rest taper attachment : In this method, a tapered guide bar is mounted over the work in the same position as the pilot bar, replacing it. The cutting tool resembles the bar turner. It can be easily adjusted for size by means for a graduated dial. The tool and the rolls recede as they progress along the work, guided by the bar, thus producing the desired taper.

The parting off operation is always done from the rear cross-slide in the case of bridge type of cross-slide.

5.7. ADVANTAGES OF TURRET LATHE

The turret lathe is adapted to quantity production work. For this, the turret lathe has got the following advantages over a simple engine lathe :

1. A large number of tools can be permanently set up in the machine in the proper sequence of their use.

2. Each cutting position is provided with a feed stop or feed trip which ensures identical cuts on successive pieces.

3. Multiple cuts and combined cuts can be taken at the same time.

4. The turret lathe can be fitted with various attachement such as for taper turning, thread chasing, and duplicating. It can be also tape controlled.

5. A less skilled operator is required.

5.8. TOOL LAYOUT

Once a turret lathe is properly tooled, an experienced operator is not required to operate it. However, skill is required in the proper selection and mounting of tools. The tool layout for a job constitutes the predetermined plan for the order and method of machining operations necessary to produce it. Accuracy and cost are largely dependent on an efficient layout and the layout is itself influenced by the number of pieces to be made. As a general rule, standard tools should be used as much as possible and for small batches of work, the layout should be simple. For large quantities and long runs, it will be more economical to use special tools which minimise machining time and retain their cutting quantities for the maximum period.

For the preparation of layout, it will be necessary to have a finished drawing of the part to be produced and if it is a forging and casting, a rough blank will show how much machining has been left on the various faces. After the preliminary list giving the order of operations has been decided upon with details as to the tools required, it may be necessary to prepare a scale layout on the drawing sheet. Thus tool layout for a turret lathe consists of two steps :

1. Preparation of operation sheet.

2. Sketching the plan showing the various tools fitted into the turret faces and on the cross-slides, in proper sequence.

5.8.1. Planning Turret Lathe Operations. Time is the most important item of cost in a turret lathe operation. As already discussed in chapter 4, the total production time for a workpiece consists of : set up time, work-handling time, machine-handling time and actual cutting time.

A major portion of the set up time is used up in mounting and adjusting the cutting tools. This can be reduced considerably by using universal tooling and maintaining a permanent set up on a turret lathe.

The work handling time is that which is consumed in mounting or removing the work from the machine. It largely depends on the type of raw material and on the type of work holding devices used. For bar-stock, this time is minimised by using bar-stock collets. Also, for average work in small or moderate quantities, standard holding devices are best. If need be, special jaws, arbors and simple fixtures can be incorporated in these. Special fixtures are justified only if the job is hard to hold in standard devices and is made in large quantities.

Machine-handling time consists of : bringing the respective tools into cutting positions, changing the feeds and speeds and time for indexing. This time can be reduced by having the tools in proper position and sequence for convenient use. Machine handling time is also reduced by taking multiple and combined cuts to save indexing.

The actual cutting time for a given operation is largely dependent on the proper use of cutting tools, proper cutting tools, speeds and feeds. Multiple and combined cutting also reduce the cutting time.

The following main rules should be followed in laying out the sequence of operations necessary to produce a workpiece :

1. Overlap working operations whenever possible and try to increase the number of tools operating simultaneously.

2. Overlap handling operations with each other and with the working operations

3. Overlap finishing and roughing cuts only in cases when this does not have a detrimental effect on the quality of the workpiece. Finishing operation should be done at the end whenever possible.

4. Arrange the cutting tools so as to counterbalance the cutting forces to a maximum extent.

5. Provide a finish cut over the full length of a surface if it was roughed in sections by several tools.

6. Do not permit substantial reduction in the rigidity of the workpiece (cutting deep grooves) until all the machining is completed. This is especially important in thread cutting.

7. Turn precise form surface in two cuts whenever possible.

8. Centre drill before drilling small diameter holes.

9. In drilling deep holes, pull out the drills several times to break up and remove chips and to facilitate drill cooling.

10. Cored holes should not be extended in diameter by using a drill but they should be bored only.

11. In drilling a stepped hole, first drill the largest hole and then the next small diameters in order. This will decrease the total travel required by all the drills.

12. Finish machine external and internal surfaces at a single position to obtain the exact alignment.

5.9. BAR STOCK FEEDING MECHANISM.

1. *For Turret Lathes.* As is known, the rotating bar is fed through the hollow spindle of the lathe. The rotating bar beyond the back of the head stock of the lathe is protected and is usually enclosed in a tube which carries the feed mechanism. The bar feed can be; By hand, pneumatically, electrically or may function automatically upon release of the bar chuck (collet) or by a separate control. As soon as the first component is machined the mouth of the collet is *opened*. The required length of the bar stock is fed through the collet and then it is closed. Electric feed is less commonly used. A small separate motor provides the power and control is usually by foot operated switches.

Peneumatic feed is interçonnected with the bar chuck so that depression of a pedal opens the chuck and feeds the bar forward. A rear pusher is unsuitable for the small sizes of the bar and the feed mechanism at the rear end of the spindle comprises a collet tube which clamps on the bar, pulls it forward from the enclosing tube and pushes it into the spindle.

Automatic bar feeding takes less time and the bar can be fed without stopping the machine. A simple and effective method, whereby the bar can be readily fed forward with the minimum loss of production time, is shown in Fig. 5.15.

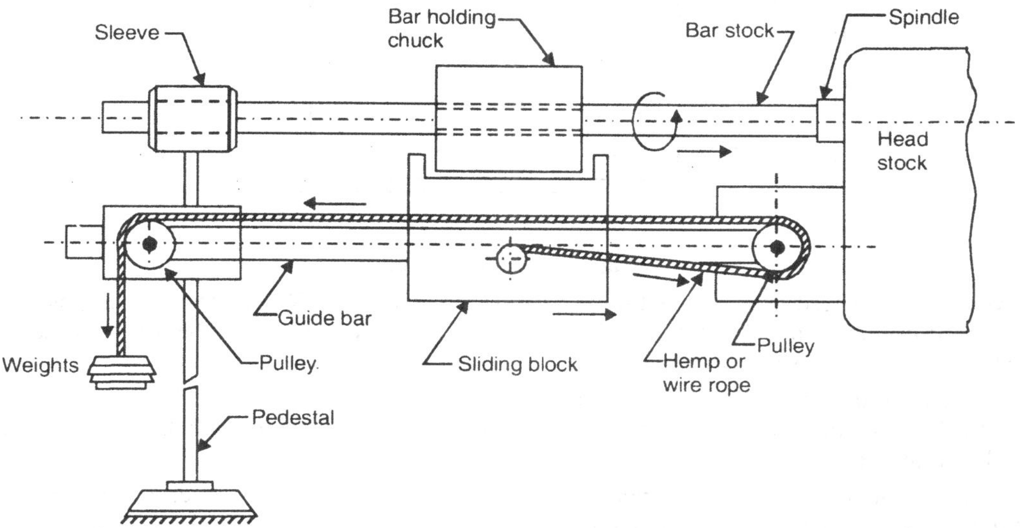

Fig. 5.15. Automatic Bar Feeding Method.

The bar stock passes through the pedestal bushing or sleeve and is held in a bar holding chuck before it enters the lathe hollow spindle and is clamped in the collet (spring chuck) of the lathe. The bar holding chuck rotates within the sliding block, when the feeding bar is rotated by the lathe spindle. A heavy weight is attached to the sliding block, by means of a hemp or wire rope, that passes over a series of groove pulleys. As soon as the bar holding collect in the lathe spindle is released by a lever, on the completion of a component, the dead weight tends to move downward. This exerts an end thrust on the sliding block. This end thrust forces the sliding block to carry the bar holding chuck and the bar forward until the bar touches the stock-stop on the face of the turret slide. As the length of the bar decreases, the sliding block and the bar holding chuck near the machine spindle and then their position on the bar is adjusted. On placing a new bar in the feed mechanism, similar adjustment becomes necessary.

In a variation of the above method, the bar is enclosed in a tube which carries the feed mechanism. The bar tube, having a slot in one side throughout its length, is carried by supports and is accurately aligned with the lathe spindle. The bar pusher inside the tube has a lug projecting through the slot. The lug is externally attached to a wire which passes over a pulley at the headstock end of the tube and returns parallel to its initial direction, to the rear support. At this point, the wire runs round two series of pulleys, one of which supports a group of weights. The pulleys are so arranged that the available descent of the weights is sufficient to feed the bar through the length of the tube. The latter is pivoted at the rear support and the front end is supported by the transverse slide. For loading purposes, the tube assembly can be pivoted by pulling out the front end along the slide after which the bar is fed in from the front end. A screw pad is clamped on to the wire, retaining the bar inside the tube until the latter has been returned to the working position, where it is locked by a handle.

The above methods of automatic bar feeding are suitable only for small-diameter bars. With the larger and heavier bars, the weight of the meterial and the impact load created, as the mass is brought to rest by the stock stop, is such that the wire feed mechanism no longer suffices. For such cases, it is essential to use the feed finger, Fig. 5.16, which also is the standard equipment for bar feeding on automatic lathes, as discussed in the next article. Upon heat-treatment, the jaws of the feed finger are closed, to form the spring grip. The action of the feed finger is to grip the bar with sufficient force, so that, when the spring collet is open, the

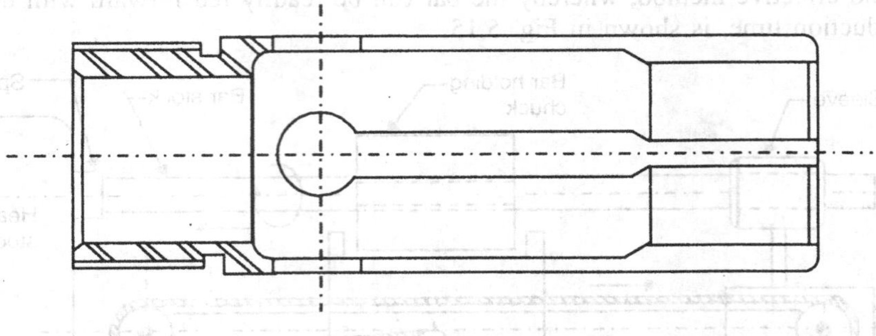

Fig. 5.16. A Feed Finger.

movement of the feed lever brings the bar against the stock stop. When the bar is in position, it is gripped by the spring collet or chuck, so that the return motion of the feed lever causes the feed finger to slide over the bar.

The large diameter bars make it necessary to use a "back end collet" which is adjusted, so that, the bar just slides through. Its aim is to support the over-hanging bar and thus take the strain off the spring collet or chuck.

Hand operated bar feeds often have an endless chain running over sprockets at the extremities of the bar tube, the front sprocket being connected to the operating handwheel. The tube is carried by two supports and pivots on the rear one. The tube is slotted along its full length and a lug on the bar pusher inside it projects through the slot and is connected to a roller chain. After the chuck is opened, operation of a hand wheel feeds out the desired length of bar. Loading is performed from the front end, the arrangement for pulling forward the tube assembly, clear of the head stock being similar to the one explained above. Hand feeding takes more time and the machine has to be stopped before feeding the bar.

Nearly all bar feeding mechanisms are equipped with extended pusher or Collet tubes, enabling the end of the bar to be fed right upto the back of the chuck to minimise wastage of stock.

5.10. SOLVED EXAMPLES

Example 1. *Draw the tool layout for the component shown in Fig. 5.17.*

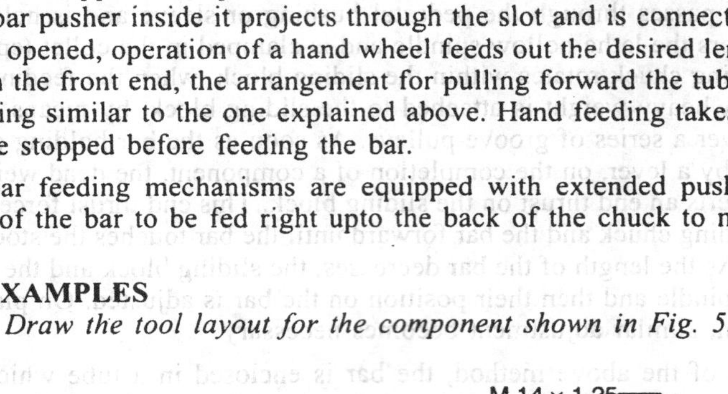

Fig. 5.17

Solution.

Material : 18-mm steel bar

Machine : Capstan lathe

Operations :

1. Feed the bar stock to stop.
2. Turn 14 mm diameter with box tool.

3. Turn 14.28 mm diameter with box tool.
4. Round end with roller steady ending tool.
5. Centre with centre drill.
6. Cut threads with die-head.
7. Form 17-mm diameter and chamfer with tool from front square slide.
8. Part off with cut of tool in rear tool post. The tool layout is shown in Fig. 5.18.

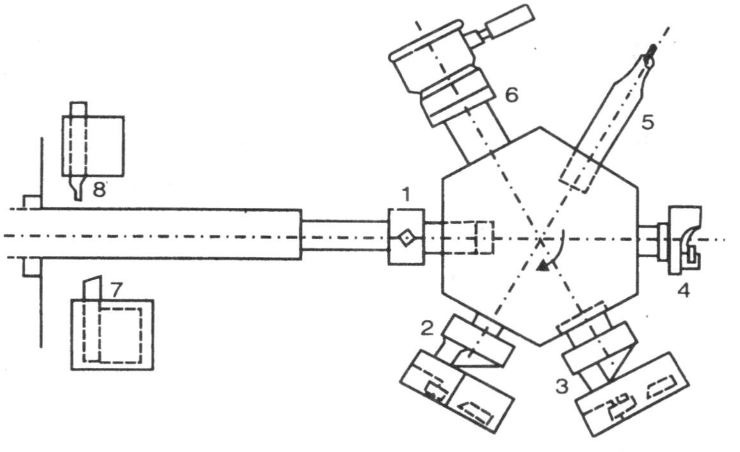

Fig. 5.18. Tool Layout.

Example 2. *Draw the tool layout for the component shown in Fig. 5.19, which is to be made on a capstan lathe.*

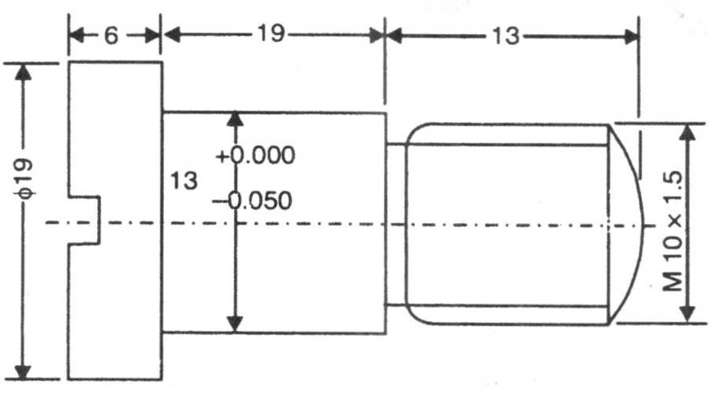

Fig. 5.19

Solution.

Operations :

1. Feed to-Turret.
2. Turn 13/12.95 mm diameter, Roller box-Turret.
3. Turn 10 mm diameter, Roller box-Turret.
4. Turn end, Roller box-Turret.
5. Undercut and Chamfer-cross slide (Rear).
6. Threading, Self opening die head-Turret.

7. Part off, Part-off tool-cross slide (Front). The tool layout is shown in Fig. 5.20.

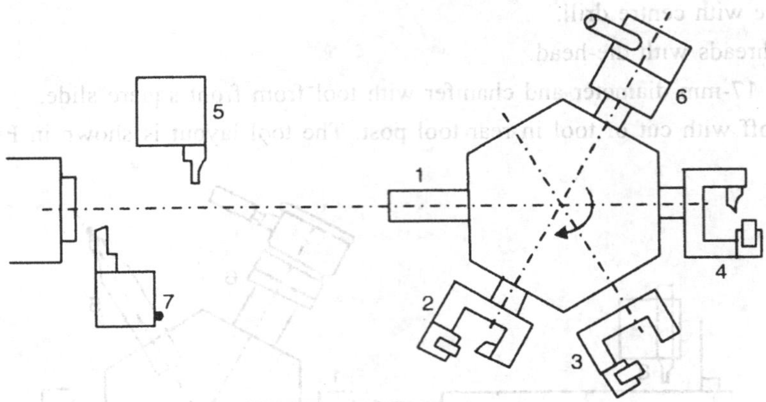

Fig. 5.20. Tool Layout

Example 3. *Draw the tool layout for manufacturing knurled screw and nut, Fig. 5.21, on capstan lathe.*

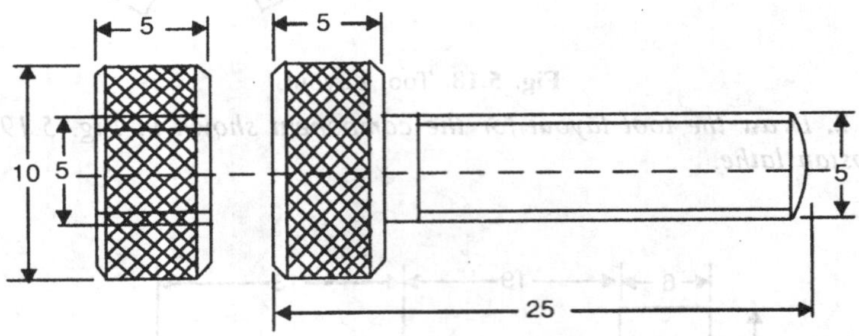

Fig. 5.21. Knurled Screw and Nut.

Solution.

Operations :

1. Feed the bar stock to stop-Turret.
2. Turn 5 mm diameter and form the end, Steady box-tool with turning and radiusing tools-Turret.
3. Cut the threads on screw, die head-Turret.
4. Knurl, Knurl tool-Turret.
5. Chamfer the four diameters, chamfer tools-Front cross slide.
6. Part off the screw, parting tool-Rear cross slide.
7. Drill and face the nut, drill and facing tool-Turret.
8. Tap the nut, Tap-Turret.
9. Part off the nut, parting tool-Rear cross slide. (The tool layout is shown in Fig. 5.22)

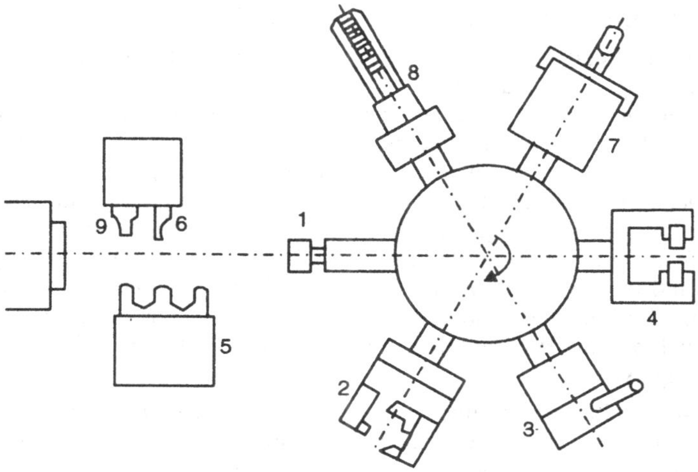

Fig. 5.22. Tool Layout for Knurled Screw and Nut.

Example 4. *Draw the tool layout for the component shown in Fig. 5.23.*

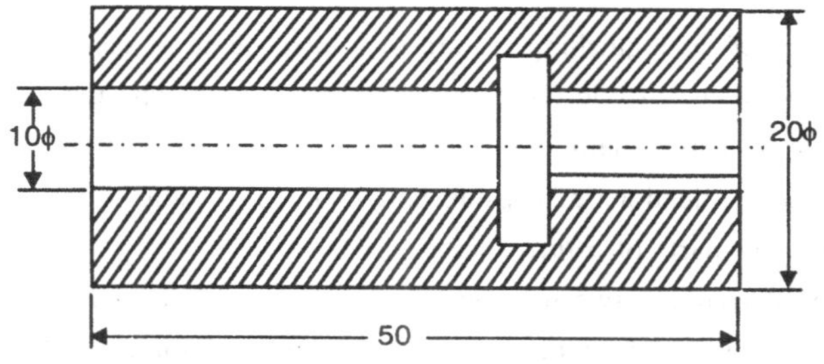

Fig. 5.23

Solution. Machine Capstan lathe

Operations :

1. Feed the bar stock to combined stock stop and start drill. Close the collect. The end of bar is then centred by advancing the start towards it—Turret.
2. Drill the internal diameter — Turret.
3. The thread diameter is bored to correct size, boring bar in a slide tool — Turret.
4. Ream the drilled hole, Reamer — Turret.
5. Make the recess, slide tool — Turret.
6. Cut the internal threads, Tap-Turret.
7. Cut-off, parting tool. — rear cross-slide. The tool layout is shown in Fig. 5.24.

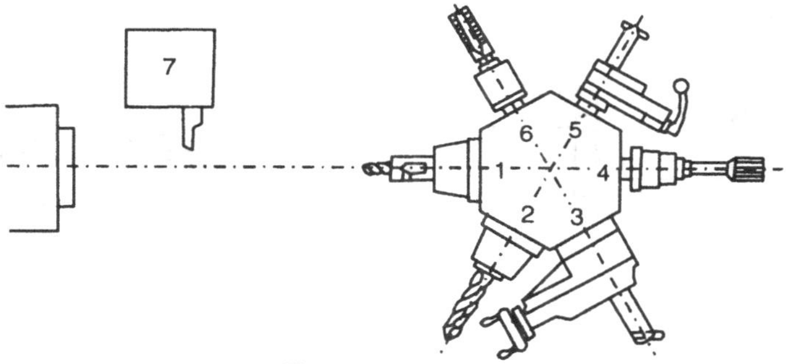

Fig. 5.24. Tool Layout.

5.11. PROGRAMME-CONTROLLED TURRET LATHES

The present day trend is towards replacing the turret lathe by the more sophisticated but costly N.C. Lathe equipped with a turret for multi-tool operation. However, where high initial and peripheral costs cannot be considered, automatic turret lathes with electronic programme control are ideal. The turret lathe operation involves, on an average, over 6000 manipulations during one working day. It is laborious for the workers to manually set the speeds and feeds, swing the tools to the required position and bring them upto the workpiece to engage the feed. The programme controlled turret lathe cuts through all the hard work of setting speeds, feeds, engaging and disengaging of feeds and indexing clamping of the turret head. The lathe operates on a fully automatic cycle and there is a saving in cycle time when compared to manually operated turret lathes. One method of having automatic controls is by electric master switches. HMT has designed and developed the Programme-Controlled Turret Lathe PTL 30 with a microprocessor based programming system.

Programme-controlled turret lathe is of great convenience and bridges the gap between the traditional turret lathe and the sophisticated and costly NC lathe. This lathe is mainly used in medium batch production industries engaged in the production of components like brake linings, bushes, bearing housings, gear blanks etc.

PROBLEMS

1. What is the chief distinguishing feature of a turret lathe as compared to an engine lathe ?
2. Distinguish between 'Saddle type turret lathe' and 'Ram type turret lathe.'
3. How the size of a turret lathe is specified ?
4. Enumerate the applications of capstan lathe and turret lathe.
5. Discuss the main parts of a turret lathe.
6. What is the difference between 'Side hung type cross-slide' and 'Reach over type cross-slide.'
7. Discuss the various work-holding equipment used on a turret lathe.
8. What is the difference between 'flange mounting tool holder', and 'Shank mounting tool holder'
9. What is meant by 'multiple cut' and 'combined cut' ?
10. Enumerate the advantages of a turret lathe.
11. What is meant by 'tool layout' of a turret lathe ?
12. How the total production time on a turret lathe be minimised ?
13. Enumerate the various rules which must be followed while laying out the sequence of operations for a turret lathe.

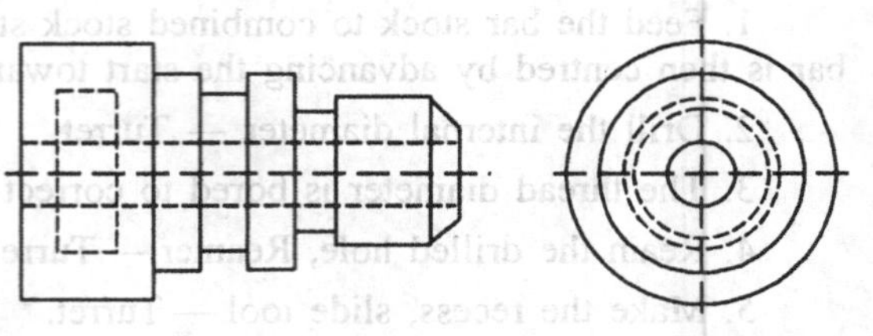

Fig. 5.25

14. Draw the tool layouts for the components shown in Fig. 5.25 to 5.27, which are to be produced on turret lathe.

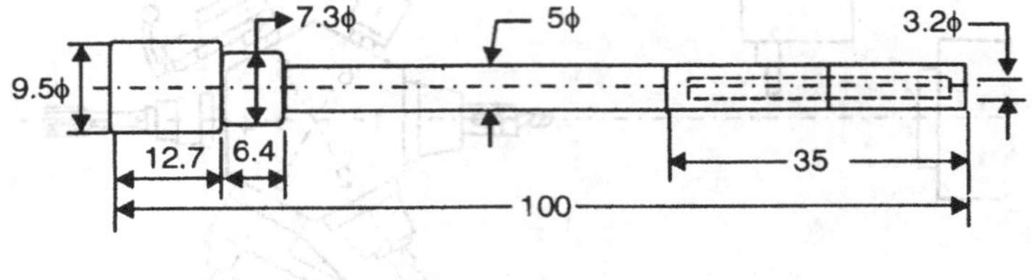

Fig. 5.26

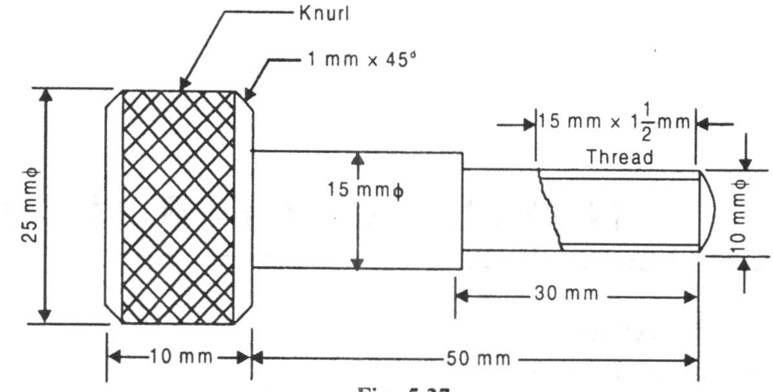

Fig. 5.27

15. Draw the tool layout for the component shown in Fig. 5.28, which is to be produced on a turret lathe.

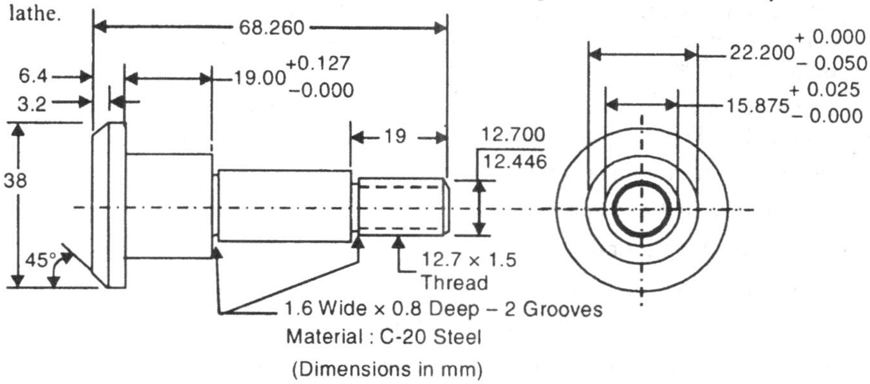

Fig. 5.28. A Stud.

16. Draw the tool layout for capstan lathe to manufacture the components shown in Fig. 5.29 and Fig. 5.30.

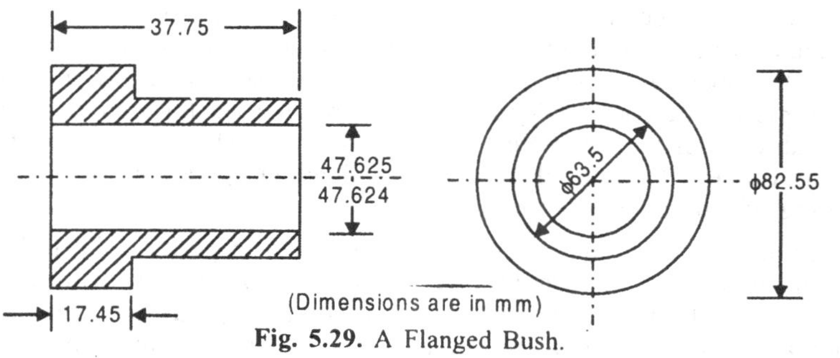

Fig. 5.29. A Flanged Bush.

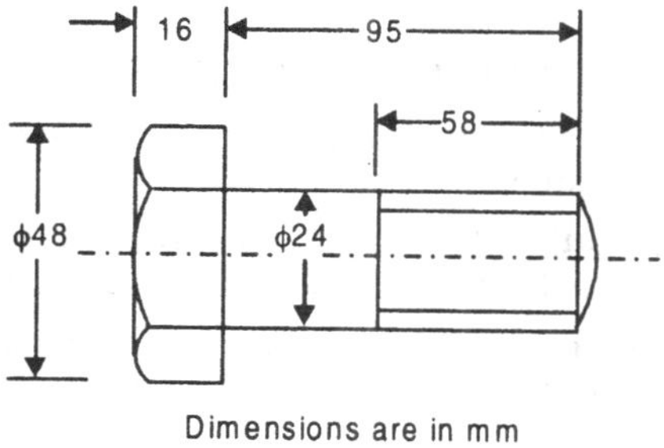

Fig. 5.30. A Hexagonal Bolt.

6

TOOL LAYOUT FOR AUTOMATICS

6.1. AUTOMATIC LATHES

Automatic lathes are machine tools in which the components are machined automatically. The working cycle is fully automatic that is repeated to produce duplicate parts without participation of the operator. All the working and idle operations are performed in a definite sequence by the control system adopted in the automatic which is set up to suit a given work. Strictly speaking the machine is not fully automatic since operator is required to load the machine for a batch of parts and start each cycle.

These machines are used when production requirements are too high for turret lathes to produce economically. Their use results in the following advantages :

1. Greater production over a given period.
2. More economy in floor space.
3. More consistently accurate work than turrets.
4. More constant flow of production.
5. Scrap loss is reduced by eliminating operator error.
6. During the automatic machining operation, the operator is free to operate another machine or inspect the completed parts.

Semi automatics. Semi-automatic machines are usually turning machines adapted to chuck work. In these machines, although the movements of workpiece or tools are automatically controlled, but the workpiece has to be loaded into and removed from the chuck at the beginning and end of each cycle of operations. The machining cycle is automated, but the direct participation of the operator is required to start each subsequent cycle, *i.e.*, to machine each subsequent workpiece. The operator loads the blank into the machine, starts the machine, checks the work size and removes the complete component by hand. This machine can be converted into a fully mechanised one if loading, clamping and unloading operations are mechanised and also magazine loading of workpieces is arranged.

Automatics and semi-automatics are chiefly designed to perform the following machining operations : centring, cylindrical turning, tapered and formed surfaces, drilling, boring, reaming, spot facing, knurling, cutting threads, facing, milling, broaching, grinding and cut off. Additional operations such as slotting and cross drilling etc. can also be performed on these machines with the help of special attachments.

6.2. CLASSIFICATION OF AUTOMATIC MACHINES

These are various ways of classifying the automatic lathes. Depending upon the type of work machined, these machines are classified as :

1. **Magazine Loaded Automatics.** These machines are used for producing components from accurate separate blanks. These machines are also called 'Automatic chucking machines'.

2. **Automatic Bar Machines.** Automatic bar machines are designed for machining components from bar or pipe stock. These machines are chiefly used for the manufacture of high

quality fasteners (screws, nuts and studs), bushings, shafts, rings, rollers, handles and other parts, usually made of bar or pipe stock, and recently of separate blanks as well.

Depending upon the number of work spindles, the automatic lathes can be classified as : single spindle automatics and multispindle automatics.

Automatics can also be classified according to : purpose and arrangement of the spindle. According to purpose, automatics are classified into : general and single purpose machines. According to the arrangement of the spindle, automatics are classified into horizontal and vertical machines.

In this chapter, we shall deal with automatics and semiautomatics in which automaticity is achieved by mechanical means using cam drives. The operating cycle is automatically controlled with the help of a camshaft carrying cams which are linked to the operative mechanisms (slides, feeding and clamping mechanisms) through a system of levers, gears etc. When the camshaft rotates, the cams actuate the various operative units, Fig. 6.1. According to their forms the cams are classified as plate cams and cylinder cams.

In accordance with their cycle control, the automatics may be grouped into three categories :

1. In the first category, automatics have a single camshaft which rotates at a constant speed for the given set up and which controls both working and idle motions. The speed of the camshaft corresponds to slow working motion and since the idle motions should be completed in shortest possible time, considerable time will be wasted for these motions for the machines of this category. However, due to its simplicity, this category is justified for small automatics with only few idle motions.

2. Automatics of the second category have also only one camshaft but it has two speeds of rotation : slow for working motions and fast for idle motions. However, the camshaft has quite a large moment of inertia (due to whole set of working and idle cams) and due to this, the change-

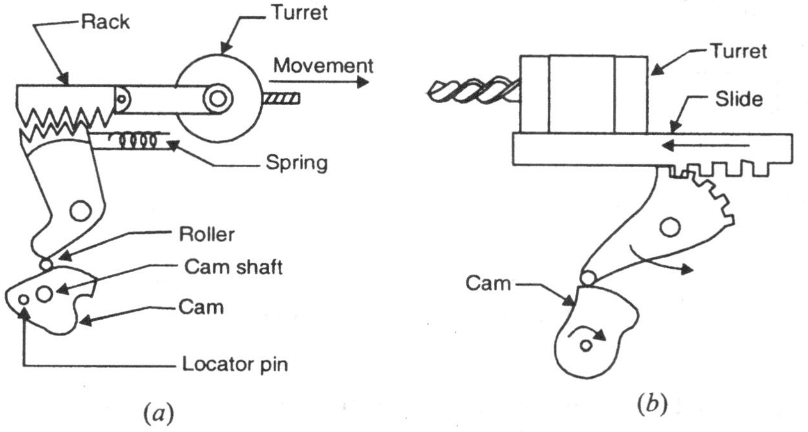

Fig. 6.1

over mechanisms will experience impact loads when switching over from one speed to the other. Thus, it is rather impossible to obtain any significant difference between the two speeds of rotation.

3. The third group of automatics have one camshaft for working motion which rotates at slow speed and a high speed auxiliary camshaft for actuating the idle motions. This group finds applications for machining complex shaped work-pieces where many auxiliary motions are required.

Automaticity can also be obtained by Hydraulic, Pneumatic, Electric and combined systems and also by punch tapes, punched cards, magnetic tapes or by a drum.

6.2.1. Type of Single Spindle Automatics. The single spindle automatic lathes are of the following types :

1. **Automatic Cutting Off Machines.** These machines produce short workpieces of simple form by means of cross-sliding tools. The machines are simple in design. The head stock with the spindle is mounted on the bed. Two cross-slides are located on the bed at the front end of the spindle. Cams on a camshaft actuate the movements of the cross-slides through a system of levers.

The principle of operation of such a machine is explained below with the help of Fig. 6.2. The required length of work (stock) is fed out with a cam mechanism, upto the stock stop which is automatically advanced in line with the spindle axis, at the end of each cycle. The stock is held in the collect chuck of the rotating spindle. The machining is done by tools held in slides operating only in the cross-wise direction. Special attachments can be employed if holes or threads are required on the simple parts. Typical simple parts (from 3 to 20 mm in diameter) machined on such a machine are shown in Fig. 6.3.

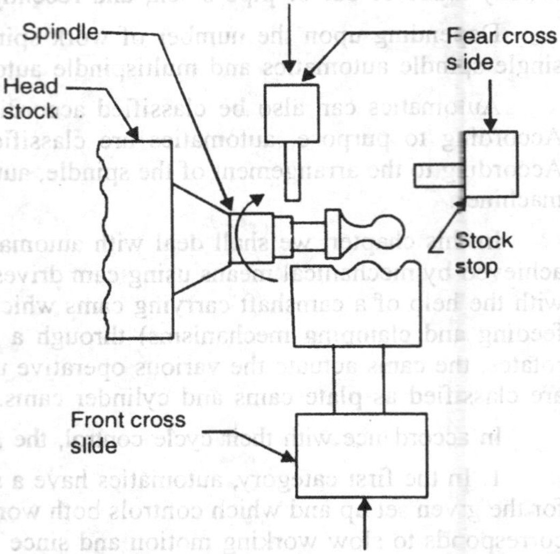

Fig. 6.2. Automatic Cutting-off Machine.

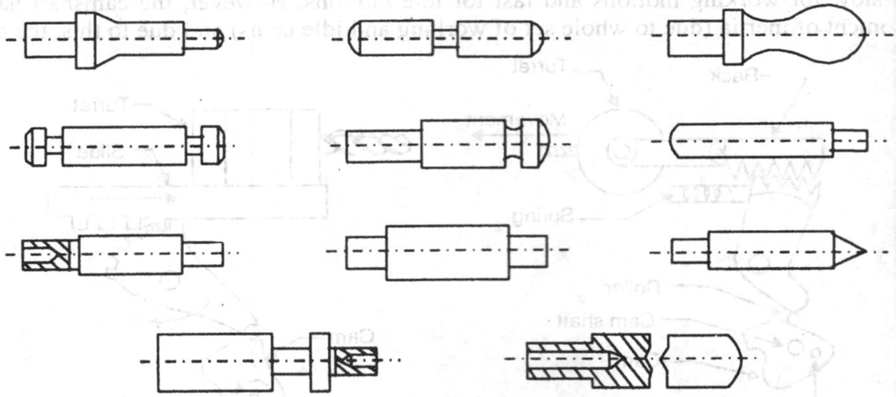

Fig. 6.3

2. **Single Spindle Automatic Screw Machine.** This machine was originally used for producing small screws, but has long since graduated to the production of all sorts of small turned parts. These machines are essentially wholly automatic bar type turret lathes. They differ from the horizontal turret lathe in that the turret revolves about a horizontal pivot pin instead of about a vertical axis. They are designed for machining complex internal and external surfaces on parts made of bar stock or of separate blanks. Upto ten different cutting tools may be employed at one time in the tooling of such a screw machine. The tools are fixed in indexing turret and in cross-slides. The turret carries six tools. Two cross-slides (front and rear) are employed for cross-feeding tools. A vertical slide for parting off operation may also be provided. It is installed above the work spindle. The head stock is stationary and houses the spindle which rotates in either direction. The bar stock is held in a collet chuck and advanced by a feed finger after each piece is finished and cut off. All movements of the machine units are actuated by cams mounted on the camshaft. These machines are made in several sizes for bar work from 12.7 mm to 60 mm diameter.

Fig. 6.4. shows a general layout of a single spindle automatic screw machine. The bar stock is pushed through stock tube in a bracket and its leading end is clamped in rotating spindle by means of a collet chuck. The bar is then fed out for the next part by stock feeding mechanism. Longitudinal turning and machining of the central hole are performed by tools mounted on turret slide. The cut off and form tools are mounted on the cross-slides. At the end of each cut, turret slide is withdrawn automatically and indexed to bring the next tool into position. The indexing motion is accompanied by an additional rapid retraction.

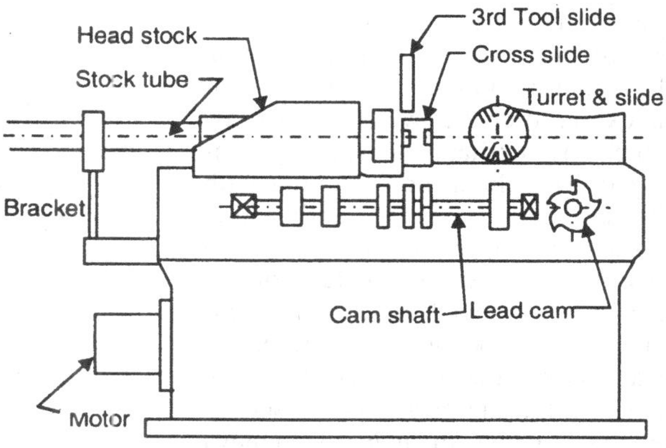

Fig. 6.4. Automatic Screw Machine.

Gearing mechanism of a single spindle automatic screw machine is shown in Fig. 6.5. Ther camshaft is at the front of the machine. It is driven from a shaft at the back of the machine (called back-shaft or auxiliary camshaft), through cycle time change gears. The backshaft rotates at a constant speed, usually 2 rev/s or 4 rev/s. One revolution of camshaft produces one component. The speed of the camshaft can be changed for each new component by means of cycle time change gears.

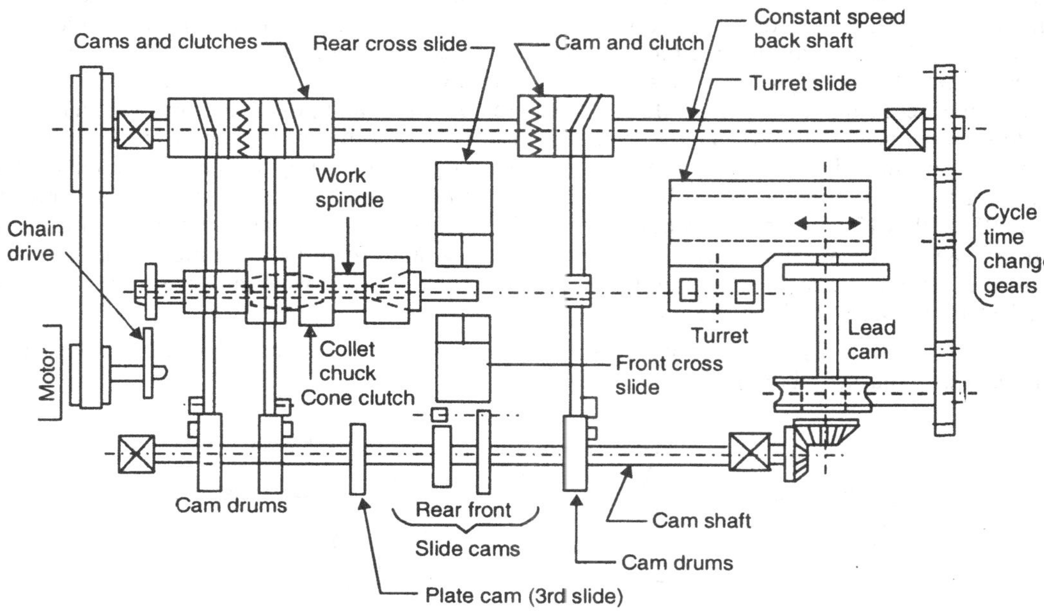

Fig. 6.5. Gearing diagram of Automatic Screw Machine.

The auxiliary camshaft carries dog clutches in three positions. These are operated through drum cams, which are actuated through levers from cam drums on main camshaft. Trip dogs are bolted in T-slots, in the sides of cam drums or dog carriers and can be adjusted circumferentially in T-slots. The trip lever, Fig. 6.6, is actuated by a trip dog (on main camshaft). The lever releases a pin in one half of the clutch on the drive shaft (auxiliary camshaft). This clutch is then caused to slide over the auxiliary camshaft until it engages the other half of the clutch which is revolving, by a dowel pin. Depending upon which trip lever is raised, the respective clutch (which then acts through a gear train) stays engaged long enough to complete : one (or two) index, one stock feed (by causing the collect chuck to open, hence the bar feeds forward to the bar stop in the turret, and the collect closes against gripping the bar), or one spindle clutch shift to cause a change in spindle speed. The dog (jaw) clutches are disengaged at the end of the cycle of auxiliary motion they control. Let us explain this for indexing, after the two halves of jaw clutch on auxiliary camshaft get engaged. The clutch now rotating with the auxiliary camshaft and acting through a gear train cause a disc to revolve. Attached to the face of the disc is a hardened roller which engages the back plate of the index head. Once the roller engages the index head, the turret is caused to revolve in the manner of Geneva motion.

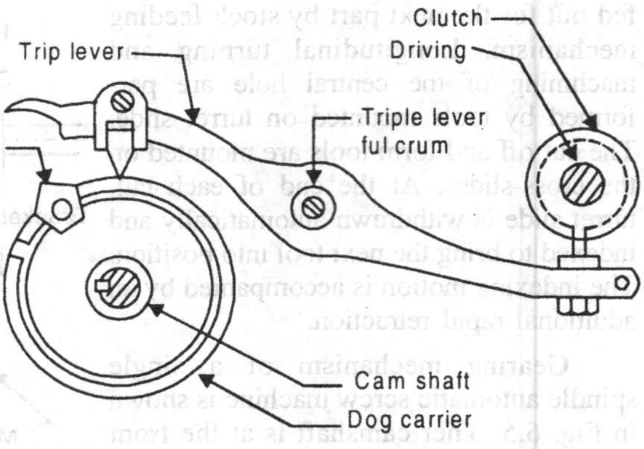

Fig. 6.6. Trip Dog and Lever Mechanism.

3. **Swiss type Automatic Screw Machine.** The machine is also known as 'Sliding head screw machine', or 'movable headstock machine', because the head stock is movable and the tools are fixed. These machines are used for machining long accurate parts of small diameter (2 to 25 mm). The stock is held by a rotating collect in the head stock and all longitudinal feeds are obtained by a cam which moves the head stock as a unit. The rotating bar stock is fed through a hard bushing in the centre of the tool head. The tool head which consists of five single point tools is placed radially around the bushing. Most diameter turning is done by two horizontal tool slides while the other three slides are used principally for such operations as knurling, chamfering, recessing and cutting off. The tools are controlled and positioned by cams that bring the tools in as needed to turn, face, form, and cut off the workpiece from the bar as it emerges from the bushing. The cutting action is confined close to the support bushing reducing the overhang to a minimum. As a result, the work can be machined to very close limits. Tolerances of 0.005 to 0.00125 mm are common. The machine does step, straight, taper, back, and form cutting. With special attachments, centering, drilling and reaming operations can also be performed. Tools are simple and adjustments easy. All tools can work at once and a piece is finished in one pass, making the time short for most work.

Fig. 6.7 shows a general layout of Swiss type automatic. It is clear that it has three main features (as discussed above) :

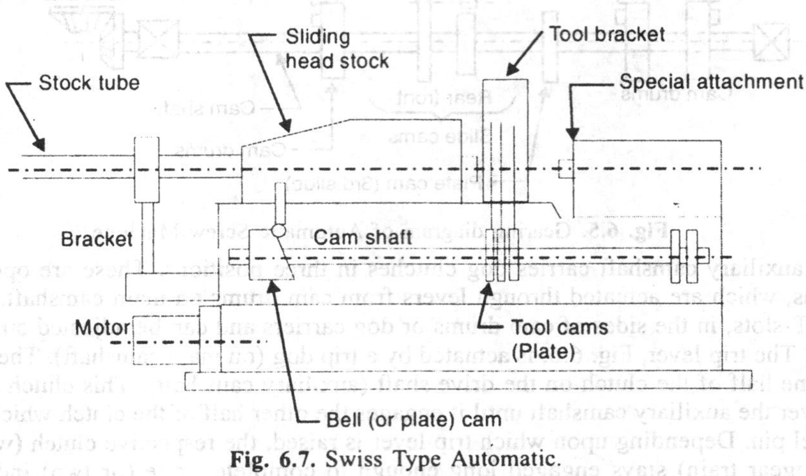

Fig. 6.7. Swiss Type Automatic.

1. A sliding head stock through which the bar stock passes.

2. A tool bracket which supports five tool slides and also guide bush to support the bar stock.

3. A camshaft at the front of the machine. Cams on the camshaft control the movement of the tool slides and headstock.

In addition to the above three main features, the machine has a special attachment mounted on the right hand side of the bed, for machining of the central hole, i.e., drilling, boring, cutting threads with taps or dies.

Fig. 6.8 shows the arrangement of tool slides viewed in the direction normal to the horizontal axis of head stock.

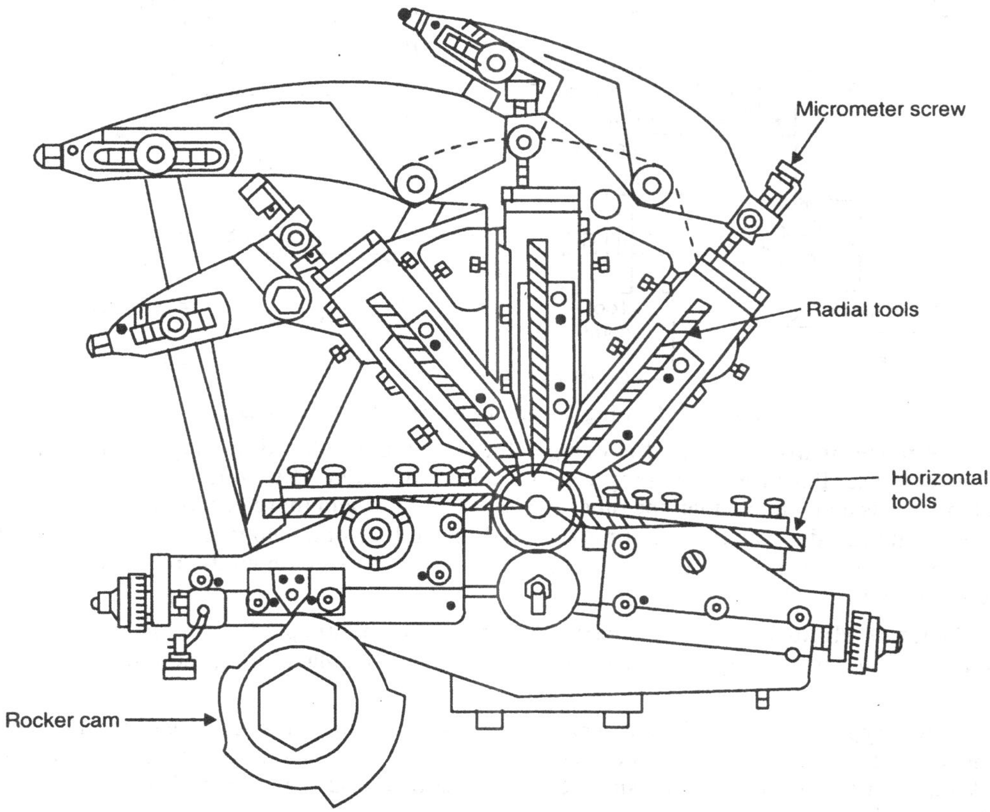

Fig. 6.8. Tool Slide Arrangement of Swiss Type Automatic.

Fig. 6.9 shows the gearing diagram of a Swiss type automatic. The machine has a single camshaft which controls both the working and idle motions (first category of automatics). The camshaft is driven from the constant speed backshaft through cycle time change gears. One revolution of the camshaft produces one component. The cycle time gears can be changed for different parts.

6.2.2. Multi Spindle Automatic Machines. The multi-spindle automatic machines are the fastest type of production machines and are made in a variety of models with two, four, five, six or eight spindles. In contrast to the single spindle machine, where one turret face at a time is working on one spindle, the multi-spindle machine has all turret faces working on all spindles at the same time. Their production capacity is higher than that of single-spindle machines but their machining accuracy is somewhat lower. The rate of production of a multi-spindle machine, however, is less than that of the corresponding number of single-spindle machines. For example, the production capacity of a four-spindle machine is not four times but only $2^1/_2$ to 3 times more than that of a single-spindle machine.

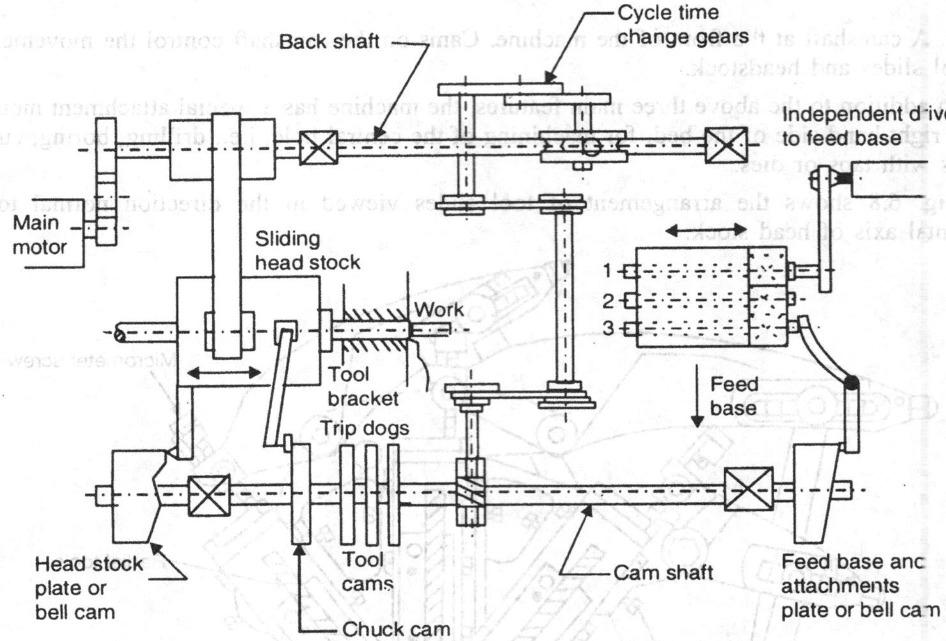

Fig. 6.9. Gearing Diagram of Swiss Type Automatic.

The range of work produced is comparable with that produced by single-spindle machines and turret lathes. Because of the longer set up time and increased tooling cost, the multi-spindle machine is less economical than the other two on short runs, and more economical on longer runs. On the single-spindle machine, the time required to produce one component is the sum of all the turret operations, whereas the time required to machine one piece on multi-spindle machine is the time of the longest cut.

Types of Multi-Spindle Automatic Machines. The multi-spindle automatic machines may be classified in various ways. In accordance with the type of blank used, these may be classified as :

1. **Bar type Machine.** The bar type machine operates on bar stock, usually round but may be hexagonal or any suitable shape. Bar type machines are rated by the largest diameter of stock that can be fed through the spindles. The size of round bars ranges from about 14.3 mm to 197 mm diameter.

2. **Magazine-loading type or Chucking type Machine.** The chucking machine is identical to bar type machine and operates in the same manner but handles several individual workpieces, such as castings or forgings held in several chucks. The capacity of this machine is the diameter of the work that can be swung over the tool slides. The common sizes are upto about 250 mm diameter.

In accordance with their principle of operation, the multi-spindle automatic machines may be calassified as :

(*i*) Parallel action automatics, and

(*ii*) Progressive action automatics.

(*i*) **Parallel action multi-spindle automatics.** Parallel action multi-spindle automatic machine is also called as 'multiple-flow' machine. In this type of machine, the same operation is performed on each spindle and a workpiece is finished in each spindle in one working cycle. It means that the number of components being machined simultaneously is equal to the number of spindles in the machine. The rate of production is thus very high, but the machine can be employed to machine simple parts only since all the machining processes are done at one position.

These machines are usually automatic cutting-off bar-type machines. They are used to perform the same work as single-spindle automatic cutting off machines. Centering or a single drilling operation can also be performed on certain models. The machine consists of a frame with a head stock at the right end. The horizontal work spindles which are arranged in a line, one above the other, are housed in this headstock (Fig. 6.10). Cross-slides are located at the right and left-hand sides of the spindles and carry the cross-feeding tools. Tool slide on each side accommodates tools for all of the spindles. All the working and the auxiliary motions of the machine units are obtained from the cam mounted on the cam-shaft.

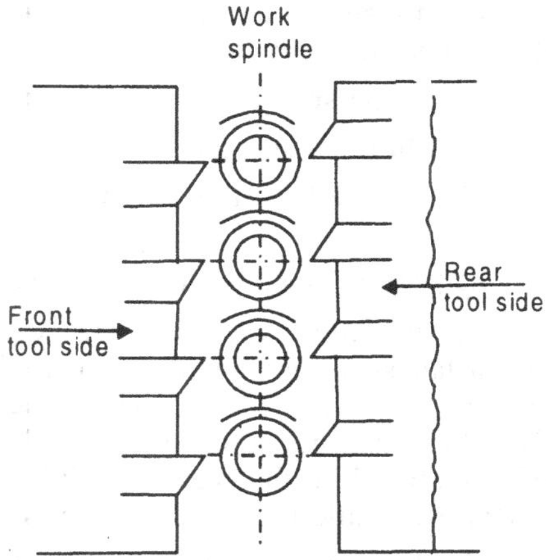

Fig. 6.10. Parallel-action multi-spindle Automatic.

(*ii*) **Progressive-action multi-spindle automatics.** In this design of multi-spindle automatic machine, the workpiece is machined in states and progressively in station after station. The head stock is mounted at the left end of the base of the machine. It contains a spindle carrier which rotates about a horizontal axis through the centre of the machine. The working spindles are mounted in this spindle carrier. The spindles carry the collets and bars from which the workpieces are machined. The bar stock is fed through each spindle from the rear. On the face of the spindle-carrier support are mounted cross-slides which carry tools for operations such as cut off,

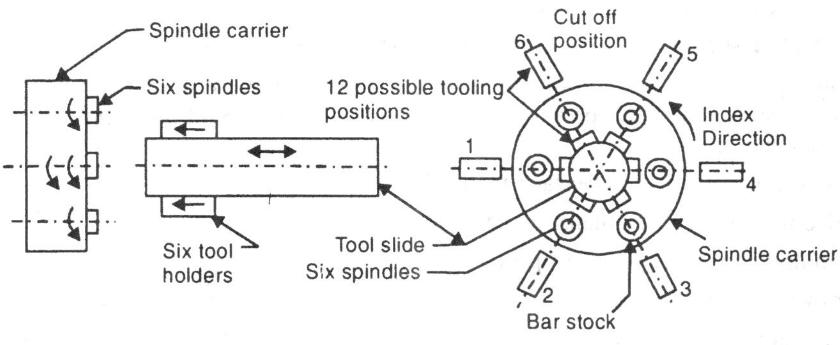

Fig. 6.11. Six Spindle Progressive Action Automatic.

turning, facing, forming and chamfering. The number of these slides equals the number of spindles. The main tool slide (end tool slide) extends from the middle of this support. Here also, the number of tool slides or faces is equal to the number of spindles. These tool slides move forward at the proper time and the proper amount and are operated by their own shaft and gears. The fed of each tool, both cross-slide tools and end-slide tools, is controlled by its own individual cam.

A six-spindle progressive action automatic is shown in Fig. 6.11. The spindle carrier indexes on its own axis by 60 (360 divided by the number of spindles) at each tool retraction. As the spindle carrier indexes, it carries the work from station to station, where various tools operate on it. The stock moves around the circle in counter clockwise direction and returns to the station number 6 for cutting off. A finished component is obtained each time the spindle carrier indexes.

6.3. CLASSIFICATION OF SEMI-AUTOMATICS

Semi-automatic lathes are employed for machining work from separate blanks. The operator loads and clamps the blank, starts the machine and unloads the finished work. Depending upon the number of work spindles, these machines are classified :

1. Single-spindle semi-automatic machine.
2. Multi-spindle semi-automatic machine.

6.3.1. Single-Spindle Semi-automatics. Single-spindle semi-automatics are of the following types.

(*a*) Centre type (*b*) Chucking type.

In the centre type machine, the workpiece is held between centres, for which a head stock and a tail stock are mounted on the bed of the machine. Usually, external stepped or formed surfaces are machined on this machine. The work is machined by two groups of cutting tools. The front tool slide holds the cutting tools which require a longitudinal feed motion to turn the steps of a shaft, while the rear tool slide carries the tools that require a transverse feed motion to perform operations such as facing, shouldering, necking and chamfering etc.

In the chucking type machine, the workpiece is held in a chuck. Such a machine may be equipped with various tool slide arrangements. In addition to longitudinal and transverse feed tool slides, these machines may also be equipped with a central end-working tool slide or a turret if internal surfaces are also to be machined in addition to the external surfaces.

6.3.2. Multi-spindle Semi-automatics. This machine may also be built in two designs.

(*a*) Centre type, and (*b*) Chucking type

These semi-automatics, in the same way as automatics, are classified as :

1. Parallel action or single station type, and
2. Progressive action or multi station type.

6.4. SETTING UP OF AUTOMATICS AND SEMI-AUTOMATICS

Setting up any machine tool involves all preparations required for manufacturing a component in accordance with the accepted manufacturing process, to the specified accuracy and at a maximum rate of production.

Before setting up, it is necessary to do the following :

1. Selecting the type of blank to be used.
2. Planning the manufacturing process.
3. Drawing up a tooling layout in accordance with the manufacturing process.
4. Calculating the time for each machine movement within the operating cycle.
5. Selecting the standard cutting tools, gauges and cams in accordance with the tooling layout.
6. Designing and manufacturing all special tools.

When the above preparatory work has been completed, the set up man proceeds to set up the machine tool. As already discussed in Chapter 5, the machining process is broken into separate operation elements, so that the work is completed at a maximum rate of production and at minimum cost. The processing breakdown between the various cutting tools must be checked several times to select the most suitable method. For this, the guidelines as discussed in Chapter 5 (Article 5.8.1) are also valid for automatics and semi-automatics.

6.5. TOOLING LAYOUT AND OPERATION SHEET

The tooling layout for an automatic contains the sketches of the work in the various stages of machining, by operation elements. The required cutting tools are indicated and slown at the end of the feed motion. The operation sheet usually has a sketch of the finished workpiece. It indicates the material of which the workpiece is to be made and lists the operations in their proper sequence. To fill up the operation sheet, it will be necessary to select :

(*a*) cutting speeds, (*b*) cutting feeds, and (*c*) to calculate the lengths of travel of all the tools in each operation element. Cutting speeds and feeds are given in standard tables in Handbooks. The following rules apply for the cutting speeds and feeds on automatics :

1. The spindle speed is the same for two tools working simultaneously and is determined by the maximum permissible cutting speed on the largest diameter machined in the given operation element.

2. Cutting tools mounted on the same slide and operating simultaneously, will always have the same rate of feed which is determined by the tool permitting the minimum feed.

The spindle speed is determined by substituting the selected cutting speed in the formula

$$N = \frac{1000\,V}{\pi D}$$

where V is the cutting speed in mpm of work or cutter, and D is the diameter of work or cutter in mm.

The nearest values of available spindle speeds are then selected from a table of speed steps and the corresponding change gears are installed in the main drive of the automatic.

The next step is to determine the travel of the cutting tools required in each operation element. This is obtained from the workpiece drawing. To this should be added about 0.5 to 3 mm (tool approach) to prevent the tool from cutting into the work during its rapid approach movement.

6.5.1. Time Required for each Operation Element. The number of spindle revolutions for each operation element is obtained by the relation.

$$n = \frac{L}{f}$$

where L is the tool travel in mm, and f is the rate of feed in mm per revolution.

The time required for each operation element is determined by the formula,

$$Tm = \frac{L}{fN}, \text{ min}$$

The total time is based on the total number of spindle revolutions to complete a piece if the spindle runs continuously at one speed. If different spindle speeds are used for different operation elements it is necessary to recalculate the number of spindle revolutions in each operation element to a single basis, by using the so called 'equivalent revolutions factor', which is the ratio of the higher to the lower speed.

6.5.2. Time Required for Idle Motions and Working Travel. The various idle motions in an automatic machine operation can be : feed stock to stop and index turret etc. The time needed for these idle motions depends on the design and type of automatic machine. In certain types, the idle motion time is constant and does not depend on the workpiece. The idle motion time is usually determined from the table given in the setting-up instructions of each automatic. The time required for the idle movements may be determined by estimating the approximate time for whole cycle, during which one workpiece is produced. This is obtained on the basis of the working travel. The total time is then increased by 25 or 30 per cent. Then, the time required for the idle movements is obtained from table corresponding to the estimated cycle time. After that the cycle time can be calculated exactly.

Cams are usually divided into either 100 or 360 parts. The time for idle movements is usually given in terms of divisions on the cam surface. For example, the following values are usually taken for the various idle movements :

Feed stock to stop = 2.5 hundredths of cam surface

Index turret = 1.5 hundredths of cam surface

Withdrawal of the cut-off tool = 3 hundredths of cam surface

To determine the time required for the working and idle movements, in terms of hundredths of cam surface, it is necessary to find the relationship between the spindle revolutions and hundredths of cam surface. If K_i is the number of hundredths required for the idle movements, then the number of hundredths of cam surface for the total working travel will be,

$$K_w = (100 - K_i)$$

Now, if 'n_w' is the number of spindle revolutions for working travel, then the number of spindle revolutions for the whole cycle will be

$$n_{cycle} = \frac{n_w \times 100}{K_w}, \text{ revolutions per cycle}$$

∴ Number of spindle revolutions for each hundredths of cam surface = n cycle/100.

6.6. CAM DESIGN

For laying out the cam, the radii of the plate cam curves are determined in the following manner :

1. The main data on the cam and the drive of the operative unit are taken from the service manual. This includes (*a*) maximum radius of cam R_{max}, (*b*) minimum radius of cam R_{min}, (*c*) height of the cam lever fulcrum above the cam centre, (*d*) ratio of the lever system.

2. The cam radii are determined for each operation element from the final position of the operative unit (turret or cross-slide).

3. For hexagonal turret slide or end working slide, R_{max}, corresponds to the minimum, distance of the slide from the spindle nose. For cross-slides, R_{max} will correspond to the position where the tool has advanced slightly beyond the spindle axis.

4. The other radii for the final positions of the operative units are determined as follows :

For first operation, $R_1 = R_{max} - (L_1 - L_{min})$

where L_1 is the distance between the turret face and the spindle for first operation and L_{min} is the minimum distance between the turret face and spindle.

5. The cam radii of the beginning of the working travel for each operation is determined by subtracting the travel of the cutting tool from the radius at the end of each corresponding travel.

After knowing the various can radii, the cam can be drawn. In doing so, the following points should be taken care of :

(*i*) The movement of the operative unit (turret, cross-slides) usually consists of : rapid approach, working travel, and rapid return.

(*ii*) Rapid approach and rapid return being idle motions, should be performed with minimum time loss. Therefore, the cam curves should rise or drop sharply at these places. Curves for these two movements are drawn to a special template in accordance with cycle time.

(*iii*) The working travel curve should provide for motion at a constant speed, *i.e.*, at a uniform rate of feed. This uniform curve is called Archimedean spiral. This curve can be drawn to a special template or geometrically as explained below :

The cam is drawn full size and its surface is divided into hundredths. The lines of the hundredths are arcs drawn with a radius equal to that of the cam roll lever. The beginnings and ends of working travel lobes are marked off on the corresponding hundredths by drawing arcs of the required radii. Then the cam roll is drawn full size, tangent to the above points. The distance between the two roll positions, *i.e.*, the cam rise is divided into as many equal parts as there are hundredths in the given working travel. The points of intersection of the arcs with Arabic numerals and Roman numerals are marked, Fig. 6.12. A uniform curve passing through these points will ensure a uniform feed of the operative unit.

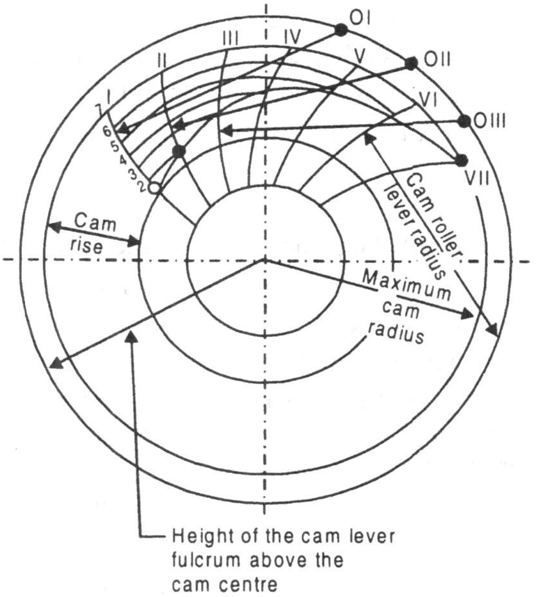

Fig. 6.12. Developing a Uniform Cam Rise Curve.

6.7. TOOL LAYOUT FOR AUTOMATIC SCREW MACHINE

In this article, the procedure for laying out the tools for an automatic screw machine to manufacture the component shown in Fig. 6.13, will be explained. The following order of procedure is recommended :

1. Determine the spindle speed for the material and the diameter of the component. Select the nearest speed in the range available.
2. Decide the proper sequence of operations.
3. Determine the travel for each tool.
4. Select the feed per revolutions for each tool.
5. Determine the number of spindle revolutions for each operation.
6. Determine the number of spindle revolutions for each of the idle movements.
7. Determine the hundredths of cam surface required for dwells and clearance.
8. Calculate the total spindle revolutions to machine one component.
9. Calculate the hundredth of cam surface for each operation.
10. Complete the operation sheet and draw the cams.

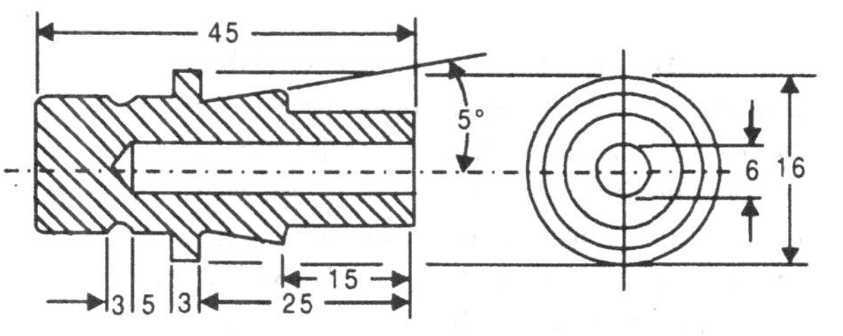

Fig. 6.13

1. **Spindle Speed.** The material of the component is Brass 16 mm diameter. The cutting speed for turning may be taken as 150 metre per min. Therefore the spindle speed is,

$$N = \frac{1000\,V}{\pi D} = \frac{1000 \times 150}{\pi \times 16}$$

$$= 2984 \text{ r.p.m.}$$

The nearest standard spindle speed of 3000 r.p.m. is selected.

2. **Sequence of Operations.** This may be fixed as follows :

(*i*) Feed stock to turret stop.

(*ii*) Centre and turn-Turret slide.

(*iii*) Drill 6 mm diameter from turret slide. The forming from the front slide, the front portion of the component and the forming of the back portion of the component from the rear slide is overlapped with the drilling operations.

(*iv*) Cut off from the top slide.

3. **Travel for each Tool (Cam Throw).** To calculate the total travel for each tool, a tool approach from 0.5 to 3 mm should be added to avoid damage to tools or component.

(*i*) **Centre and Turn.** Here, the length to be turned will be taken for calculation purposes. Therefore,

$$\text{Tool travel} = 15 + 1 = 16 \text{ mm}$$

(*ii*) **Drilling and Forming.** The travel of the drilling tool exceeds that for forming, therefore,

$$\text{Tool travel} = 33 + 2 = 35 \text{ min.}$$

(*iii*) **Cut Off.** As is clear from Fig. 8.13, the tool travel for cut off tool will be 6 mm plus tool approach and tool travel, say 7 mm.

4. **Feed per Revolution per each Tool.** The feed per revolution depends upon the type of material, the rigidity of the component, the finish required and the tool used. The following feeds are taken for the present problem :

$$\text{Turing} = 0.18 \text{ mm/rev.}$$

$$\text{Drilling} = 0.10 \text{ to } 0.12 \text{ mm/rev.}$$

$$\text{Forming} = 0.018 \text{ to } 0.050 \text{ mm/rev.}$$

$$\text{Cut off} = 0.025 \text{ to } 0.075 \text{ mm/rev.}$$

The throw of the parting tool is divided into a low feed to cut off the component, followed by an increased feed to clean up the end of the next component.

5. **Spindle Revolutions for each Working Operation.** Having determined the tool travel for each operation and selecting the suitable feed, the revolutions of the spindle are calculated as :

$$\text{Revolutions per operation} = \frac{\text{Travel of tool}}{\text{Feed per revolution}}$$

6. **Dwells and Clearances.** After the tool has reached the end of its travel, it should be allowed to dwell for a short time to ensure a clean face and a uniform diameter. Usually, one hundredth on cam surface is allowed for dwell. Under certain conditions, the cross-slides cannot be advanced whilst the turret is operating, and lapse of time whilst the turret is receding and the cross-slide is advancing is expressed as clearance idle time. Also, after cutting off the workpiece, the parting tool must drop back to clear the bar on the next feed out. The interval necessary for this clearance is usually 2 hundredths of cam rotation.

7. **Tool Spindle Revolutions to Machine one Component.** In table 6.1, the spindle revolutions for working travels shown within parentheses are for overlapped operations and are not to be counted for calculating production. Similarly, cam hundredths for idle motions, shown within parenthesis are for overlapped operations and are not to be considered. From table 6.1.

Total number of spindle revolutions for working travel = 528

Number of hundredths for idle motions, $K_i = 30$

Number of hundredths for working travel,

$$K_w = 100 - 30 = 70$$

Table 6.1. Operational Sheet for Automatic Screw Machine

Material : Brass 16 mm diameter **Production Time : 15 seconds**

Cutting Speed for turning : 150 m/min **Spindle R.P.M. : 3000**

Operative unit	Operation element	Tool travel, mm	Feed mm/rev	Spindle revolution for each Operation element	Accepted for Calculation		Number of hundredths on Cam	Hundreths on Cam for			
					Working travel	Idle motion		Calculating Production		Cam design	
					Spindle revolution	Hundredths on Cam		Start	Finish	Start	Finish
1	2	3	4	5	6	7	8	9	10	11	12
	1. Feed Stock to stop	-	-	-	-	4.0	4.0	0	4.0	0	4.0
Turret											
	Double Index	-	-	-	-	4.0	4.0	4.0	8.0	4.0	8.0
	2. Centre and turn	16	0.19	84	84	0	11.0	8	19	8	19
	Dwell	—	—	—	—	1.5	1.0	19	20	19	20
Turret	Drop back	—	—	—	—	5.0	5.0	20	25	20	25

	Double Index	–	–	–	–	4.0	4.0	25	29	25	29
	3. Drill 6 mm diameter	22	0.13	183	183	–	24	29	53	29	53
	Clear drill	–	–	–	–	8.0	8	53	61	53	61
Turret	Drill	13.3	0.10	133	133	–	18	61	79	61	79
	Dwell	–	–	–	–	1.0	1.0	79	80	79	80
	Double Index	–	–	–	–	(4.0)	(4.0)	–	–	–	80
Cross-Slide	3. Form	3.3	0.05	(66)	(66)	–	9	–, –	–	49	50
Front	Dwell	–	–	–	–	(1.0)	(1.0)	–	–	49	50
Rear	3. Form	3.5	0.017	(203)	(203)	–	27	–	–	50	77
Cross											
Slide	Dwell	–	–	–	–	(1.0)	(1.0)	–	–	77	78
	4. Cut off	1.5	0.027	(37)	(37)	–	5	–	–	75	80
	Cut off	5.5	0.043	128	128	–	17	80	97	80	97
	Dwell	–	–	–	–	1.0	(1.0)	97	98	97	98
	Clear	–	–	–	–	2.0	(2.0)	98	100	98	100

Number of spindle revolution to make one workpiece

$$= \frac{528 \times 100}{70} = 750$$

Cycle time (seconds to make one piece)

$$= \frac{60 \times 750}{3000} = 15 \text{ secs}$$

$$\text{Gross production} = \frac{3600}{15} = 240 \text{ pcs/hr}$$

Actual production (assuming 15% losses)

$$= 240 \times 0.85$$

$$= 204 \text{ pcs/hr.}$$

8. **Number of hundredths on Cam Surface for each Spindle Revolution.** For drawing the cam, it is necessary to determine the portion on cam surface devoted to each operation. The cam surface is divided into 100 divisions round its circumference.

In the present example,

Total spindle revolution per piece = 750

$$\therefore \quad \text{Revolution per hundredth} = \frac{750}{100} = 7.5$$

Hence, by dividing the spindle revolutions for each working travel with 7.5 the equivalent hundredths on cam surface can be calculated.

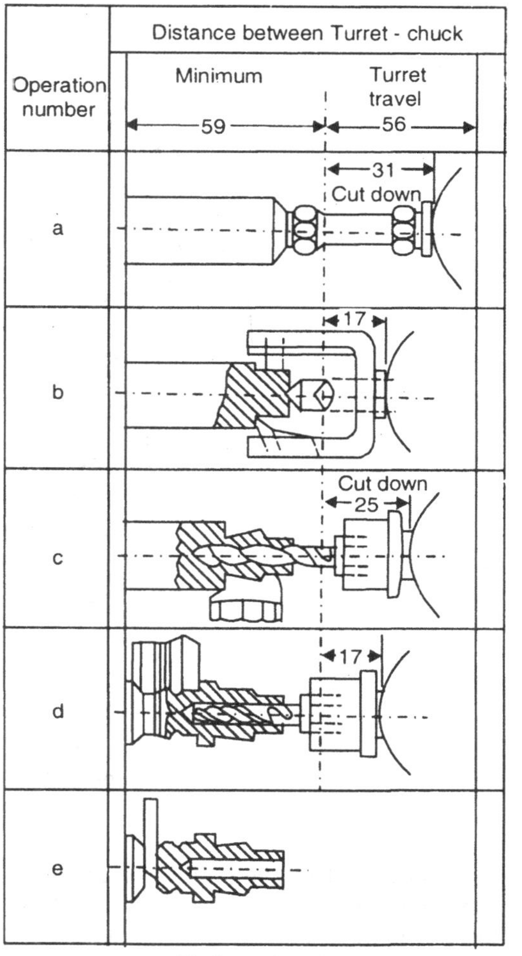

All dimensions in mm

Fig. 6.14. Tool Layout for Cam Design.

9. **Drawing the Cams.** To draw the cam layout, the following procedure is adopted. Let us take the case of turret slide cam :

(*a*) A circle is drawn for the maximum diameter of cam blank and this is divided off into 100 equal parts.

(*b*) To know the maximum radius of the cam lobes for each operation, we ought to determine the "cut down". For this, the accurate scale drawings for each operation (at the end of tool travel) have to be made. A vertical line is drawn touching the face of the collect chuck. Another line to the right is drawn which represents the forward-most position of the turret. A third line is drawn on the extreme right to represent the turret face at its maximum backward position, Fig. 8.14. It will now be observed that the turret face is at varying distances behind the vertical line drawn for the turret forward position. These distances represent the "cut drawn", that is, the amounts by which the cam lobes must be reduced from the original outside radius of the cam blank.

(*c*) After determining the "cut down" for each operation (at the end of the tool travel), moving clockwise from the vertical line (zero division), Fig. 6.15, the hundredths of cam are divided off in the order and amounts set out in column 9 to 12 (Table 6.1). The radial lengths from the outer circle of the cam may now be set out by taking into account the "cut downs" at the end of the operation plus the cam throw for that operation. The cam surfaces joining the start and end of cutting operations may now be drawn as explained under article 6.6 or by means of a template. The cam surfaces for dwells and feeding stock to stop are drawn as

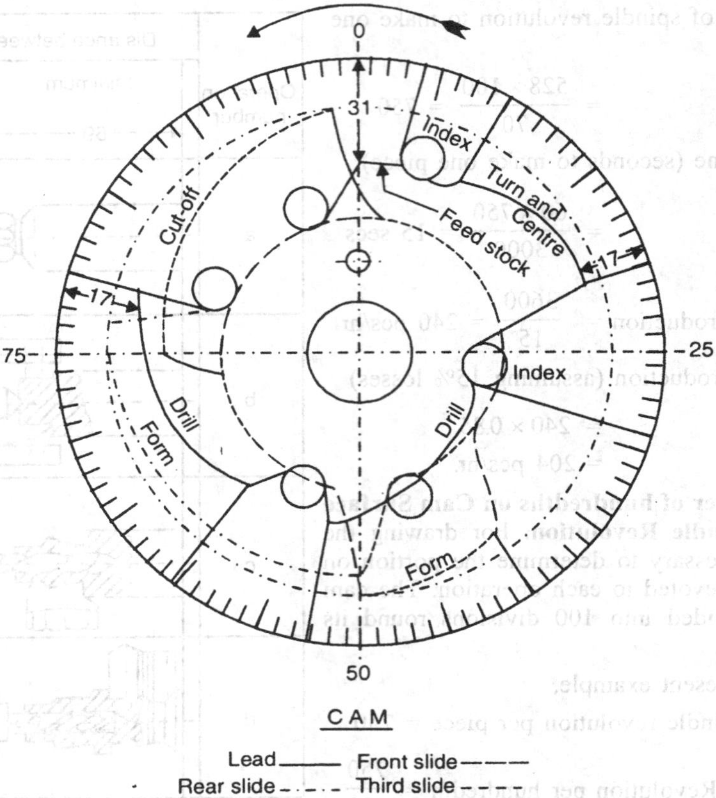

Fig. 6.15. Cam Layout.

concentric circles. The radii between the lobes, where the indexing occurs, are swung from the cam centre, the radius drawn being slightly less than the start of the next throw. This radius and the rise or fall adjacent to it are then joined by an arc equal to the cam roller radius

The cams for the front, rear and top cross-sides may be set in the same manner as the turret cam.

6.8. PROGRAMMED AUTOMATIC LATHES

The standard automatic lathe is programmed to produced parts by means of Cams, stops or other mechanical methods. However, cams are suited to the production of a specific part and cam controlled automatic lathes are not referred to as programmed lathes. Complete versatility in programming is provided by Numerical Control. In a N.C. lathe, the programme is provided by the punched tape and no Cams are required.

6.9. BAR STOCK FEEDING

Bar stock may be fed by one of the following methods :

(*a*) By gravity. The devices are simple in design but are inconvenient in renewing long bars. Such devices are seldom used and then mainly in multi-spindle vertical automatic bar machines for small diameters and comparatively short bar stock.

(*b*) By driven rolls. This method is used for machining long components.

(*c*) By advancing the bar stock together with the head stock as in Swiss type machine.

(*d*) By the use of a feeding finger (stock pusher) and tube. This is the main method applied in general purpose single spindle automatic screw machines and multi spindle automatic bar machines.

The spring feeding finger is carried in the feeding tube which is arranged inside the collet tube. The stock pusher is screwed on the front end of the feeding tube. The feeding finger is a slitted spring bushing in which the jaws were compressed or closed before heat treatment. The jaws maintain a constant grip on the stock sufficient to feed the bar forward the moment it is released by the collet chuck.

The spring action of the jaws should not be excessive, however, so that the feeding finger may be drawn back over the stock to the initial position when the bar is clamped.

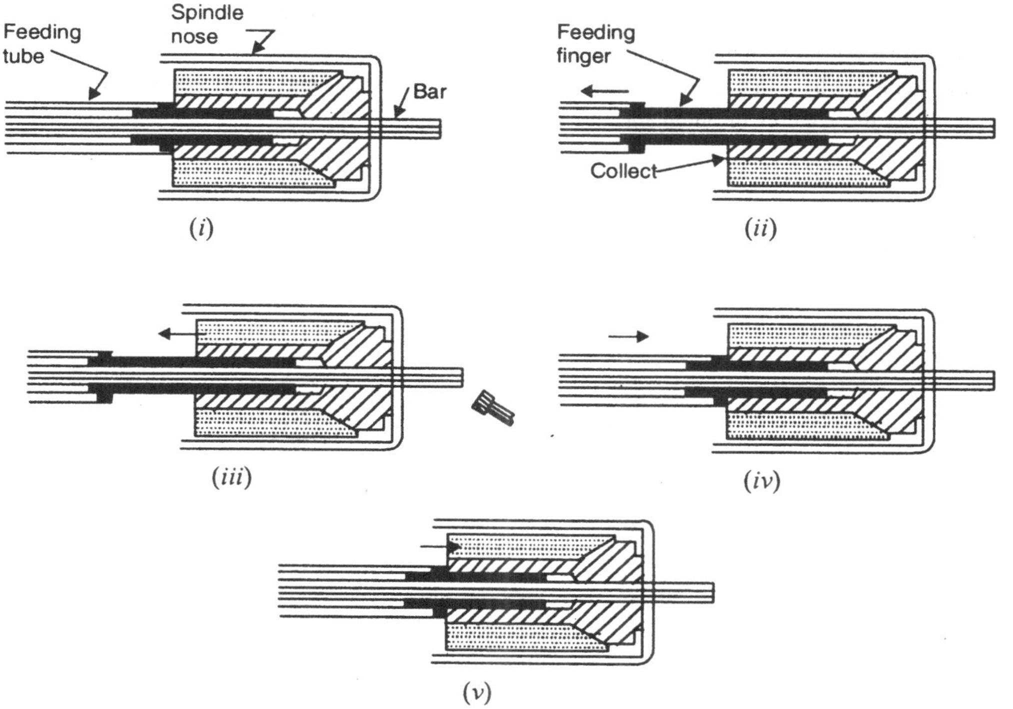

Fig. 6.16. Bar Stock Feeding for Automatics.

Bar stock feed and chucking proceed in the following order, Fig. 6.16.

(*i*) bar stock is clamped.

(*ii*) feeding finger is retracted.

(*iii*) work is cut-off and collet opens.

(*iv*) feeding tube and finger feed the bar out (to the right) upto the stock stop. After this, the stock is clamped by the movement of the collet tube. When the collet grips the bar, the feeding tube is returned to its initial left position and, the feeding finger slides along the clamped bar.

The feeding tube is operated by permanent cylinder cams through a lever system.

PROBLEMS

1. Define an automatic and a semi-automatic lathe.
2. Differentiate between an automatic and a semi-automatic lathe.
3. Discuss the various types of automatic lathes.
4. Describe a typical single-spindle automatic chucking machine.
5. Describe a typical single-spindle automatic bar machine.

6. Describe a Swiss-type automatic and tell how it differs from other single-spindle automatics.
7. Describe a single spindle cutting off automatic machine.
8. What are the relative advantages and disadvantages of single-spindle and multi-spindle automatics ?
9. Discuss the various types of multi-spindle automatics.
10. Differentiate between parallel action and progressive action multi-spindle automatics.
11. Discuss the various types of semi-automatics.
12. Contrast the automatic and capstan lathes, giving a general description of the two machines as required in a comparison of the two machines and their relative merits for a particular job.
13. Explain the procedure for drawing cams for turret type automatic.
14. What are programmed automatic lathes ?
15. With the help of a neat diagram, explain the bar feed mechanism for automatics.
16. List the features of automatic screw machines.

7

GEAR MANUFACTURING

7.1. INTRODUCTION

Gears are used extensively for transmission of power. They find application in : Automobiles, gear boxes, oil engines, machine tools, industrial machinery, agricultural machinery, geared motors etc. To meet the strenuous service conditions the gears should have : robust construction, reliable performance, high efficiency, economy and long life. Also, the gears should be fatigue free and free from high stresses to avoid their frequent failures. The gear drives should be free from noise, chatter and should ensure high load carrying capacity at constant velocity ratio. To meet all the above conditions, the gear manufacture has become a highly specialised field. Below, we shall discuss the various materials and manufacturing processes to produce gears.

7.2. MATERIALS

The various materials used for gears include a wide variety of cast irons, steels, non-ferrous materials (Bronzes, Aluminium) and non-metallic materials such as : leather, plastic, nylon, texolite etc. The selection of the gear materials depends upon :

(*i*) Type of service

(*ii*) Peripheral speed

(*iii*) Degree of accuracy required

(*iv*) Method of manufacture

(*v*) Required dimensions and weight of the drive

(*vi*) Allowable stress

(*vii*) Shock resistance; and

(*viii*) Wear resistance.

1. Cast iron is popular due to its good wearing properties, excellent machinability, and ease of producing complicated shapes by the casting method. However, it is suitable where stresses on gear are not high and where large gears of complicated shapes are needed.

2. Steel is sufficiently strong and highly resistant to wear by abrasion. The teeth can be machined to the required degree of accuracy and also it can transmit heavy loads.

3. Cast steel is used where stress on gear is high and it is difficult to fabricate the gears.

4. Plain carbon steels (medium carbon steels and high carbon steels, C = 0.3 to 0.6%) find application for industrial gears where high toughness combined with high strength, fatigue limit in bending, long life and accuracy of teeth are required.

5. Where high tooth strength and low tooth wear are required, for example, industrial and automobile gears alloy steels (Mn, Ni, Cr, Mo) are used.

6. To combine properly the toughness, gears made of steels are heat treated. The various heat treatment processes employed are : case hardening, through hardening, nitriding etc. High hardness heat treated steels give smaller gears for a given load and also the service life is increased by about ten times without increasing the size or weight of the gears.

(*a*) Case hardened steel with carbon less than 0.2 per cent and alloyed with Ni, Cr and Mo is used for industrial gears where the conditions are of varying loads and shocks.

(*b*) Surface hardening steel (C = 0.4 to 0.6%) and alloyed with Mn, Cr and Mo and flame/ induction hardened, is used where tooth surface has to be very hard and core soft, for example, gears used for transmitting shocks loads.

(*c*) Nitrided steel is used where hardness has to be very high (750-800 BHN), for example, aircraft gears, marine propulsion gears and industrial gears etc.

7. Bronzes are very popular in worm gear drives because of their ability to withstand heavy sliding loads and ability to "wear in" to fit hardened steel worms. They are also very useful for corrosive conditions. Like C.I., they can also be easily cast into complex shapes.

8. Aluminium is used where low inertia of rotating mass is desired, for example, instrument gears and appliances.

9. Gears made of non-metallic materials give noiseless operation at high peripheral speeds, but their load carrying capacity is low, for example, gears for instruments and appliances etc.

7.3. METHODS OF MANUFACTURE

The various manufacturing methods used for gears can be listed under the following headings :

1. Casting
2. Metal forming
3. Metal removal.

Out of the above methods, metal removal or machining is the best and most widely used method for mass production of high quality gears.

7.3.1. Casting. Gears can be produced by the various casting processes. Sand casting is economical and can take up large size and module, but the gears have rough surfaces and are inaccurate dimensionally. These gears are used in machinery where operating speed is low and where noise and accuracy of motion can be tolerated, for example, farm machinery and some hand operated devices. Sand casting is suitable for one off or small batches. Large quantities of small gears are made by "Die-Casting". These gears are fairly accurate and need little finishing. However, the materials used are low melting ones, such as alloys of zinc, aluminium and copper. So, these gears are suitable for light duty applications only (light loads at moderate speeds), for example, gears used in toys, cameras and counters etc. Gears made by "Investment Casting" may be accurate with good surface finish. These can be made of strong materials to withstand heavy loads. Moderate-size gears are currently being steel cast in metal moulds to produce preforms which are later forged to size. Light gears of thermoplastics are made by "Injection Moulding". This method is suitable for large volume production. However, gear tooth accuracy is no high and initial tool cost is high. These gears find use in instruments, household appliances etc.

For phosphor bronze worm wheel rims, "centrifugal casting" is used far more extensively than any other method. Centrifugal casting is also applied to the manufacture of steel gears. Both vertical and horizontal axis spinners are used. After casting, the gears are annealed or nomalized to remove cooling stresses. They may then be heat treated, if required, to provide the needed properties. Centrifugally cast gears perform as well as rolled (discussed ahead) gears and are usually less expensive. "Shell moulding" is also sometimes used to produce small gears and the product is a good cast gear of sowewhat lower accuracy than one made by investment casting but much superior to sand casting.

7.3.2. Forming Method. Methods included under this heading are : Roll forming, extrusion and cold drawing.

1. Roll forming. In roll forming of gears, the gear blank is mounted on a shaft and is pressed against hardened steel rolling dies. The rolls are fed inward gradually during several revolutions which produce the gear teeth from the solid. The forming rolls are very accurately made and the roll formed gear teeth usually have excellent accuracy. Roll forming is done both hot and cold. In hot roll forming, the hot-rolled gear is usually cold-rolled which completes the gear with a smooth mirror finish. In cold roll forming, higher pressures are needed as compared to hot rolling. Many of the gears produced by this process need no further finishing. Since the material is strained plastically during the process, it becomes stronger against both tension and fatigue. Since this method can be called as chipless method (no metal removal), there is lot of material saving. Also, there is saving in machining operation, may be 80%. The machinery is very expensive and so the method is suitable only for mass production. Both spur and helical gears are being made by this process.

2. Extrusion. Small sized gear can also be made by extrusion process. Other operations like piercing the hole upsetting or heading etc. are also needed to produce the complete gear. At the end, only finish grinding is required to complete the job. There is saving in material and machining time. The method can produce any shape of tooth form and is suitable for high volume production. Gears produced by extrusion find application in watches, clocks, typewriters, etc. Previously, only brass, bronze, aluminium and magnesium alloys were extruded. But now steel bars upto about 60 mm diameter are cold extruded.

7.3.3. Metal Removal. In this method, the metal is removed from the gear blanks to produce the gears. Under this, the main gear cutting methods are :

(*a*) Profiling or Form Cutter Method

(*b*) Generation.

Profiling or Form Cutter Method. In this method, the spaces between the teeth are cut by means of a cutter conforming to the gear tooth form. This method necessitates the use of special cutter for each application and so is not very suitable for large volume production. The various methods under this category are :

1. Gear cutting on milling machine with formed disc cutter or end-milling cutter.
2. Gear cutting on planer or shaper with single point formed tool.
3. Cutting external or internal teeth with formed cutter in broaching machine.
4. Gear cutting with form tool blades by shear speed process.

Gear generating process. This process is based on the principle that any two involute gears, or any gear and a rack, of the same module will mesh together properly. If one of the gears (or the rack) is made into a cutter by proper sharpening and it meshes with the gear blank, then the teeth on blank are developed or generated by the relative rolling motion of the cutter and the blank. This process in which the cutter is similar to a mating pinion or a rack is called "Gear Shaping". If the generating tool (cutter) is threaded and gashed, it is called a "hob" and the process is called as "Gear Hobbing".

The major advantages of gear generation is that the same cutter of a particular module can cut gears of different number of teeth without profile deviation. Due to this, this is the most popular method of producing gears in mass.

7.4. GEAR CUTTING BY MILLING

For cutting a gear on a milling machine, the gear blank is mounted on an arbor which is supported between a dead centre and a live centre in the indexing head. The cutter is mounted on the arbor of the milling machine. The geometric centre of the cutter must be aligned exactly vertically with the centre line of the indexing head spindle. The table of the machine is moved upward until the cutter just touches the periphery of gear blank. The vertical feed dial is set to zero. The table is then moved horizontally until the cutter clears the gear blank. The table is then moved upwards by an amount equal to the full depth of the gear tooth. The vertical movement may be less if the gear is to be cut in two or more passes. After this, the longitudinal feed of the table is engaged. The gear blank moves under the rotating cutter and a tooth space is cut. After this, the movement of the table is reversed so that the cutter again clears the gear blank. The gear blank is then indexed to the next position for cutting the second tooth space. This procedure is repeated until all the teeth have been milled.

This method is shown in Fig. 7.1(*a*). The cutter is flat circular disc type cutter. The plane is rotation of the cutter is radial with respect to the blank. The second method, shown at (*b*), makes use of an end-cutter or miller. The cutter rotates about an axis which is set radially with respect to the blank and at the same time the cutter is traversed parallel to the axes of the blank. The cutting edges lie on a surface of revolution, so that any axial cross-section of the cutter corresponds to the shape required for the space between two adjacent teeth on the finished wheel. The milling machine used in this method is vertical milling machine.

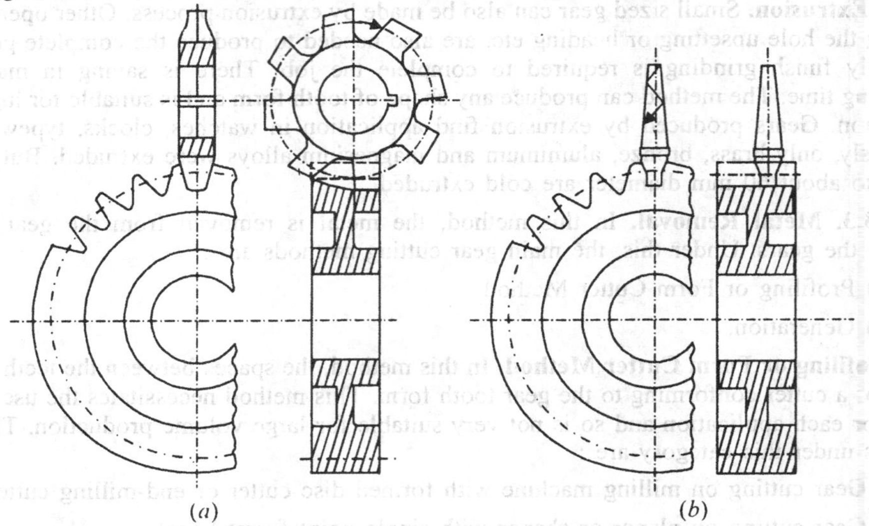

Fig. 7.1. Gear Milling.

The end mill cutter is mounted straight on the milling machine spindle through a chuck. The disk type of cutter is used to cut big spur gear of large pitch, whereas, the end mill type of cutter is employed for the manufacture of pinion of large pitch.

The quality of gear produced by this method is not high since the operation of the indexing device is not precise. Also, this method is very slow since only one tooth is cut at a time. To overcome these two drawbacks, multiple tool shaping cutter head, Fig. 7.2, is used to cut all the tooth spaces of the gear at the same time. The form tools are arranged radially in the cutter head

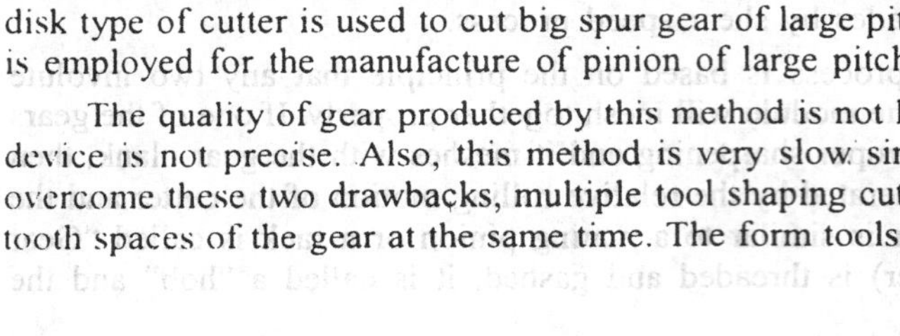

Fig. 7.2. Multiple-tool Shaping Cutter Head.

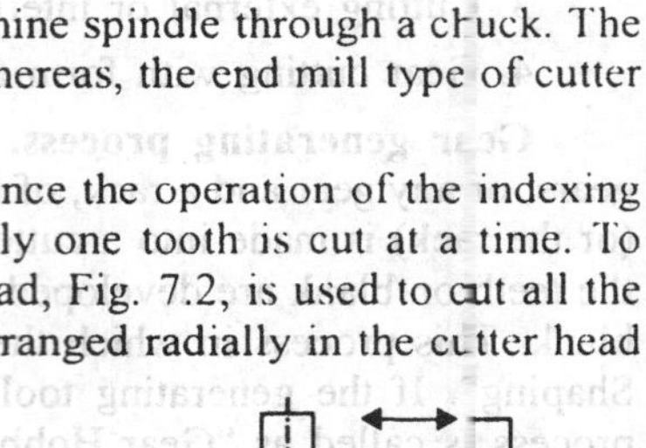

Fig. 7.3

and their number is equal to the gear-tooth spaces desired. All the tools are fed radially towards the centre of blank by an amount equal to the tooth depth, prior to each cutting stroke. After the cutting stroke (into the plane of paper), all the tools are simultaneously retracted.

Gear milling is a simple, economical and flexible method of gear making. Spur, helical, bevel gears and racks can be produced by this method. The method is versatile in that tooth forms, that are not possible by generation methods, can be produced by milling.

The major disadvantage of this method is that a separate cutter must be used not only for every pitch but for every number of teeth. If this method is followed, the cost of producing a gear will be very high. To reduce the cost, the usual procedure is to use the same cutter for 8 to 10 different numbers of teeth at a given pitch. However, obviously, one out of 8 or 10 gears will be properly cut. Due to this, the milling cutter method is seldom used and that only for small quantity production and for gears finding application in machinery where operation speed is low and where deviation in angular motion is not the major concern.

7.5. GEAR CUTTING WITH SINGLE POINT FORMED TOOL ON SHAPER/PLANER

Gear can also be cut on a shaper or a planer by using a dividing head. The ordinary dividing head on a milling machine can be bolted directly on to the table of a shaping machine. However, a simplified dividing head can also be used. The gear blank is held in a chuck mounted on the spindle of dividing head. If the job necessitates, a tail stock can also be mounted on the table of shaper. The teeth are cut and the gear blank is indexed in the same manner as on the milling machine, but the cutter is single point and its motion is reciprocating instead of a multi-point rotary cutter used on the milling machine. This method is suitable only for small batch production. The method is very slow and the cutters are not universal.

This method is shown in Fig. 7.3. The tool is ground so that its cutting edge corresponds to the shape of the tooth space and it reciprocates parallel to the axis of the blank. When one space has been cut, the gear blank is turned about its own axis, through an angle corresponding to the pitch of the teeth in order to cut the second space. This process is continued until all the tooth spaces have been cut.

7.6. BROACHING

Broaching produces highly accurate gears with an excellent surface finish. Internal gears, racks and gear sectors have been produced by broaching. By an operation, called "pot broaching", external gears can be made in one pass by a circular broach having in ward-facing teeth. Broaching is a highly productive method and is suitable for high volume production. The method has the drawbacks : tool is not universal and a separate broach must be used for each size of gear. Also, the parts with blind end cannot be machined.

7.7. SHEAR SPEED PROCESS

In shear speed process or "shear speed gear shaper", all the teeth on a gear are cut simultaneously by a ring of form cutting tools or blades. Single point tools, one for each tooth space, are arranged in a circle and point radially. For cutting external gear, the cutters are mounted around a hollow head. The gear blank is mounted on a fixture below and is reciprocated into the head. The tools are fed radially a predetermined amount each stroke and are retracted slightly for clearance, on the return stroke. All tooth spaces are cut at the same time. The method is very rapid. Because of tooling cost, the method is economical only for large quantities of parts. Gears with unsymmetrical or unequally spaced teeth can be conveniently produced by this method. The method can be used to cut internal and external spur gears, splines, clutch teeth and many special shapes.

7.8. GEAR PLANING

This is one of the oldest methods of gear production but is still extensively used. It employs rack type cutters for generation of spur and helical gears. Involute rack has straight edges and sharp corners and hence can be manufactured easily and accurately. The cutters generate as they cut and as the name implies, the machine cuts the teeth by the reciprocating planing action of the cutter. This is a true generating process since it utilizes the principle that an involute curve can be formed by a straight generator (rack type cutter) when a circle (gear blank) is made to roll without slip relative to the generator, Fig. 7.4.

There are two types of gear planing machines, one based on "The Sunderland Process" and the other on "The Maag Process". Both the methods are identical in principle but differ in machine configuration and detail.

The Sunderland process. In this method the work (gear blank) is mounted with its axis horizontal, and the cutter slide is carried on a saddle that moves vertically downwards as cutting proceeds. For cutting spur gears, the cutter reciprocates parallel to the work axis but because it can be swivelled in the vertical plane to any desired angle, the machine is also used for cutting single helical gears. The cutter is gradually fed to the desired depth of teeth after which the depth remains constant. Simultaneously, the gear blank is rotating and the rack is traversed at a tangent, the motion of rack and blank being geared to act in unison on their respective pitch lines. This relative motion brings fresh part of the blank and rack into contact and thus causes the teeth of the cutter to generate wheel teeth of exactly the correct form. Since a long rack to cut a full gear is not practicable, the machine is arranged to index automatically after the rolling motion has occurred for a distance of one or more pitches. The indexing really consists in stopping the rotation of the blank and causing the rack to move back the amount it has advanced and the process is repeated until the blank has completed one revolution.

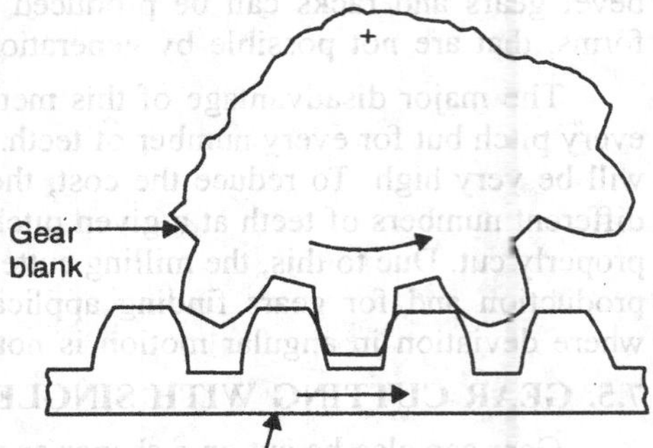

Fig. 7.4. Generation by rack.

Fig. 7.5 explains the principle of gear planing. Thus, while the cutter during its cutting stroke is in contact with several teeth at the same time, and with different part of each tooth, it planes comparatively a narrow strip on each tooth at each stroke, and a different part of each tooth is submitted to the action of the cutter at the next stroke.

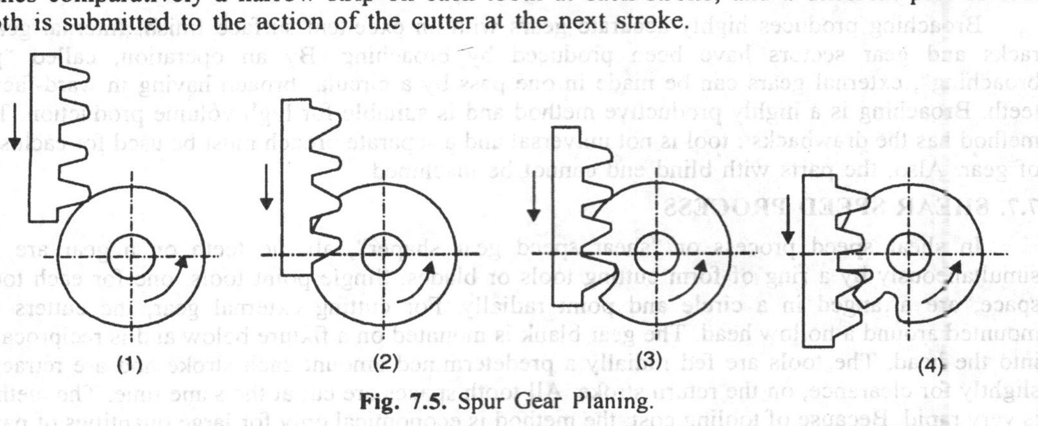

Fig. 7.5. Spur Gear Planing.

To improve the above method regarding the output and quality, two modifications are incorporated :

(*a*) In double cutting method, two blanks are mounted side by side and two cutters are mounted on a duplex cutter holder. The cycle is so arranged that the cutting action of each cutter takes place on alternate strokes.

(*b*) In the double action method, a special box carries two cutters back to back. One cuts during forward stroke and the other during return stroke. One cutter works at the bottom of the tooth space while other finishes the sides. This avoids idle stroke and maintains the cutter condition over a long period.

The Maag Process. In this method, the work is mounted on the machine table with its axis vertical. The rack cutter is carried in a cutter head that is made to move in a vertical plane but the actual direction of motion of the enter in its slide can be set at any desired angle.

7.9. GEAR SHAPERS

In gear shapers, the cutter reciprocates rapidly. The teeth are cut by the reciprocating motion of the cutter and because of this, these machines are called "gear shapers". The cutter can either be a 'rack-type cutter', or a 'rotary pinion type cutter', Fig. 7.6. The main drawback of 'rack-type cutter' is that once the length of rack is covered by the gear blank, the cutting process is interrupted to index the blank back to starting point. In the case of 'rotating pinion type cutter', such an indexing is not required, therefore, this type is more productive and so common. The principle of a gear shaper is shown in Fig. 7.7. The cutter is pinion shaped with a clearance on the tooth face. The gear blank is mounted on a vertical spindle and the cutter on the end of a second, vertical, reciprocating spindle. Cutting takes place either on the downward stroke or on the upward stroke of the cutter (depending upon the design of the machine). During each return stroke, either the table with the blank is withdrawn from the cutter, or the cutter is withdrawn from the blank. Its purpose is to prevent rubbing and resulting wear of the cutting edges of the cutter and damage to the tooth profiles of the gear being cut.

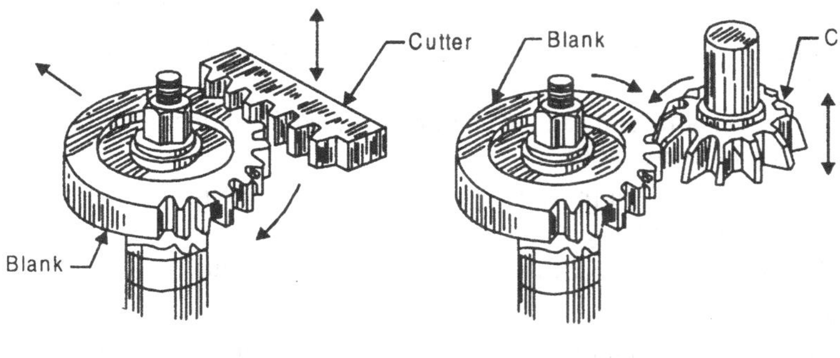

Fig. 7.6. Gear Shaper Cutters.

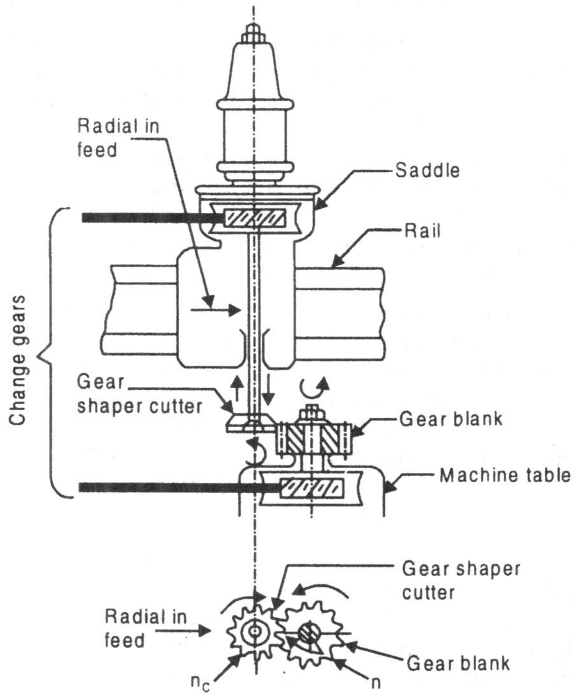

Fig. 7.7. Principle of a Gear Shaper.

To start cutting a gear, the tool is fed into the gear blank before each cutting stroke. When a required depth is reached, the inward feed stops and the cutter and blank slowly start rotating as if they were in mesh. The gear is cut in one complete rotation of the blank. The rotary motions of the cutter and the blank are co-ordinated through change gears so that the surface speeds of the cutter and the blank are the same. The cutter reciprocates at about 100 strokes per minute for an average job and the strokes can be upto 2000 per minute for fine tooth gears. Both straight and helical-tooth gears can be cut on gear shapers. For cutting helical-tooth gears, both the tool and the gear blank are given an oscillating rotational motion during each stroke of the cutter, turning in one direction during the cutting stroke and in the opposite direction during the return stroke.

Advantages

1. One cutter can be used for cutting all spur gears of the same module, irrespective of number of teeth on the gear.

2. It can cut teeth, up close to a shoulder with only a slight clearance recess.

3. Because the cutting stroke can be adjusted, gear shapers are particularly suitable for cutting cluster gears.

4. The method is versatile and can cut spur, helical, herring-bone, internal and cluster gears, racks, splines, pawls and many others.

5. The cutter has a generated profile which is more accurate in shape than some cutters, so, the gears produced are accurate.

6. The method is suitable for medium and large batch production.

Limitations

1. The cutting action is reciprocating, and cutting takes place only during one half of the stroke, so, only about half the machine time is spent in metal removal.

2. Separate helical guide is required for cutting helical gear of particular helix angle in a particular direction.

Applications. Gears produced by gear shaper find applications in automobiles, machine tools, instruments, machinery, clocks and other equipment.

The above method of gear shaping, in which the work is mounted on a rotating table with its axis vertical and the cutter is carried on a ram that is given a reciprocating vertical motion, is known as 'The Fellows Process'. Another technique of cutting helical gears on such a machine is to use a cutter made with helical teeth and the ram carrying the tool is fitted with helical guides. During the working stroke, the cutter teeth move in a helical path corresponding to the angle of the tooth spirals required on the work.

The Sykes Process. In this type of gear shaping machine, the work is mounted with its axis horizontal. The cutter is carried on a ram that reciprocates horizontally and also rotates uniformly. Spur, single helical and double helical gears can be cut on this machine. For cutting double helical gears, two helical guides are required, one right- and the other left-handed, and one forms a sleeve over the other.

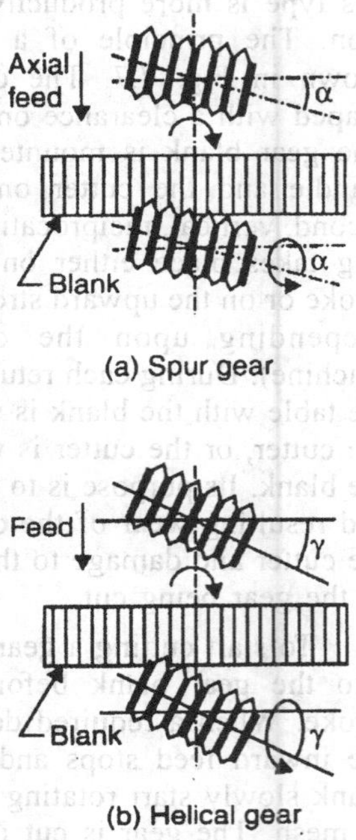

Fig. 7.8. Hobbing Gear Teeth.

7.10. GEAR HOBBING

Hobbing is the process of generating gear teeth by means of a rotating cutter called a "hob". A hob resembles a worm, with gashes made parallel to its axis to provide cutting edges. Relief is provided behind each of the helically arranged cutting faces, for clearance. Gear hobbing is a continuous cutting operation. The hob and the gear blank are connected by means of proper change gears. The ratio of hob and blank speeds is such that during one revolution of the hob, the blank turns through as many teeth as there are starts (threads) on the hob. To start cutting a thread, the hob is made to clear the blank. It is then moved inwards to obtain the required tooth depth. After the tooth depth is reached, the hob is fed in a direction parallel with the axis of rotation of the gear (Axial hobbing). As the gear blank rotates, the teeth are generated and the feed of the hob across the face of the blank extends the teeth to the desired tooth face width. One rotation of the blank completes the cutting unless the blank has a wide face.

To cut spur or helical gears, the hob is set in relation to the axis of the gear blank, so that the threads of the hob facing the gear blank are directed along the axes of the tooth spaces for cutting spur gears and at the helix angle of the teeth while cutting helical gears. This is achieved by setting the hob axis at an angle to the horizontal equal to the helix angle of the hob α, (for cutting spur gears) and at angle $\gamma = \beta \pm \alpha$ for cutting helical gears, where β is the helix angle of the helical gear, Fig. 7.8. Plus sign is to be used if the hands of the helical gears and hob are different (one right hand and the other left hand) and if the hands are the same, minus sign is to be used.

Types of Hobbing. Depending upon the direction of feed of the hob, the gear hobbing may be classified as :

Axial hobbing, radial hobbing and tangential hobbing.

1. **Axial Hobbing.** In this type, firstly the gear blank is brought towards the hob to get the desired tooth depth. The table slide is then clamped. After that, the hob, while rotating, moves along the face of the blank to complete the job, Fig. 7.9 (*a*). Axial hobbing, which is used to cut spur and helical gears can be "climb hobbing" or "Conventional hobbing".

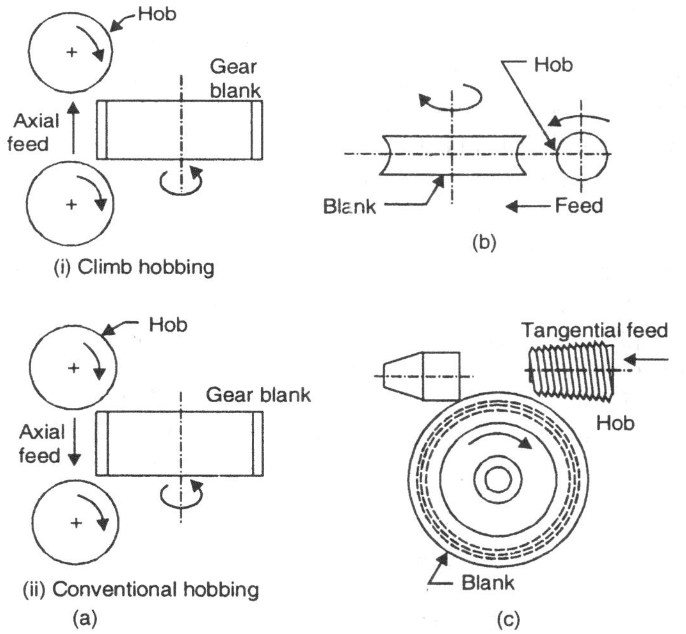

Fig. 7.9. Types of Hobbing.

2. **Radial Hobbing.** In this type, the feed of the hob is radial towards the centre of blank. Radial infeed stops, when the full depth of cut is reached, Fig. 7.9 (*b*).

Radial hobbing has got the following characteristics :

(*i*) Only a small portion of hob length is doing cutting at any time, so, the hob wears non-uniformly. This can affect the tooth profile accuracy of the cut gear.

(*ii*) Again, a comparatively few hob teeth cut each tooth on gear, producing more or less visible flats which also reduces the tooth profile accuracy.

Radial hobbing has a higher production capacity and is used for producing worm wheels whose helix angle does not exceed 6° to 7°.

3. **Tangential Hobbing, Fig. 7.9. (*c*).** In this method, the hob is set at the start of cutting to the full tooth depth and is then fed into the gear blank by an axial feed motion. This method is generally used for generating worm wheels.

Axial hobbing is most commonly used but it has the drawback that there should be enough space for the hob approaching length. When such a condition is not available, the combined "Radial Axial Hobbing" is used. In "Diagonal hobbing", both the axial and the tangential feeds are combined. This method produces a crossed pattern of cutting on the cut teeth, imparting better rolling characteristics to the gear. The wear is uniform all along the hob length which gives longer hob life.

This method is popular for generating tapered gears.

Advantages of Gear Hobbing

(*a*) The method is versatile and can generate spur, helical, worm and worm wheels.

(*b*) Since gear hobbing is a continuous process, it is rapid, economical and highly productive.

(*c*) The method produces accurate gears and is suitable for medium and large batch production.

(*d*) The cutter is universal, because it can cut all gears of same module, irrespective of number of teeth on the gear.

Disadvantages

(*a*) Gear hobbing cannot generate internal gears.

(*b*) Enough space has to be there in component configuration for hob approach.

Applications. The gears produced by gear hobbing are used in automobiles, machine tools, various instruments, clocks and other equipments.

7.11. BEVEL GEAR GENERATING

For generating straight bevel gears, the rolling motions of two pitch cones are employed instead of pitch cylinders (used for cutting spur and helical gears). In cylindrical gears, we have seen that the teeth are generated by tools that simulate or are derived from the basic rack, whereas bevel gear teeth are produced by cutters representing the basic crown wheel (A crown wheel is a bevel gear in which the tooth faces lie in one plane. It is equivalent to the rack in spur gearing).

7.11.1. Straight bevel gear planing. In the production of straight bevel gears by planing, use is made of two straight sided reciprocating tools which shape the profile of the teeth being cut, while the workpiece is given the appropriate rolling motion.

The arrangement to generate straight bevel gears is shown in Fig. 7.10. Tool slides with tools reciprocate along ways provided on the face of the cradle of the machine.

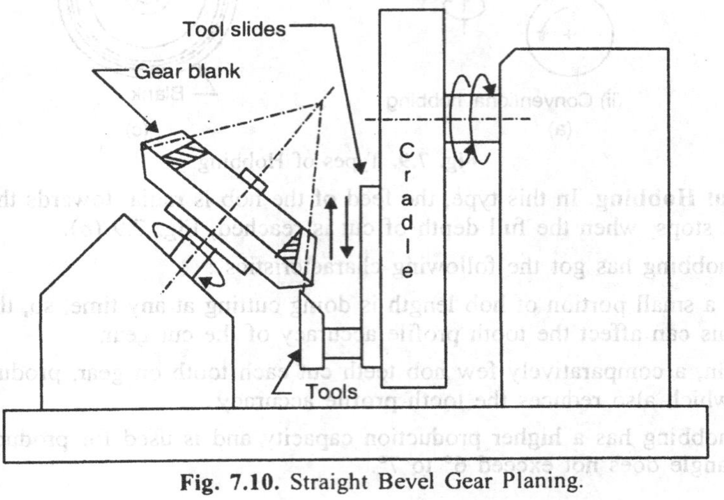

Fig. 7.10. Straight Bevel Gear Planing.

The gear blank is rotated. Also rotated is the cradle with the reciprocating tools. The tools cut with their motion towards the apex of the gear pitch cone. The return stroke is idle, when the tools are withdrawn from the blank to avoid damaging the machined surfaces of the teeth. The two tools represent only one tooth space of the imaginary crown gear.

The complete cycle of generating a tooth is illustrated in Fig. 7.11. At the start, the tools for machining one of the side surface of teeth begins to cut into the blank (positions 1 and 2). Then the second tool, designed to generate the other side surfaces of the teeth, starts to cut the gear (positions 3 and 4). At this point, the first tooth has been completely cut. As the cradle rolls further the tools run out of mesh with the gear blank. So, both gear blank and cradle are reversed. At the end of the reverse rotation, the blank is indexed for the next tooth (positions 5 and 6). This cycle is repeated for each tooth of the gear.

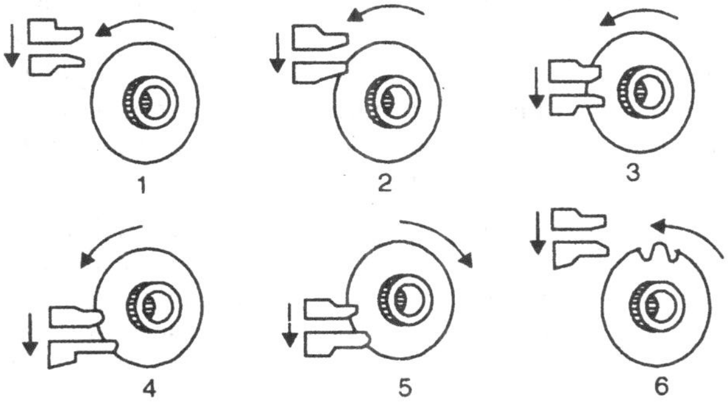

Fig. 7.11. Complete Cycle of Generating a Tooth.

7.11.2. Gleason Method. In this method, two disc milling cutters are employed, Fig. 7.12. The tools form the blanks of a tooth simulating the basic crown wheel. Cutter teeth are inter-meshing and the discs are inclined to each other at the required pressure angle (usually 20°). The following motions are involved while cutting a tooth :

1. The rotating cutters revolve about their axes to provide the cutting action.

2. They travel in planes passing through the sides of the teeth on the imaginary crown gear to shape the teeth along their teeth.

3. At the same time, they participate in the relative rolling motion between the cutters and blank to obtain the required tooth profile.

Indexing takes place after each tooth space has been completed and the machine is fully automatic in its motions. When gear has been completed, the machine stops, the cutters withdraw and the workpiece can be changed with little delay. This type of machine is a high production rate machine and very useful for dealing with large batches of identical gears.

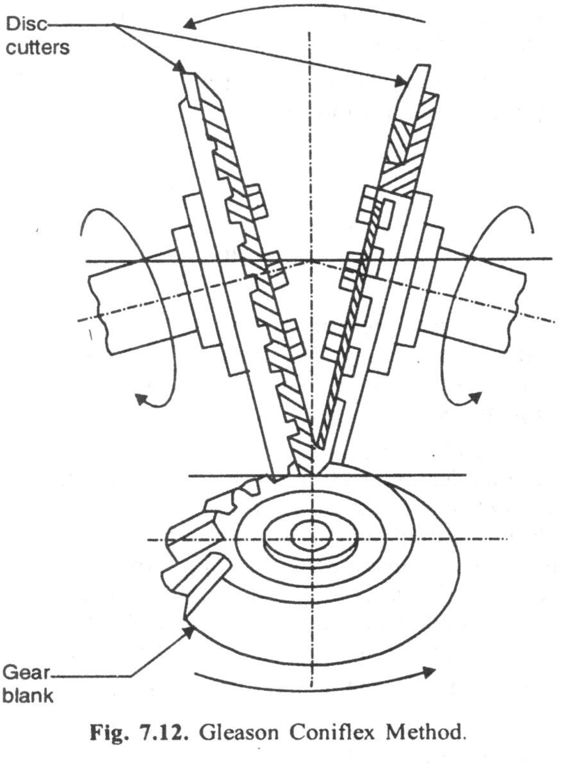

Fig. 7.12. Gleason Coniflex Method.

7.12. MISCELLANEOUS GEAR CUTTING METHODS

Some other methods for making gears can be : Gear stamping, cold drawing, powder metallurgy, electro-chemical matching, electrical discharge machining and HERF.

1. **Cold drawing.** In this method, the material is passed through a number of dies. The shape of the final die corresponds to the desired shape of the tooth. During the passage of the material through the dies, it is squeezed into the shape of the die. Since the material is displaced by pressure, the tooth surface is quite hard and smooth. The materials which can be employed for cold drawing are : low to medium carbon steels, brass, bronze and aluminium. Gears produced by this method find application in : watches, electric clocks, spring wound clocks, typewriters and so on.

2. **Stamping.** Large quantities of gears are made by the method known as "stamping", "blanking" or "fine blanking". The gears are made in a punch press from sheet metal, upto about 12.7 mm (maximum) thick. Such gears find applications in : toys, clocks and timers, watches, household gadgets, water and electric meters and some business equipment. If, after blanking, the gears are shaved, they give best finish and accuracy. The materials which can be stamped are : low, medium and high carbon steels, stainless steel, all brass alloys and aluminium alloys. The method is suitable for large volume production.

3. **Powder metallurgy.** High quality gears (both dimensional and surface quality) can be made by powder metallurgy (P/M) method. The metal powder is pressed in dies conforming to the tooth shape, after which the product is sintered. After sintering, the gear may be coined to increase density and surface finish. This method is usually used for small gears (upto about 25 mm in diameter). Large gears can be made by powder metal forging (P/M forging) in which performs made by powder metallurgy are forged to the final shape. The method is suitable for large volume production.

Sintered gears can be held to close limits of dimensional accuracy with a good surface texture and they require little or no subsequent machining. Sintered gears can be impregnated with a plastic to impart additional wear resistance or with another metal of lower melting point than the sintered material, to give them increased strength.

An example of a suitable powder for producing gears by P/M technique can have the following composition :

Nickel	1.75%
Molybdenum	0.50%
Copper	1.50%
Carbon	0.50%
Iron	Remainder

Accurate gear pump rotors with wear and corrosion resistant surfaces can be made from stainless steels by this method, since stainless steels are very tough to be cut easily by conventional means. Gears made by P/M method find applications in : toys, instruments, appliances, small motor drivers, lawn and garden tractor transmission gears etc.

4. **Hot forging of gears.** In this method, the whole gear is forged from a billet, usually of case hardening steel, in a pair of accurate dies split on the axial plane. It is necessary, therefore, to control the size of the fully machined billet to a fairly close tolerance to reduce the forging flash to a minimum. The dies are produced by a copying technique from a master gear. Dies are expensive to manufacture but the forging process is rapid and is fit for production. This method followed by a cold coining operation is attractive for bevel gear manufacture and is also applicable to spur-gear manufacture. A good grain flow pattern in the teeth, an accuracy in pitch and profile within 0.012 mm and a high surface finish can be obtained by this method.

7.13. PRACTICAL ASPECTS OF GEAR MANUFACTURING

Here, we will discuss the various steps involved in the cutting of gears by different machining methods.

1. Spur Gears : Gear cutting by milling has been dealt with under Art. 7.4. Fig. 7.1 (*a*) shows the milling of spur gears on a horizontal milling machine with the help of a flat circular disc type cutter and Fig. 7.1 (*b*) shows the gear milling on a vertical milling machine by making use of an end-cutter or miller.

(*a*) **Spur gear cutting on Horizontal Milling Machine :** The following steps are involved in cutting spur gears on Horizontal Milling Machine with involute disc gear cutter :

(*i*) For every module, a set of involute disc gear cutters are available as given below :

Cutter Number	No. of gear teeth (z) for which cutter is used
8	12–13
7	14–16
6	17–20
5	21–25
4	26–34
3	35–54
2	55–134
1	135 and above

From module, the number of teeth of the gear is determined as below :

$$\text{Outer diameter} = \text{P.C.D.} + 2x \text{ addendum}$$

$$\therefore \quad D_o = m.z. + 2m = m\,(z + 2)$$

$$\therefore \quad z = D_o/m - 2$$

∴ For the given outside diameter of the gear blank, D_o, the number of gear teeth, z, is determined for the desired module.

(*i*) For z number of teeth, the gear cutter of suitable number is selected from the above table.

(*ii*) The cutter is mounted on the arbor of the horizontal milling machine.

(*iii*) The gear blank is mounted on the mandrel or between centres of a dividing head fixed on the machine table, and made true.

(*iv*) By trial and error, the cutter is brought just over the gear blank which is checked by equal cutting of the blank on both the sides.

(*v*) The table is now moved back to clear the blank. It is then moved vertically up (by elevating screw) by 2.25 m (addendum + dedendum). The longitudinal feed of the table is engaged for automatic cutting.

(*vi*) When the cutting of one tooth space is completed, the table is moved back.

(*vii*) The gear blank is indexed $\frac{40}{z}$ rotation by means of the dividing head, using simple indexing procedure.

(*viii*) Following the above steps, the complete gear in milled.

Gears are cut on milling machines only when required in short quantities. We know tht the involute curve or form of gear tooth depends upon the gear diameter or the module and number of teeth of the gear (as noted above). Therefore, a different cutter is needed if the number of gear teeth changes. This requires the inventory of a very large number of cutters, which is not economical. Therefore, the whole range of number of teeth is divided into 8 groups (as in Table above) and there is one cutter for each group. This cutter has the true profile for the lowest number of teeth in the group. For example, cutter number 3 (Table above) is meant to cut gears with 35 to 54 number of teeth. This cutter will cut true profile only for gears with 35 number of teeth. For others, it would be approximate. This is the biggest drawback of cutting the gears on milling machines. In order to have greater accuracy for gears with number of teeth towards the higher end of the range, a half number cutter is provided for each group. For example. a cutter number 3½ is there and it would give higher accuracy for number of teeth ranging from 42 to 54.

(*b*) **Spur gear cutting on a Gear Shaper :** See Art 7.9.

As explained, the principal motions involved in the gear shaping are as follows :

(*i*) Reciprocating and Rotary motion of the cutter.

(*ii*) Rotary motion of the work spindle on which the gear blank is mounted.

(*iii*) Infeed cutter motion into the blank by cam rotation, to get depth of cut.

(*iv*) Gear blank relief movement during return stroke of the cutter (to avoid their rubbing with each other).

(*v*) Indexing motion : Slow speed continuous rotation of the cutter spindle and work spindle to provide circular feed, the two speeds being regulated through the change gears such that against each rotation of the cutter, the blank rotates through z_c/z rotation, where,

z = No. of teeth to be cut or the gear blank

z_c = No. of teeth on the cutter.

The cutter head moves towards right (Fig. 7.6) to achieve radial infeed, with the help of infeed cam. At least 2 or 3 passes are needed to obtain the required depth of cut. The indexing and reciprocating motions (of spindle) continue until the required number of teeth are cut all along the periphery of the gear blank.

On completion of cycle, the cutter is retracted back completely, the cut gear in unloaded and a new gear blank mounted on the work spindle.

(*c*) **Spur gear cutting on a Hobbing machine :** See Art. 7.10.

It involves the following steps :

(*i*) The gear hob of the desired module is mounted on the machine spindle.

(*ii*) The angle of hob, as punched or written on the hob is noted, say α_h^0 (right). The cutter head with the hob is rotated by α_h^0 to the left.

(*iii*) The gear blank is mounted on the table and centred.

(*iv*) The table and hob speed ratio is adjusted so that for every single rotation of the hob, the table turns by n/z rotation, where n is the number of starts of the hob cutter.

Note : The hobs are commonly 75 to 1500 mm ϕ. Two and three start hobs are also made. They cut faster but are less accurate.

(*v*) Suitable feed and speed are selected.

(*vi*) The hob is advanced to the blank, so that it just starts cutting.

(*vii*) The tool head is now moved up, above the blank so as to clear it.

It is them moved towards the gear blank by (2.25 m). For larger module, several cuts may be required.

(*viii*) The machine is now started on automatic feed and gear cut. See Fig. 7.9 *a* (*ii*).

(*ix*) Several gears can be cut at a time is one setting.

2. Helical Gears

(*a*) **Gear cutting on a Hobbing Machine :** The method differs from that for spur gear cutting, in the following two aspects :

(*i*) The tool head alongwith the hob is turned by $\beta \pm \alpha_h$ where β is the helix angle of the helical gear. Plus sign is to be used, if the hands of the helical gear and hob are opposite (one R.H. and the L.H.) and if the hands are the same, then, minus sign is to be used.

(*ii*) Table speed, Rev/min $= \frac{n}{z}$ (Hob speed, Rev./min.) $\pm \frac{\text{Tan}\ \beta}{\pi . m . z} \times$ vertical feed of the hod in mm/min.

If the table speed comes out to be equal to (whole numbers + a fraction), the extra rotation is achieved with the help of differential gearing.

(*b*) **Gear cutting on Gear Shaper :** See Art. 7.9.

(*c*) **Helical gear cutting on Universal Horizontal Milling Machine :** To facilitate the cutter to cut helical grooves, the workpiece must be given two simultaneous motions normal to each other : the linear motion along its axis and a rotary motion. The workpiece is held in the dividing hear (mounted on the machine table) which is connected by means of gears to the table feed screw of the machine. When the table feed screw rotates, the spindle of the dividing head and hence the work will rotate through change gears, worm and worm wheel of the dividing head. Also, when the table feed screw rotates in the nut, the machine table and hence the dividing head has linear motion. The milling cutter is mounted on the arbor of the machine. To cut the helix, it is necessary to swing the machine table through an angle equal to the helix angle, on the vertical axis. Due to this, the spiral grooves are cut on a universal milling machine, Fig. 7.13.

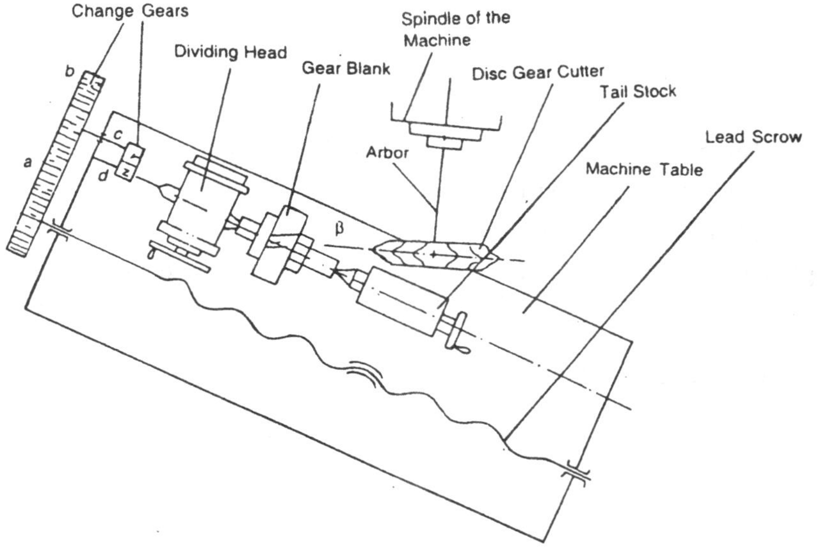

Fig. 7.13. Milling of spiral grooves.

The helix angle of the groove on the job is given as.

$$\tan \beta = \frac{\pi d}{L}$$

where d is the diameter of the job and L is the lead of the helix. The helix angle is the angle between the tangent to the thread helix on the pitch cylinder and the axis of the job.

Supposing the gear ratio between the table feed screw and the worm shaft (of the dividing head) is 1 : 1. Then, one revolution of table feed screw will rotate the worm shaft through one revolution. The dividing head spindle and hence the work will rotate through $\frac{1}{40}$ of a revolution. That is, for the work to complete one revolution, the worm shaft and hence the table feed screw will make 40 revolutions. If "p" is the pitch of the table feed screw, then, the distance travelled by the table feed screw for one revolution of the work will be $40p$. This is called the "Lead of the machine". Hence, the lead of the machine is defined as the distance travelled by the machine table corresponding to one complete revolution of the dividing head spindle (and hence of the work) when the gear ratio between the table feed screw and worm shaft is 1 : 1. Thus,

$$\text{Lead helix will be} = \text{Lead of machine} = 40p.$$

The diameter of the job and lead of the helix to be cut will determine the setting of the machine and the gear ratio. The hand of the helix will determine whether an idler is needed or not in the gear train. The hand of the helix will determine whether an idler is needed or not in the gear train. The work must rotate left handed for left hand helix and right handed for right hand helix. Thus, for right handed helix, the table feed screw and the worm shaft should rotate in the same direction and for left handed helix, they should rotate in the opposite directions.

$$\text{Gear ratio } \frac{a}{b} \times \frac{c}{d} = \frac{\text{Drivers}}{\text{Driven}}$$

$$= \frac{\text{Lead of the machine}}{\text{Lead of the helix to be cut}}$$

$$= \frac{40\,p}{L} \times n$$

$$= \frac{40\,pn}{L} \times \beta = \frac{40\,pn \tan \beta}{\pi . z . m}$$

where n = Number of starts of the milling table lead screw

z = Number of teeth on helical gear

m = Module of the teeth

Example : *A right hand helix is to be cut on a job with diameter 40 mm. Lead of helix to be cut is 450 mm, and the pitch of the table screw is 6 mm. Determine is the setting of the machine and the change gears.*

Solution. Helix angle of the groove to be cut,

$$\beta = \tan^{-1} \frac{\pi d}{L} = \tan^{-1} \frac{\pi \times 40}{450}$$

$$= \tan^{-1} 0.2792$$

$$= 15.6°$$

Gear ratio,

$$i = \frac{\text{Drivers}}{\text{Driven}} = \frac{\text{Lead of the machine}}{\text{Lead of the helix}}$$

$$= \frac{40p}{450} = \frac{40 \times 6}{450}$$

$$= \frac{8}{15} = \frac{2 \times 4}{5 \times 3}$$

$$= \frac{40 \times 32}{100 \times 24}$$

It is a compound gear drive.

Use a gear of 40 teeth on table feed screw, gear of 24 teeth on worm, 100 teeth gear on Fist gear on stud and Second gear of 32 teeth on stud, and no idler gear.

After mounting the gears, the job and the cutter, the machine feed table should be swung at an angle of 15.6°.

The following steps are involved in cutting helical gears on Universal Horizontal Milling Machine :

(*i*) The disc gear cutter number corresponding to virtual number of teeth ($zv = z/\cos^3 \alpha$: α pressure angle) and of given module is selected from the table given under Art. 7.13. It is then mounted on the arbor of the milling machine.

(*ii*) The gear blank is mounted on the mandrel or between the centres of the dividing head fixed on the machine table. The motion of the dividing head and of the machine lead screw are connected through suitable changeable gears (a, b, c, d) as shown in Fig. 7.13, so that for one rotation of the gear blank, the table advances by an amount equal to the pitch of the helix.

(*iii*) The machine table or the spindle head of the universal machine is rotated through angle β of the gear helix, as shown in Fig. 7.13.

(*iv*) The remaining steps are exactly similar to the ones for milling spur gears.

2a. Herringbone Gears/Double Helical Gears : are two helical gears, having the same helix angle but of opposite hand.

Milling each gear is done either on a special machine or more commonly on a gear hob/ universal horizontal milling machine. The two gears are separately turned and milled, and then assembled together.

For the purpose of proper assembly, one section of the gear is turned with a projection the other section with the corresponding locating surface. Both the sections are then locked with locking screws, Fig. 7.14.

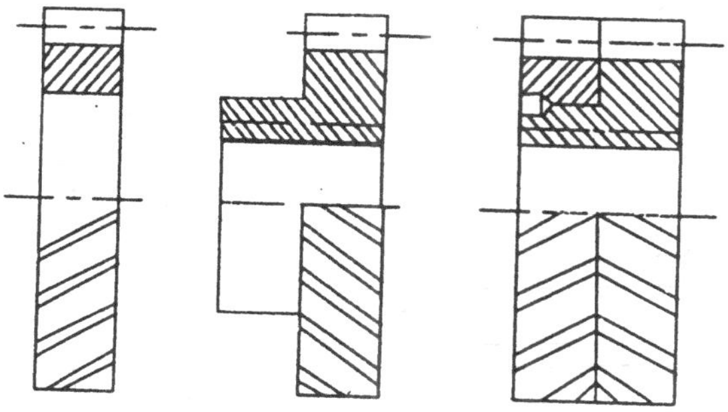

(*a*) (*b*)

Fig. 7.14. Herringbone Gears

This was the original method of manufacturing herringbone or double helical gears. But nowadays, herring bone gears are produced as a single unit on special machines which cut the teeth in two directions at one time.

3. Manufacturing of Bevel Gears : See Art. 7.11.

4. Worm and Worm - Wheel Cutting :

4.1 Worm Cutting : The worm can be cut on a thread milling machine, or more frequently cut on a lathe. The method involves the following steps :

(*i*) After the blank has been turned, the machine is set for the axial thread pitch of the worm ($p = \pi.m. n$; m = module, n = number of starts of the worm).

(*ii*) The tool is ground to the required profile and the worm cut with the usual thread operation for the given number of starts. After the first thread has been completed, second thread is cut by advancing the compound slide by π. m. For more number of starts, the compound slide is advanced by $\pi. m.$ after the completion of each thread.

(*iii*) During the process, the thread thickness at a certain depth is checked to see whether the process in under control or not.

4.2. Worm-Wheel Cutting : On gear Hobbing Machine, Fig. 7.15.

The steps are :

(*i*) The hob cutter should have the same number of starts as that of the worm and diameter equal to or slightly greater (upto 5 mm) than that of the worm. This cutter is mounted on the machine.

(*ii*) If the angle of the hob is αh (R.H.) and helix angle is β (L.H.), then the tool head must be rotated to the left by an angle = $\beta + \alpha h$, and so on.

(*iii*) The hob cutter and table speeds are adjusted such that,

$$\frac{\text{Hob speed}}{\text{Table speed}} \text{ (Rev./min.)} = \frac{n}{z}$$

(*iv*) The hob cutter is adjusted centrally to the gear blank (by trial and error) so that it cuts equally both the edges.

(*v*) The hob is now advanced till it just starts cutting in the centre. From this position, the hob is advanced by (2.2 m) in steps. At every one complete rotation of the table, the cutter is fed till the required depth of cut is obtained.

(*vi*) During the process, the axial distance 'A' should be checked, which serves as a control for the proper cutting.

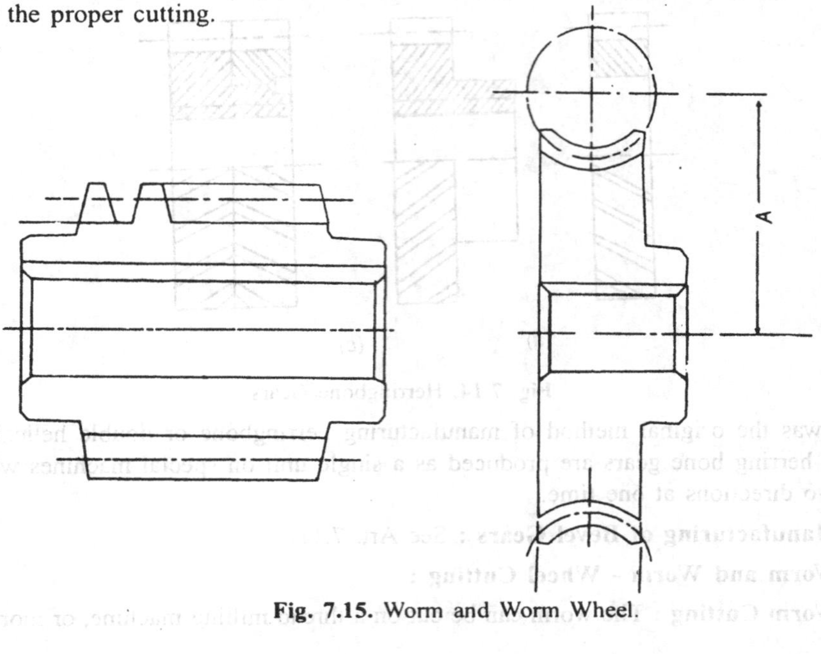

Fig. 7.15. Worm and Worm Wheel.

7.14. GEAR FINISHING OPERATIONS

For the gears to operate efficiently and have satisfactory life, the tooth profile must be very accurate and the teeth should be hard and smooth. Gear made by cold rolling method may not need further finishing, but, gear made by other methods (particularly made by metal removal) often needs further finishing operations. The finishing operations can be : shaving and burnishing for untreated gears and grinding, lapping and honing for heat treated gears (hardened).

7.14.1. Gear Shaving. Gear shaving is the most common method for gear finishing. In this method, a very hard gear (gear-shaving cutter) is used to remove fine chips from the gear-tooth profile. The shaving cutter can be : Rotary type or rack type. In rotary shaving, the cutter and the gear are run in mesh. As they rotate, the gear is traversed longitudinally across the shaving cutter or vice versa. The rotary shaving cutter, Fig. 7.16 has a number of peripheral gashes, serrations or grooves, to form a series of cutting edges. The cutter and gear are set up in gear shaving machine with crossed axes in the form of spiral gearing. The usual angles are 10° to 15°. In rack shaving, the cutter is in the form of a rack. During the operation, the gear is rolled in mesh with the cutter. The cutter is reciprocated and at the end of each stroke is fed into the gear.

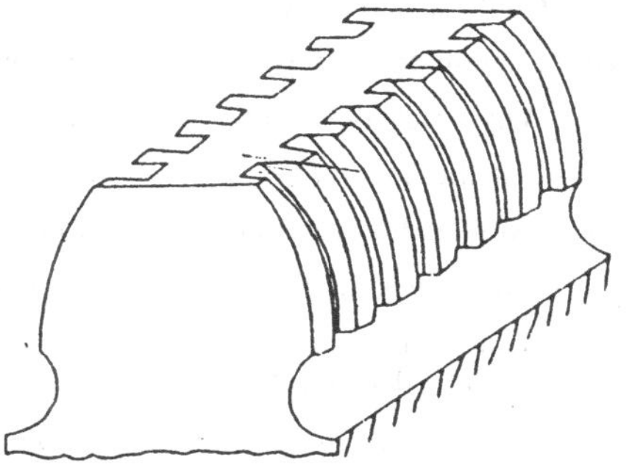

Fig. 7.16. Gear Shaving Cutter.

7.14.2. Gear burnishing. This is not a common method of gear finishing. In this method, the gear is rolled under pressure with three hardened, accurately formed, burnishing gears. This is a cold working process in which any high points on the tooth surface are plastically deformed to get accurate and finished tooth profile. However, the operation may have a smeared metal surface.

7.14.3. Gear grinding. Grindling is the most accurate method of gear finishing. The operation can be either by : form grinding or generation grinding. In form grinding, the wheel is dressed to the form of an involute and is similar to the formed cutter used on a milling machine. The cutter is plunged into a tooth space of the gear. The gear is completely finished by indexing after each tooth has been finished by the rotary cutter (disc grinding wheel).

In the generating type, the gear is rolled past the revolving wheel and the finishing is done by the flat face of the wheel. The wheel is wide enough to cover the entire tooth width, so that it does not have to be traversed parallel to the axis of the gear. The drawbacks of the method are : low production capacity and the grinding wheels are expensive and need skilled labour for operation.

7.14.4. Lapping. The lapping process only corrects minute heat treatment distortion errors in hardened gears. The gear to be finished is run in mesh with a gear shaped lapping tool or another mating gear (cast-iron). An abrasive lapping compound is used in between them. Only the tooth contact is improved by this method.

7.14.5. Honing. Honing is employed mainly to remove burrs and nicks and thus for improving the surface finish. In this method, the honing tool (made of a plastic material with abrasive material embedded in it) is run in mesh at crossed axes, with the gear to be finished. The gear is driven at high speeds by the tool and is also reciprocated across the tool. During the cycle, the gear is run in both directions.

PROBLEMS

1. List the desirable characteristics of a gear.
2. List the factors on which the selection of material for gear depends.
3. List the various gear materials and give the field of application of each.

4. Give the advantages and limitations of casting for the manufacture of gears.
5. Differentiate between forming and generating.
6. Explain the "Roll forming" method of gear manufacture. Give the advantages of roll forming.
7. Explain the method of cutting gears by milling.
8. Why are not more gears made by broaching ?
9. Could a helical gear be cut on a plain milling machine ? Why ?
10. Explain the process of gear shaping.
11. Why "rack-type cutter" is not used so commonly in gear shaper ?
12. List the advantages and limitations of gear shaping.
13. Explain the principle of gear hobbing.
14. Why a gear hobbing machine is more productive than a gear shaper ?
15. What is the relation between the axes of a hob and the gear blank for cutting (*a*) spur gears, (*b*) helical gears ?
16. Give the advantages and limitations of gear hobbing.
17. Write about the following methods of gear manufacture : extrusion, cold drawing, stamping and power and metallurgy.
18. Why gear finishing is required ?
19. Discuss the various gear finishing operations.

8

NUMERICALLY CONTROLLED MACHINE TOOLS

8.1. GENERAL

In the modern age of business competition, the aim of manufacturing activity is to manufacture the parts of desired quality at the minimum possible cost resulting in profit to the organisation. To compete in the market, the quality of the product will have to be reliable and consistent with delivery of the product offered on schedule.

For consistency in quality, especially in the manufacture of complex components, human intervention has to be eliminated. This necessitates the use of special purpose or automatic machine tools. These machine tools are highly specialised and have high rates of production. However, their initial costs are high. Due to these factors, these machine tools are suitable and economical for mass production only. Moreover, these machine tools are inflexible (that is, the complete set up will have to be changed for a new product). Due to this inflexibility also, these machine tools are not suitable for job or small batch production. Thus, when the main requirement is : quantity, consistency in quality and delivery schedule, mass production or automation is the answer. Mass production machine tools include : automatic unit machine tools and automatic transfer machines etc.

However, the mass production system of manufacturing accounts for only about 20% of the total of manufactured parts.

The remaining demand is met by jobbing and the batch production systems. More than half of the total machine tools are engaged in small lot and piece production. These include general purpose machine tools, hydraulic tracer controlled machine tools and programmed operation cycle machine tools. The general purpose machine tools, eventhough highly flexible (can easily change from one product to another), are not suitable for mass production, because of longer set up times, machine and tool adjustments, and very low machine utilization time (maximum upto about 30% of the total time). All this results in low rates of production and delays in the delivery schedule. Also, they require too highly skilled operators. Because of the intervention of operators, (eventhough skilled) the consistency in quality cannot be assured. Therefore, when consistency in quality and delivery schedule are of prime importance, these machine tools are not suitable. Similarly, hydraulic tracer controlled machine tools etc. (which are mainly used for small batch productions) require longer set up times while changing over to new jobs as these machines require cams, templates, stops, electrical trip dogs etc.

Because of the above factors, a great need was felt for machine tools that could bridge the gap between highly flexible general purpose machine tools (which are not economical for mass production) and highly specialized, but inflexible mass production machines. Numerically controlled machine tools have taken up this role very well. These machines are highly flexible and are economical for producing a single or a large number of parts.

Numerical control, NC, can be defined simply as control by numbers. Electronics Industries Association defines NUMERICAL CONTROL as "a system in which actions are controlled by the direct insertion of numerical data at some point. The system must automatically interpret at least some protion of these data." In numerically controlled machine tools, the input information for controlling the machine tool motion is provided by means of punched paper tapes, plastic tapes(Mylar), floppy disk, hard disk or magnetic tapes in a coded language. Thus, with numerical control, the operation and motions of a machine tool are controlled electronically. NC machine tools are thus automated production machines.

The key difference in this system is that of "reprogrammability". In NC, a changeover to a new product, does not require extensive physical changes in the machine set up. Only a change in the control program itself is required. Thus, the flexibility of general purpose machine tools can be combined with the precision and accuracy of special-purpose machine tools. The human intervention, as an interface translating the design information into machine activities, is replaced by some form of information processing device, such as a computer.

The principles of NC were established in 1950s. However, with the availability of low-cost programmable control, based around the microprocessor in 1970s, this technology diffused widely. This opened up the development of the concept of Computer Numerical Control (CNC), and other functions such as automatic tool change or part manipulation. All this led to highly sophisticated, multipurpose machining centres, complete with a range of support functions such as tool change, head change, transport and manipulation and so on–all under computer control. The next development was the idea of Direct (or distributed) Numerical Control (DNC), in which more than one machine tool (alongwith associated functions) could be grouped into a manufacturing cell under the overall control of a larger, supervisory computer. The early numerically controlled machine tools were milling and profiling machines that provided faster and less expensive means for producing aircraft parts. Now, NC, has been applied to other machine tools, such as lathes, drilling and boring machines, welding and flame cutting machines, punching machines and inspection devices

8.2. WORKING OF NC MACHINE TOOL

To understand the working of a NC machine tool, let us refer to Fig. 8.1. The first two

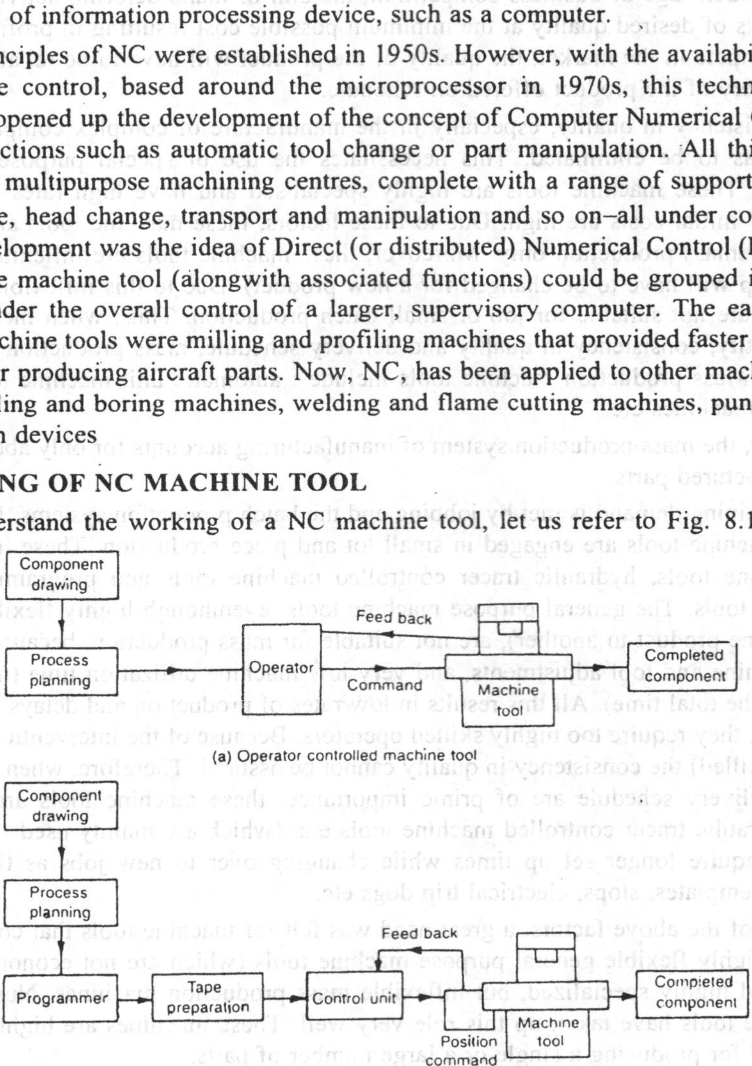

Fig. 8.1

steps, component drawing and process planning are similar in both operator and NC machine tools. In the operator controlled machine tools, the operator controls the cutter position during machining. He also makes the necessary adjustments and corrections to produce the desired component. However, in numerically controlled system, the place of the operator is taken by the data processing part of the system and the control unit. In the data processing unit, the co-ordinate information regarding the component is recorded on a tape by means of a teleprinter. Tape is fed to the control unit which sends the position command signals to slide-way transmission elements of the machine tool. At the same time, the command signal is constantly compared with the actual position achieved, with the help of position feedback signal derived from automatic monitoring of the machine tool slide position. The difference in the two signals, if any, is corrected until the desired component is produced.

The conventional NC system consists of basically two parts : Machine Tool and Machine Control Unit (MCU).

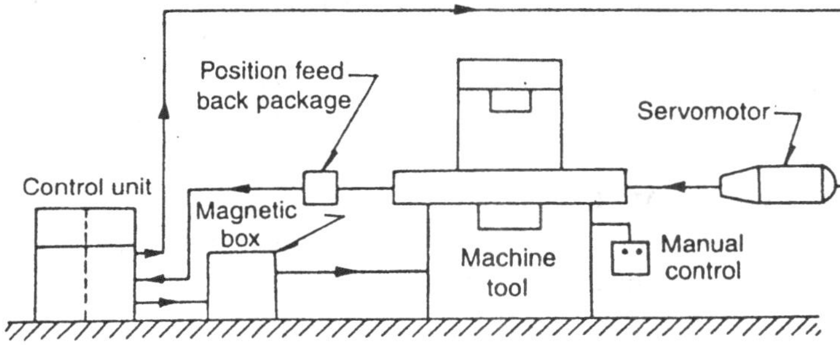

Fig. 8.2. Main Elements of a NC Machine.

The main elements of a NC machine tool are shown in Fig. 8.2. These are :

1. The machine control unit (MCU), also known as NC Console or Director.
2. The drive units, or actuators.
3. The position feedback package.
4. Magnetic box.
5. Manual control.

The hardware also includes the associated circuits.

The functional elements involved in MCU are :

1. Data Input. The instructions for manufacturing the component are written in a Coded language (Paper tape is the most commonly used device) are read by a tape reader.

2. Data Processing. The instructions undergo electronic processing resulting in information in the form of electrical signals (pulsed commands).

3. Data Output. The control unit sends command signals to the drive units of machine tool and also to the Electrical control cabinet called "Magnetic Box". Command signals sent to the drive units of the machine tool, control the lengths of travel and the feed rates, while the command signals sent to the magnetic box control other functions such as : spindle motor starting and stopping, selecting spindle speeds, actuation of tool change, coolant supply etc.

It is clear from above that MCU basically consists of two elements : Data Processing Unit (DPU) and the Control Loops Units (CLU).

The DPU processes the data (in coded form) read from the tape or anyother source and passes the information regarding the various controls to CLU. NC is used to provide the following controls:

- all cutting speeds
- complete path and feed rates of a cutter in relation to the workpiece or fixture
- all auxiliary functions such as turning cutting fluids off and on.

The CLU operates the drive mechanisms of the machine tool, receives feedback signals regarding the actual position and velocity of each of the axes and signals the completion of the operation.

The DPU reads and processes the data sequentially. When a line has completed execution as signalled by CLU, the DPU reads and processes the next line of the programme and so on.

To check whether the required lengths of travel have been obtained, a feedback transducer is provided. The feedback transducer sends the information of the actual position achieved to the control unit. If there is any difference between the input command signal and the actua position achieved, the drive unit is actuated by suitable amplifier from the error signal.

Manual control or operator control helps the operator to perform some functions manually such as : motor start-stop, coolant supply control, axes movements, speed change, feed change etc.

8.3. CLASSIFICATION OF NC MACHINES

NC machine tools can be classified in different ways. Based on the type of power drive or the actuation system used, there are three common NC systems :

1. Electro-mechanical
2. Hydraulic, and
3. Pneumatic

The range of the machine's performance capabilities and its applications will normally depend upon the type of the power source.

Hydraulic power provides the largest power. The movement of the machine slides is more uniform in speed. However, the drawbacks are : Higher cost, noise, hydraulic contamination from leaking fluid and the need for additional equipment such as reservoir, intensifier, valves etc.

Pneumatic power is least expensive, because the shop air can be tapped for this purpose. However, the motion of the worktable is not uniform.

Electro-mechanical drives are the most precise and the most commonly used power sources.

The group of electro-mechanical, hydraulic or pneumatic equipment used to control the motion of an NC machine is called the "Servo".

Based on control system features, NC machines are classified as :

1. Point-to-Point system
2. Straight-line system
3. Contour system

These three systems are illustrated in Fig. 8.3.

In point-to-point system, the machining is done at specific positions. The working-piece remains unaffected as the tool moves from one position to the next. This system is the simplest. In Fig. 8.3 (*a*), after drilling the hole at position A, the tool moves to position B, along the dotted line. A drilling machine is the best example of point-to-point system.

Straight-line system, Fig. 8.3 (*b*) is an extension of point-to-point system. Here the tool moves at a controlled feed rate in one axis direction at a time. The examples of straight-line system are: stepped turning on lathe, pocket milling etc.

In contour or continuous path system, Fig. 8.3 (*c*), there are continuous, simultaneous and co-ordinated motions of the tool and the workpiece along different co-ordinate axes. The various profiles, contours and curved surfaces are machined by this system.

Depending on the feedback, NC machine systems are of two types :

1. Open loop system
2. Closed loop system.

As discussed above, a command signal is sent to the machine tool to carry out a certain operation. In the open loop system, there is no 'feedback' and no return signal to indicate whether the tool has reached the correct position at the end of the operation, or not. Hence, there are no means of knowing whether there is an error or not between the input command signal and the result achieved. An example of open loop system is a co-ordinate drilling machine.

Fig. 8.4 (*a*) illustrates an "**open loop system**". The command signal from the MCU is given to the servo-motor (here a stepper motor). The motor is driven a precise angular rotation for every pulse issued by CLU. So, the response of the motor is in incremental steps. An incremental step of 1.8° is common. This will result in a corresponding linear movement of the lead screw (depending upon its lead) and hence of the machine slide.

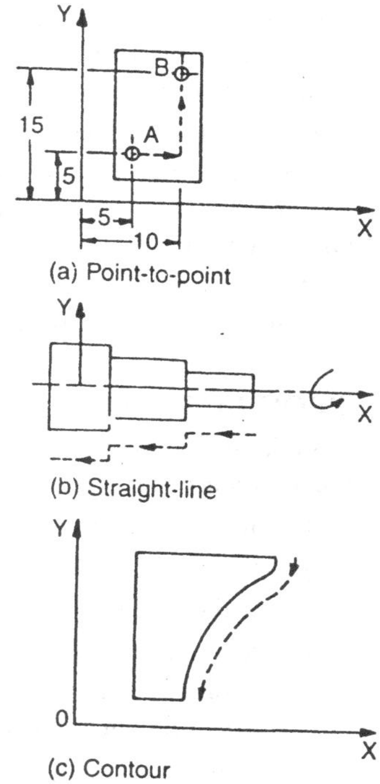

Fig. 8.3. NC Control System.

In **closed loop system**, a 'feedback' is built into the system, which automatically monitors the position of the tool. From the feedback signal to the control unit, it is known whether the tool has moved correctly to its new position or not. If not, its position is automatically controlled until it is in the right position. Such a device is known as a position feedback transducer. Common position feedback devices used on modern NC machines are : shaft encoders, linear scales or

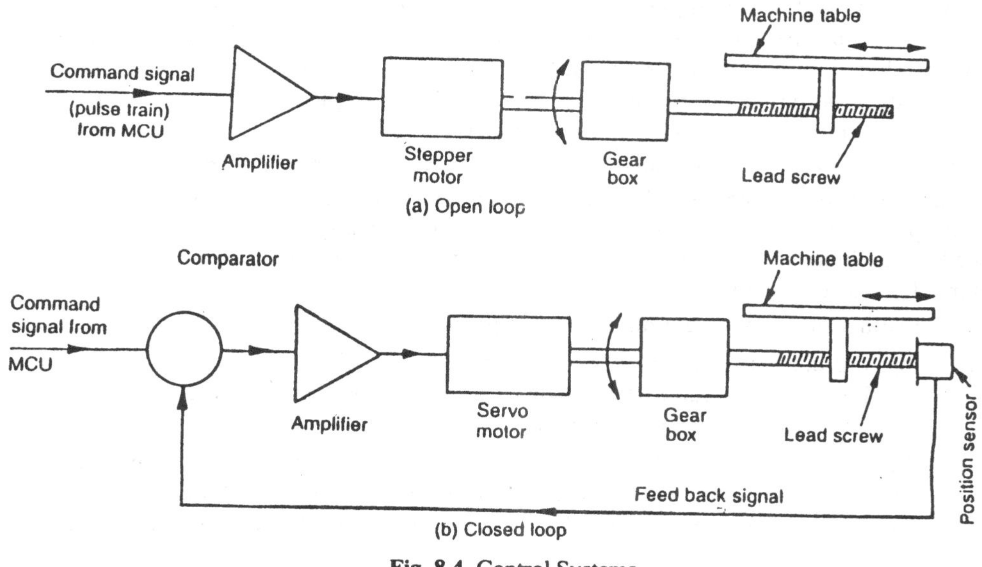

Fig. 8.4. Control Systems.

Inductosyns. In the contour or continuous path system, velocity control is also essential to ensure that the cutter path is as required by the profile. In this system, a velocity feedback is provided by a transducer known as *tacho-generator* attached to the feed drive system.

A closed loop system is more expensive than an open loop system.

A block diagram of a closed-loop system is shown in Fig. 8.4 (*b*). Since, there is a continuous monitoring of the positions, the drive is a continuous-position device. These servo-motors provide a smoother and continuously controllable movement of the work-table.

AC servo-motors have better reliability, better performance-to-weight ratio and lower power consumption. So, these are rapidly replacing the DC servomotor on new NC machine tools.

8.4. PROGRAMMING FOR NC MACHINES

As discussed above, the complete information for producing a component on an NC machine, is punched on a paper tape. The information punched on the paper tape includes :

1. the co-ordinate values of the entire tool path.
2. the co-ordinate values are prefixed with certain codes indicating the type of movement of the tool (point-to-point, straight-line, contour) from one co-ordinate to the next.
3. the co-ordinate values are also suffixed with certain codes indicating the various machine functions, such as, start/stop, spindle coolant etc. The co-ordinate values are also supplemented with other functions such as feed rates, spindle speeds, etc.

Programming to obtain the punched paper tape can either be done manually or with the help of a computer. Simple point-to-point programs can be easily developed manually, but more complex ones, as well as almost all contouring programs are developed with the help of computers.

Before making the part programme, the programmer first studies the part drawing and decides upon : the proper sequence of operations, the cutting tools, the path of the cutter/tool, speeds and feeds at various points and the other related information, such as starting and stopping of the machine etc.

8.4.1. Manual Programming. The first step is to establish the zero reference axes on the part drawing and determine the co-ordinate dimensions for each operation. Also, the set up instruments to establish the workpiece properly on the machine table with respect to the tool, are established. Then, all the data and the instructions are entered in a 'program sheet' in a particular format acceptable to the machine tool-control unit combination. This sheet includes the following information : the co-ordinate dimensions for each operation, the spindle traverse that determines the depth of cut, the spindle speed and feed, tool change. After preparing the 'programme sheet', the programmer uses it to prepare the punched paper tape on a typewriter-like tape punching machine.

8.4.2. Computer Programming. To develop the program with the help of a computer, specialized programming languages, such as APT (Automatically Programmed Tool) have been developed to facilitate the production of punched paper tapes with relatively little knowledge of computers. The various steps in computer programming are :

1. Writing the 'part programme manuscript' using the simplified programming language.
2. Copying the part programme manuscript onto a deck of punched cards with the help of a keypunch operator. This is known as the 'part programme'. On each card, a specific machine-tool instruction is punched.
3. The 'part programme' alongwith 'computer programme' and 'post-processor programme' are then fed to a general-purpose computer. The 'computer programme' is a previously prepared deck of punched cards, or tape, that contain instructions for

the computer for executing the part-programme instructions, and also those from 'post-processor programme'. The computer translates the 'part programme' into numbers and computing instructions and performs the calculations. The output of the computer, which is translated with the aid of 'post processor programme' into a form acceptable to the particular machine tool, may be in the form of a magnetic tape or a punched paper tape. If the output is in the form of a magnetic tape, it is converted into punched tape with the help of conversion equipment. The punched tape is now ready to be fed into machine tool control. The complete scheme of computer programming is shown in Fig. 8.5.

APT, the first programming language used for NC machine tools, was developed in 1950.

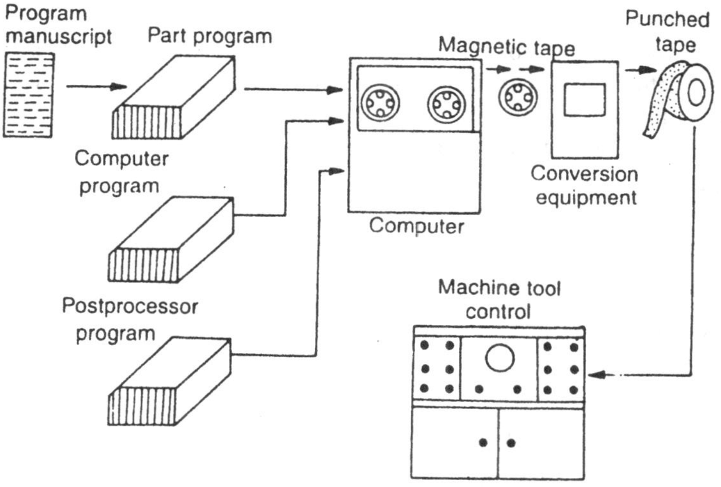

Fig. 8.5. Computer Programming.

It used English type statements. After that, many subsets of APT and new languages have been developed. In APT-II, the complete job of part program preparation from the part drawing was undertaken by the computer.

Then came the APT Long Range Program (ALRP). In 1965, ALRP was changed to Computer-Aided Manufacturing-I (CAM-I).

The subsets of APT and its expanded forms and other new languages are given below :

ADAPT (Adaptation of APT) : for smaller components.

UNIAPT : handles full power of APT on a smaller computer. It differs from APT only in its internal design of processor.

NELAPT : developed by National Engineering Laboratory in U.K.

This language contains advanced features not available in APT. EXAPT (Extended APT) : was developed in Germany. It is both geometrically and technologically oriented. Its various versions are available :

(*i*) EXAPT 1 : for positioning machine tools, *e.g.*, drilling machines,

(*ii*) EXAPT 2 : for turning centres.

(*iii*) EXAPT 3 : for milling operations.

AUTOSPOT (Automatic System for Positioning Tools) : was developed by IBM in 1962 for three axes, point-to-point motion control.

COMPACT : This language was developed for simultaneous servicing of multiple users from a remote computer over telephone lines.

COMPACT - II : This is the latest version of COMPACT. This is available to the users on a time-sharing basis.

SPLIT (Sunstrand Processing Language Internally Translated) : This language can handle upto 5-axis positioning and possesses contouring capability.

8.5. METHODS OF LISTING THE CO-ORDINATES OF POINTS IN NC SYSTEM

There are two methods of mentioning the co-ordinates of points while programming for an NC system :

1. Absolute System.
2. Incremental System.

In the absolute system, the co-ordinates of points are mentioned with respect to one reference point, that is, datum. In the incremental system, the co-ordinates are written in the programme with respect to the previous point. Referring to Fig. 8.3 (*a*), the co-ordinates of points *A* and *B*, in the two systems, are given as below :

Absolute system	*X–Co-ordinate*	*Y–Co-ordinate*
Point A	5	5
Point B	10	15
Incremental System		
Point A	5	5
Point B	5	10

8.6. APPLICATIONS OF NC MACHINES

The major applications of NC machines are :

1. For parts which are complex and it will not be possible to manufacture them very accurately on conventional machines, due to human error involved.
2. For parts which are frequently subjected to design changes.
3. Repetitive and precision quality parts which are to be produced in low to medium batch quantity.
4. In situations where the investment on tooling and fixture inventory will be high if the parts are made on conventional machine tools.
5. To cut down 'lead time' in manufacture.

8.7. ADVANTAGES AND DISADVANTAGES OF NC MACHINES

Advantages :

1. Greater accuracy.
2. Lesser production cost per piece due to reduction in lead time and also set up time.
3. Improved product quality and provision of high order of repeatability.
4. High production rates as the machining conditions (feeds and speeds) are optimised and the non-machining times are reduced to a minimum.
5. Less scrap due to consistent accuracy and absence of operator errors.
6. Reduced inventory of 'parts in process' because parts can be made economically in

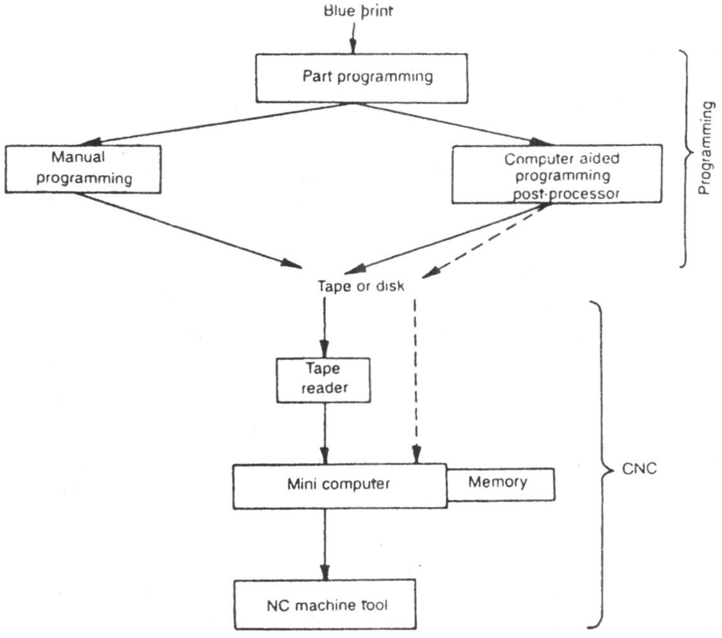

Fig. 8.6.

smaller quantities.

7. Less operator skill is required to run NC machine.
8. Due to reduced idle time, the machine utilization is better.
9. Changes in part design can be incorporated very easily and at a low cost by simply changing parts of tape program.
10. Excellent reliability as the control equipment now is virtually all made of solid state modulus.
11. Lower tooling costs as expensive jigs and fixtures are not required.
12. Reduced cycle time and increased tool life.

The major *disadvantage* of NC machines are their costs. Therefore, the machines must have sufficient use to justify the investment. Programming of NC machines has been greatly simplified and machine availability above 95% is common.

8.8. COMPUTER NUMERICAL CONTROL (CNC) AND DIRECT NUMERICAL CONTROL (DNC)

Computer has been associated with NC right from the beginning, and over the years this association has resulted in many new conceptions like : Computer Numerical Control (CNC). Direct Numerical Control (DNC), Computer Aided Design (CAD), Computer Aided Manufacture (CAM), adaptive Control (AC), and Integrated Manufacturing System.

It is clear from the above discussion that in conventional NC systems, the instructions (program) stored on a paper tape (in a coded language) are read by tape reader. The instructions are processed by DPU and the command signals are controlled by CLU. There is no provision for storage or memory in MCU. For every new component, the tape needs to be loaded again, since the program is not stored in MCU and so can not be called back. The entire data input and data handling sequence including control functions are determined only by the fixed circuit interconnections of DPU and CLU. Any modifications will need a corresponding change in the circuitry of the system. Thus, the conventional NC system, which is also known as "Hard-wired system" is a rigid one. Reloading of tape each time a component is to be manufactured results in loss of time and sometimes errors in reading. Also, modifications and editing in the program is not possible.

To overcome the above mentioned drawbacks of conventional NC systems, Computer Numerical Control (CNC) has been developed. Here, a bare minimum of electronic hardware is used and the functions of MCU of the conventional NC system are partly or wholly taken over by a dedicated micro-computer of the machine tool.

The control program (called Executive programme) is stored which performs all the basic NC functions. Thus, in this system, the machine control data comes directly from a micro-mini-computer, instead of from a continuously read tape (as in conventional NC system). By changing or modifying the executive programme (through paper tape), the system can be used for another machine tool. Thus, this system is flexible or 'soft wired system'. For repetitive production, the memory of the dedicated mini-computer can be utilized for temporary storage of the part programme being run. However, tape or disc is still the medium of transmitting the part programme from the programming computer to the dedicated micro-mini-computer of the machine tool, Fig. 8.6.

However, the instructions (program) can also be entered in the memory of the micro-computer with the help of a key-board (known as direct programming). As computer-Aided Design (CAD) systems become more common generating data bases, the use of tape for storing program will decrease with the corresponding increase in the use of floppy disk, hard disk and other computer storage devices for this purpose. Again, computer-aided programming is being replaced by computer-aided manufacturing (CAM), where the programs for the NC machine tools can be generated on the graphic terminal directly from the data base created by CAD (CAD file), with the help of CAD/CAM software.

Advantages of CNC systems over Conventional NC Systems :

1. Because the computer can be readily and easily reprogrammed, therefore, the system is very flexible. The machine can manufacture a part followed by other parts of different designs.
2. More versatility. Editing and debugging programs, reprogramming; and plotting and printing part shapes are simpler.
3. Program to manufacture a component can be easily called. This saves time and eliminates errors due to tape reading.
4. Greater accuracy.
5. Ease of operation.
6. The micro-processors used in place of hard-wired NC circuits are very reliable and have self diagonistic features. This makes trouble shooting extremely easy.
7. The dedicated computers need less memory storage and even small computers possess large memory.

In DNC system, several NC machines can be controlled by a large central computer. Here, direct link may be established between the programming computer and the large central control computer. But, this system is very expensive and a highly skilled software knowledge is needed and can usually be justified only in large corporations or if the system is expanded to include process management and control (CAM and CAD) Fig. 8.7.

The NC units used in the conventional DNC system may be :

1. of the conventional hard-wired type, with the tape reader being replaced by a direct

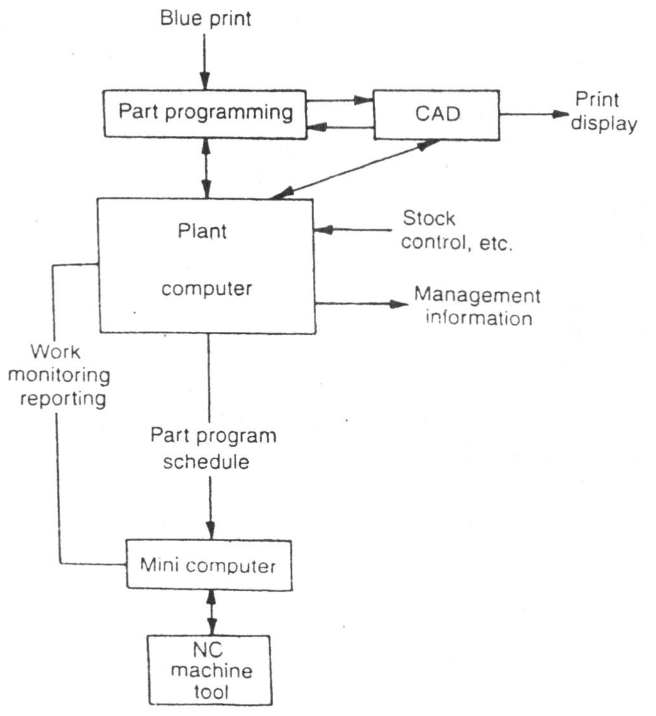

Fig. 8.7

communicating link to the Central Computer, or

2. Speicialized units which like CNC system, use a micro/mini computer as the MCU.

DNC system has the main drawback that if the Central mainframe computer goes down, all the machine tools become inoperative. This drawback has been overcome by a more recent definition of DNC as "Distributed Numerical Control" in place of "Direct Numerical Control", still abbreviated as DNC. In this system, a Central large computer is still used as in conventional DNC. But, the individual NC machines are not directly controlled by this central computer. Each NC machine has its own dedicated on-board micro-computer just like a CNC system. All these invidual micro-computers are linked to the central large computer, which serves as the Central control system. This arrangement overcomes the main drawback of the conventional DNC system, in that if the central computer goes down, the NC machines will not become inoperative. Moreover, this system provides a larger memory and computational capabilities and increased flexibility.

DNC, system appeared in the market earlier than CNC system. but with the development of dedicated mini-computers, the benefits of DNC system can be realised in CNC system.

Moreover, with the availability of small computers with large memory, micro-processors, and program editing capabilities, CNC machines are widely used at present. Also, the availability of low cost-programmable logic controllers (PLCs) has helped in the successful implementation of CNC systems.

8.9. MACHINING CENTRE

A machining centre or work centre consists of a single, but sometimes, two machine tools with the specific feature of an automatic tool changer and capable of performing a number of operations (drilling, tapping, milling, boring and turning etc.) on a workpiece. Most machines are numerically controlled, but other types of controls will work as well. The major advantage is that the job needs clamping on the workholding surface only once; the machine then performs a variety of machining operations on all the job's faces except the base. Work handling time is thus decreased because there is no movement of the workpiece from one machine to another. Some machines are equipped with two work-tables that can be moved into or out of the machine. While the work is going on one works table, the next component can be set up on the other work-table. When the job is complete on the first work-table, it is moved out of the way and the second work-table is moved into position.

For storing the preset tools, various magazine systems are used. These preset tools are removed from their slots by a hand-arm mechanism. When a particular machining operation is completed, the tool is removed from the spindle and returned to its storage slot, then another tool is picked up and mounted in the tool spindle.

A machining centre is mainly used for batch production of main components of a product. The main components of a product are usually small (about 10%) but are expensive (they represent about 50% of the product value) because they have considerable material value and usually require a large amount of machining. For such components, the machining centre is generally most economical.

Machining centres have high metal removal rate capabilities. The high degree of accuracy

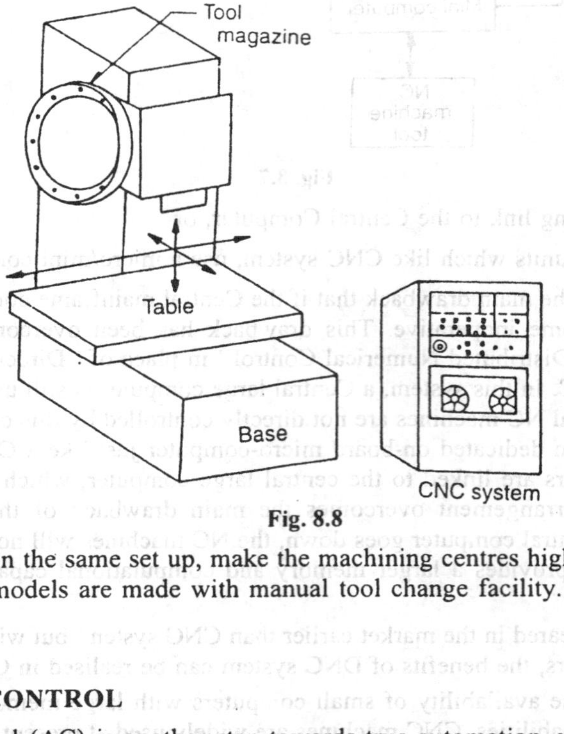

Fig. 8.8

and multi-operation in the same set up, make the machining centres highly versatile and increase productivity. Some models are made with manual tool change facility. Fig. 8.8 shows a typical machining centre.

8.10. ADAPTIVE CONTROL

Adaptive control (AC) is another step towards true automation, wherein the vital element of decision making is added. As the component is being manufactured, the important variables are measured and then if need be, certain variables are altered within programmed limits, to get as accurate the finished part as possible. For example, in drilling, the torque on the drill is measured and speed and feed or both are adjusted within programmed limits. Similarly, in milling process, feed and/or speed and depth of cut are controlled in response to signals from torque, force, vibration and dimensional transducers. The limits set by the process, *e.g.*, surface finish, maximum feeds, speeds and depth of cut, cutting force and torque; and those set by the machine tool, *e.g.*, vibration and elastic deformation, are programmed.

Adaptive control is still in its infancy since, the effect of the various process variables on the finished part is still relatively unknown. For the adaptive control to grow, process reliability has to be improved and also a more quantitative understanding of the process has to be gained. Another major problem is the on line measurement of the dimensional, without which the tool wear cannot be compensated for with any great accuracy.

PROBLEMS

1. Justify the need for NC machines.
2. Define NC
3. With the help of a diagram, explain the working of a NC machine tool.
4. List the main component of a NC machine tool and explain their functions.
5. Compare a closed loop system with an open loop system.
6. Classify the NC machines.
7. Discuss the programming of NC machines.
8. Give the applications of NC machines.
9. List the advantages and disadvantages of NC machines.
10. Define CNC and DNC.
11. List the advantages of CNC systems over the conventional NC systems.
12. List the main advantages of DNC.
13. What is "Distributed Numerical Control" ?
14. What is a machining centre ?
15. List the main advantages of a machining centre.
16. What is "Adaptive Control" ?
17. Write briefly, how numerical control works.
18. Explain the main difference between point to point and continuous path type of numerically controlled machine tools.
19. Under what conditions of production the numerically controlled machine tools are employed ?
20. What is point-to-point type of path motion of NC ? For which machines this system is employed ?
21. What is straight line NC ?
22. What is continuous path NC ?
23. How manual programming of a NC machine is done ?
24. How computer programming of a NC machine is done ?
25. Discuss the control system of a NC machine.
26. Write the functions of MCU, DPU and CLU, in a NC machine.
27. Write on some of the programming languages used for NC machine tools.
28. Write on the methods of listing the co-ordinates of points in NC system.

Appendix

PROBLEMS FROM COMPETITIVE EXAMINATIONS

1. Cylindrical bars of 100 mm diameter and 576 mm length are turned in a single-pass operation. The spindle speed used is 144 rpm and the total feed is 0.2 mm/rev. Taylor's tool life relationship is $VT^{0.75} = 75$, where V is the cutting speed (m/min.) and T is the tool life (min.). Calculate:–
   (*i*) Time for turning one piece.
   (*ii*) The average total change time per piece given that it takes 3 minutes to change the tool each time, and
   (*iii*) The time required to produce one piece given that the handling time is 4 m n.
   [*GATE 1993*]

Sol. (*i*) $$T_m = \frac{L}{fN} \text{ (See eqn. 4.10)}$$

$$= \frac{576}{0.2 \times 144} = 20 \text{ min.}$$

(*ii*) Taylor equation is,

$$VT^{0.75} = 75$$

Now $$V = \frac{\pi DN}{1000} = \frac{\pi \times 100 \times 144}{1000} = 45.24 \text{ m/min}$$

∴ From here, T = 1.962 min

∴ Number of tool change during the cutting of one piece = $\frac{20}{1.962} \approx 10$ times

∴ Total tool change time = 10 × 3 = 30 min.

∴ Total time to machine one job= 20 + 30 = 50 min

Now handling time = 4 min.

∴ Total time needed to produce one piece

= 50 + 4 = 54 min.

2. If under a given condition of plain turning, the life of the cutting tool decreases by 50% due to increase in the cutting velocity by 20%, then by what percentage will the life of that tool increase due to reduction in the cutting velocity by 20% from its original value?
   [*GATE 1995*] (**Ans.** 44%).
3. While turning a C-15 steel rod of 160 mm diameter at 315 rpm, 2.5 mm depth of cut and feed of 0.16 mm/rev. by a tool of geometry 0°, 10°, 8°, 9°, 15°, 75°, 0 (mm), the following observations were made:–
   Tangential component of the cutting force = 500 N
   Axial component of the cutting force = 200 N
   Chip thickness = 0.48 mm.
   Draw schematically the Merchant's circle diagram for the cutting forces in the present case.
   [*GATE 1995*]

4. In a turning trial using orthogonal tool geometry, a chip length of 84 mm was obtained for an uncut chip length of 200 mm. The cutting conditions are: V $\approx$ 30 m/min, uncut chip thickness = 0.5 mm, Rake angle = 20°. Cutting tool is of H.S.S. Estimate the shear plane angle ϕ, chip thickness and the shear plane angle for minimum chip strain.

[*GATE 1997*] (**Ans.** 24.7°, 1.19 mm).

5. In an orthogonal machining operation, the chip thickness and the uncut thickness are equal to 0.45 mm. If the tool rake angle is 0°, the shear plane angle is

(*a*) 45° (*b*) 30° (*c*) 18° (*d*) 60°

[*GATE 1998*] (**Ans.** a)

6. What is approximate percentage change in life of a tool with zero rake angle used in Orthogonal cutting when its clearance angle is changed from 10° to 7°?

[*GATE 1999*] (**Ans.** b).

(*a*) 30% increase (*b*) 30% decrease (*c*) 70% increase (*d*) 70% decrease

Hints. Wear rate is proportional to tan γ.

7. In an orthogonal cutting experiment with a tool of rake angle of 7°, the chip thickness was found to be 2.5 mm when the uncut chip thickness was set to 1 mm

(*a*) Find the shear angle.

(*b*) Find the friction angle assuming that Merchant's formula holds good.

[*GATE 1999*] (**Ans.** 22°–39'; 51°– 42')

8. The lives of two tools governed by equation $VT^{0.125} = 2.5$ and $VT^{0.25} = 7$ respectively in certain machining operation where V is the cutting speed in m/s and T is the tool life in seconds.

(a) Find out the speed at which both the tools have the same life. Also calculate the corresponding tool life.

(*b*) If you have to machine at a cutting speed of 1 m/s, which one of these tools will you choose in order to have less frequent tool changes? [*GATE 1999*]

Sol. Let V_0 = speed at which both the tools have the same life.

Now for the first tool,

$$V_1 T_1^{0.125} = 2.5$$

$$\therefore \quad T_1 = \left(\frac{2.5}{V_1}\right)^8$$

Similarly, for the second tool,

$$T_2 = \left(\frac{7}{V_2}\right)^4$$

Now $T_1 = T_2$ at $V_1 = V_2 = V_0$

$$\therefore \quad \left(\frac{2.5}{V_0}\right)^8 = \left(\frac{7}{V_0}\right)^4$$

From here, $V_0 = 0.893$ m/s.

(*b*) At $V_1 = V_2 = 1$ m/s

$T_1 = (2.5)^8 = 1525.88$ seconds

and $T_2 = (7)^4 = 2401$ seconds

Since $T_2 > T_1$

$\therefore$ Second tool will be selected

9. Side cutting edge angle (according to American System) of a single point tool is 30° *and* this cutting edge lies in a plane parallel to the base of the tool. The effective rake angle in a plane perpendicular to both the base and the cutting edge is 20°. Find the back rake and side rake angles. [*GATE 2000*]

Hint:
$$\lambda = 90 - 30 = 60°$$
$$i = 0,\ \alpha = 20°$$
$$\tan\alpha_b = \cos\lambda\tan\alpha + \sin\lambda\tan i$$
$$\tan\alpha_s = \sin\lambda\tan\alpha - \cos\lambda\tan i$$

10. A 15 mm diameter HSS drill is used at a cutting speed of 20 m/min. and a feed rate of 0.2 mm/rev. Under these conditions, the drill life is 100 min. The drilling length of each hole is 45 mm and the time taken for idle motions is 20 s. The tool change time is 300 s. Calculate
(*a*) Number of holes produced using one drill
(*b*) Average production time per hole. [*GATE 2000*]

Sol.
$$V = \frac{\pi DN}{1000}$$
$$\therefore \quad N = \frac{1000 \times 20}{\pi \times 15},\ \text{rev/min}$$
$$\therefore \quad T_m = \frac{L}{fN} = \frac{45 \times 15\pi}{0.2 \times 1000 \times 20} = 0.53\ \text{min.}$$
$$= 31.8\ \text{s.}$$

(*a*) $\therefore$ Number of holes produced using one drill $= \dfrac{300}{31.8}$
$$= 9.43 \text{ say } 9$$

(*b*) Average production time per hole
= machining time + idle motion time + tool change time
$$= 31.8 + 20 + \frac{300}{9} = 85.1\text{s}$$

11. (*a*) Schematically draw a single point turning tool which is designated as: 10°, 12°, 7°, 5°, 20°, 50°, 0 (mm) and show the angles in that sketch.
(*b*) Also determine the values of normal rake and normal clearance of the above mentioned tool. [*GATE 1995*]

Hint: (*a*) The ASA designation of single point cutting tool is:
$$\alpha_b, \alpha_s, \theta_s, C_e, C_s, R$$
$$\therefore \quad \alpha_b = 10°,\ \alpha_s = -12°,\ \theta_e = 7°,\ \theta_s = 5°$$
Ce = 20°, Cs = 50°, R (nose radius) = 0 mm.
$\therefore$ Tool can be drawn as shown in Fig. 1.6

(*b*) Normal rake angle is given as:
$$\tan\alpha = \tan\alpha_s\sin\lambda + \tan\alpha_b\cos\lambda$$
Now λ = Approach angle
$= 90° - \text{Cs} = 90 - 50 = 40°.$

12. The time (in minutes) for drilling a hole is given by
$$t = \frac{\text{Depth of hole} + h}{\text{Feed} \times \text{RPM}}$$ where '*h*' is the

(*a*) length of the drill (*b*) drill diameter
(*c*) flute length of the drill (*d*) cone height of the drill

[*IAS 1994*] (**Ans.** *a*)

13. Stroke of a shaping machine is 250 mm. It takes 30 double strokes per minute. Overall average speed of operation is
(*a*) 3.75 m/min (*b*) 5.0 m/min (*c*) 7.5 m/min (*d*) 15.0 m/min
[*IAS 1994*] (**Ans.** *c*)

14. A drill bit of 20 mm diameter rotating at 500 rpm with a feed rate of 0.2 mm/rev. is used to drill a through hole in an M.S. plate of 20 mm thickness. The depth of cut in this drilling operation is
(*a*) 100 mm (*b*) 20 mm (*c*) 10 mm (*d*) 0.2 mm
[*IAS 1995*] (**Ans.** *b*)

15. A grinding wheel of 150 mm diameter is rotating at 3000 rpm. The grinding speed is
(*a*) 7.5π m/s (*b*) 15π m/s (*c*) 45π m/s (*d*) 450π m/s
[*IES 1996*] (**Ans.** *a*)

16. A multispindle automatic performs four operations with times 50, 60, 65 and 75 seconds at each of its work centres. The cycle time (time required to manufacture one workpiece) in seconds will be
(*a*) 50 + 60 + 65 + 75 (*b*) (50 + 60 + 65 + 75)/4
(c) 75/4 (*d*) 75 [*IAS 1994*] (**Ans.** *a*)

17. In a single spindle automatic lathe, two tools are mounted on the turret, one form tool on the front slide and the other, a parting tool on the rear slide. The parting tool operation is much longer than form tool operation and they operate simultaneously (overlap). The number of cams required for this job is
(*a*) one (*b*) two (*c*) three (*d*) four
[*IES 1994*] (**Ans.** a)

18. If the chip-tool contact length is reduced slightly by grinding the tool face, then
(*a*) both cutting force and interface temperature would decrease.
(*b*) both cutting force and interface temperature would increase.
(*c*) the cutting force would decrease but the interface temperature would increase
(*d*) the cutting force would increase but the interface temperature would decrease.
[*IES 1993*] (**Ans.** *c*)

19. Tool life in the case of a grinding wheel is the time
(*a*) between two successive regrinds of the wheel
(*b*) taken for the wheel to be balanced
(*c*) taken between two successive wheel dressings
(*d*) taken for a wear of 1 mm on its diameter. [*IES 1993*] (**Ans.** a)

20. In ASA system, if the tool nomenclature is 8–6–5–5–10–15–2 mm, then the side rake angle will be
(*a*) 5° (*b*) 6° (*c*) 8° (*d*) 10°
[*IES 1993*] (**Ans.** *b*)

21. Consider the following characteristics :
1. The cutting edge is normal to the cutting velocity
2. The cutting forces occur in two directions only
3. The cutting edge is wider than the depth of cut.
The characteristics applicable to orthogonal cutting would include
(*a*) 1 and 2 (*b*) 1 and 3 (*c*) 2 and 3 (*d*) 1, 2 and 3
[*IAS 1994*] (**Ans.** a)

22. Thrust force will increase with increase in
(*a*) Side cutting edge angle (*b*) tool nose radius
(*c*) rake angle (*d*) end cutting edge angle
[*IAS 1995*] (**Ans.** *a*)

23. In an orthogonal cutting, the depth of cut is halved and the feed rate is doubled. If the chip thickness ratio is unaffected with the changed cutting conditions, the actual chip thickness will be
(*a*) doubled (*b*) halved (*c*) gradrupled (*d*) unchanged
[*IAS 1995*] (**Ans.** *a*)

24. In a single point turning operation with a cemented carbide and steel combination having a Taylor exponent of 0.25, if the cutting speed is halved, then the tool life will become
(*a*) half (*b*) two times (*c*) eight times (*d*) sixteen times
[*IAS 1995*] (**Ans.** *d*)

25. The angle between the face and the flank of the single point cutting tool is known as
(*a*) rake angle (*b*) clearance angle (*c*) lip angle (*d*) point angle
[*IES 1995*] (**Ans.** *c*)

26. Single point thread cutting tool should ideally have
(*a*) zero rake (*b*) positive rake (*c*) negative rake (*d*) normal rake
[*IES 1995*] (**Ans.** *a*)

27. Consider the following statements about nose radius
1. It improves tool life 2. It reduces the cutting force.
3. It improves the surface finish. Of these statements
(*a*) 1 and 2 are correct (*b*) 2 and 3 are correct
(*c*) 1 and 3 are correct (*d*) 1, 2 and 3 are correct
[*IES 1995*] (**Ans.** *d*)

28. Match list I with List II and select the correct answer using the codes given below the lists

	List I (*Wear type*)	List II (*Associated mechanism*)
A	Abrasive wear	1. Galvanic action
B	Adhesive wear	2. Ploughing action
C	Electrolytic wear	3. Molecular transfer
D	Diffusion wear	4. Plastic deformation
		5. Metallic bond

Codes: A B C D
(*a*) 2 5 1 3 (*b*) 5 2 1 3 (*c*) 2 1 3 4 (*d*) 5 2 3 4
[*IES 1995*] (**Ans.** *a*)

29. Crater wear is predominant in
(*a*) carbon tool steel (*b*) tungsten carbide tools
(*c*) high speed steel tools (*d*) ceramic tools [*IES 1995*] (**Ans.** *b*)

30. Consider the following work materials:
1. Titanium 2. Mild steel 3. Stainless steel 4. Grey cast iron
The correct sequence of these materials in terms of increasing order of difficulty in machining is
(*a*) 4, 2, 3, 1 (*b*) 4, 2, 1, 3 (*c*) 2, 4, 3, 1 (*d*) 2, 4, 1, 3
[*IES 1995*] (**Ans.** *a*)

31. The primary tool force used in calculating the tool power consumption in machining is the
(*a*) radial force (*b*) tangential force (*c*) axial force (*d*) frictional force
[*IES 1995*] (**Ans.** *b*)

32. Chip equivalent is increased by
(*a*) an increase in side cutting edge angle of tool
(*b*) an increase in nose radius and side cutting edge angle of tool

(c) increasing the plan area of cut
(d) increasing the depth of cut [*IES 1996*] (**Ans.** *a*)

33. Notch wear at the outside edge of the depth of cut is due to
(a) abrasive action of the work hardened chip material
(b) oxidation
(c) slip-stick action of the chip
(d) chipping [*IES 1996*] (**Ans.** *c*)

34. Specific energy requirements in a grinding process are more than those in turning for the same metal removal rate because of the
(a) specific pressures between wheel and work being high
(b) size effect of the larger contact areas between wheel and work
(c) high cutting velocities
(d) high heat produced during grinding [*IES 1996*] (**Ans.** *d*)

35. In orthogonal cutting, the depth of cut is 0.5 mm at a cutting speed of 2 m/s. If the chip thickness is 0.75 mm, the chip velocity is
(a) 1.33 m/s (b) 2 m/s (c) 2.5 m/s (d) 3 m/s
[*IES 1997*] (**Ans.** *a*)

36. Consider the following elements:
1. Nose radius 2. Cutting speed 3. Depth of cut 4. Feed
The correct sequence of these elements in decreasing order of their influence on tool life is
(a) 2, 4, 3, 1 (b) 4, 2, 3, 1 (c) 2, 4, 1, 3 (d) 4, 2, 1, 3
[*IES 1997*] (**Ans.** *a*)

37. The rake angle in a twist drill
(a) varies from minimum near the dead centre to a maximum value at the periphery
(b) is maximum at the dead centre and zero at the periphery
(c) is constant at every point of the cutting edge
(d) is a function of the size of the chisel edge. [*IES 1997*] (**Ans.** *c*)

38. Consider the following forces acting on a finish turning tool:
1. Feed force 2. Thrust force 3. Cutting force
The correct sequence of the decreasing order of the magnitudes of these forces is
(a) 1, 2, 3 (b) 2, 3, 1 (c) 3, 1, 2 (d) 3, 2, 1
[*IES 1997*] (**Ans.** *c*)

39. In turning operation, the feed could be doubled to increase the metal removal rate. To keep the same level of surface finish, the nose radius of the tool should be
(a) halved (b) kept unchanged (c) doubled (d) made four times
[*IES 1999*] (**Ans.** *b*)

40. The radial force in single-point tool during turning operation varies between
(a) 0.2 to 0.4 times the main cutting force (b) 0.4 to 0.6 times the main cutting force
(c) 0.6 to 0.8 times the main cutting force (d) 0.5 to 0.6 times the main cutting force
[*IES 1999*] (**Ans.** *a*)

41. Consider the following criteria in evaluating machinability:
1. Surface finish 2. Type of chips
3. Tool life 4. Power consumption
In modern high speed CNC machining with coated carbide tools, the correct sequence of these criteria in decreasing order of their importance is
(a) 1, 2, 4, 3 (b) 2, 1, 4, 3 (c) 1, 2, 3, 4 (d) 2, 1, 3, 4
[*IES 1998*] (**Ans.** *c*)

42. Internal gears are made by
(a) hobbing (b) shaping with pinion cutter
(c) shaping with rack cutter (d) milling [*IES 1993*] (**Ans.** *b*)

43. While cutting helical gears on a non-differential gear hobber, the feed change gear ratio is
(*a*) independent of index change gear ratio
(*b*) dependent on speed change gear ratio
(*c*) interrelated to index change gear ratio
(*d*) independent of speed and index change gear ratio [*IES 1995*] (**Ans.** *c*)

44. Consider the following processes of gear manufacture
1. Milling with form cutter 2. Rack type gear shaper (gear planer)
3. Rotary gear shaper (gear shaper) 4. Gear hobbing
The correct sequence of these processes in increasing order of accuracy of involute profile of the gear is
(*a*) 3, 2, 4, 1 (*b*) 2, 3, 4, 1 (*c*) 3, 2, 1, 4 (*d*) 2, 3, 1, 4
[*IES 1996*] (**Ans.** a)

45. Gear cutting on a milling machine using an involute profile cutter is a
(*a*) gear forming process (*b*) gear generating process
(*c*) gear shaping process (*d*) highly accurate gear producing process
[*IES 1996*] (**Ans.** *a*)

46. For the manufacture of full depth spur gear by hobbing process, the number of teeth to be cut = 30, module = 3 mm and pressure angle = 20°. The radial depth of cut to be employed should be equal to
(*a*) 3.75 mm (*b*) 4.50 mm (*c*) 6.00 mm (*d*) 6.75 mm
[*IES 1996*] (**Ans.** *d*)

47. Which of the following motions are not needed for spur gear cutting with a hob?
1. Rotary motion of hob
2. Linear axial reciprocatory motion of hob
3. Rotary motion of gear blank
4. Radial advancement of hob
Select the correct answer using the codes given below:
(*a*) 1, 2 and 3 (*b*) 1, 3 and 4 (*c*) 1, 2 and 4 (*d*) 2, 3 and 4
[*IES 1997*] (**Ans.** *d*)

48. Match List I (Gear component) with List-II (Preferred method of manufacturing) and select the correct answer using the codes given below the lists:

List I	*List II*
A. Gear for clocks	1. Hobbing
B. Bakelite gears	2. Stamping
C. Aluminium gears	3. Powder Compacting
D. Automobile transmission gears	4. Sand Costing
	5. Extrusion

Codes: A B C D A B C D A B C D A B C D
(*a*) 2 3 5 1 (*b*) 5 3 4 2 (*c*) 5 1 2 3 (*d*) 2 4 5 3
[*IES 1997*] (**Ans.** *a*)

49. A straight teeth slab milling cutter of 100 mm diameter and 10 teeth rotating at 200 r.p.m. is used to remove a layer of 3 mm thickness from a steel bar. If the table feed is 400 mm/min, the feed per tooth in this operation will be:
(*a*) 0.2 mm (*b*) 0.4 mm (*c*) 0.5 mm (*d*) 0.6 mm
[*IES 1999*] (**Ans.** *a*)

50. Consider the following processes for manufacture of gears:
1. Casting 2. Powder metallurgy
3. Machining from bar stock 4. Closed die forging

The correct sequence is increasing order of bending strength of gear teeth is

(*a*) 1, 2, 3, 4 (*b*) 1, 2, 4, 3 (*c*) 2, 1, 4, 3 (*d*) 2, 1, 3, 4

[*IES 1999*] (**Ans.** *d*)

Directions : The following questions (51 to 56) consist of two statements, one labelled the 'Assertion A' and the other labelled the 'Reason R', you are to examine the two statements carefully and decide if the Assertion A and Reason R are individually true, and if so, whether the Reason is a correct explanation of the Assertion. Select your answers to these questions using the codes given below and mark your answer sheet accordingly.

(*a*) Both A and R are true and R is the correct explanation of A
(*b*) Both A and R are true and R is not a correct explanation of A
(*c*) A is true but R is false
(*d*) A is false but R is true.

51. *Assertion (A):* In a multispindle automat, the turret is indexed to engage each of the cutting tools mounted on it.
Reason (R): Turret is a multiple tool holder so that the machining can be continued with each tool without the need to change the tool. [*IAS 1995*] (**Ans.** *d*)
52. *Assertion (A):* In a Swiss-type automatic lathe, the turret is given longitudinal feed for each tool in a specific order with suitable indexing.
Reason (R): A turret is a multiple tool holder to facilitate machining with each tool by indexing without the need to change the tools. [*IES 1995*] (**Ans.** *b*)
53. *Assertion (A):* For a negative rake tool, the specific cutting pressure is smaller than for a positive rake tool under otherwise identical conditions.
Reason (R): The shear strain undergone by the chip in the case of negative rake tool is larger. [*IES 1993*] (**Ans.** *a*)
54. *Assertion (A):* An increase in depth of cut shortens the tool life.
Reason (R): Increase in depth of cut gives rise to relatively small increase in tool temperature. [*IAS 1995*] (**Ans.** *a*)
55. *Assertion A:* Tool wear is expressed in terms of flank wear rather than crater wear.
Reason R: Measurement of flank wear is simple and more accurate. [*IES 1994*] (**Ans.** *a*)
56. *Assertion A:* Special purpose machine tools and automatic machine tools are quite useful for job shops.
Reason R: Special purpose machine tools can do special types of machining work automatically. [*IES 1996*] (**Ans.** *d*)
57. Given that average cutting speed = 9 m/min, the return time to cutting time ratio = 1 : 2, the feed rate = 0.3 mm/stroke the clearance at each end of c ut = 2 mm and that the plate is fixed with 700 mm side along the direction of tool travel, the time required for finishing one flat surface of size 700 × 30 mm in a shaper will be
(*a*) 10 min (*b*) 12.5 min (*c*) 15 min (*d*) 20 min
[*IES 1994*] (**Ans.** *b*)
58. The following parameters determine the model of continuous chip formation:
1. True speed 2. Cutting velocity
3. Chip thickness 4. Rake angle of the cutting tool.
The parameters which govern the value of shear angle would include:
(*a*) 1, 2 and 3 (*b*) 1, 3 and 4 (*c*) 1, 2 and 4 (*d*) 2, 3 and 4
[*IES 1994*] (**Ans.** *b*)
59. What is the correct sequence of the following parameters in order of their maximum to minimum influence on tool life?
1. Feed rate 2. Depth of cut 3. Cutting speed

Select the correct answer using the codes given below:
Codes:
(*a*) 1, 2, 3 (*b*) 3, 2, 1 (*c*) 2, 3, 1 (*d*) 3, 1, 2
[*IES 1994*] (**Ans.** *d*)

60. Tool geometry of a single point cutting tool is specified by the following elements:
1. Back rake angle 2. Side rake angle 3. End cutting edge angle
4. Side cutting edge angle 5. Side relief angle
6. End relief angle.
7. Nose radius.
The correct sequence of these tool elements used for correctly specifying the tool geometry is
(*a*) 1, 2, 3, 6, 5, 4, 7 (*b*) 1, 2, 6, 5, 3, 4, 7 (*c*) 1, 2, 5, 6, 3, 4, 7 (*d*) 1, 2, 6, 3, 5, 4, 7
[*IES 1994*] (**Ans.** *b*)

61. The ratio of cutting force to thrust force is nearly 2.5 in
(*a*) turning (*b*) broaching (*c*) grinding (*d*) plain milling
[*IES 1994*] (**Ans.** *c*)

62. Which one of the following operations is carried out at the minimum cutting velocity if the machines are equally rigid and the tool work materials are the same?
(*a*) turning (*b*) grinding (*c*) boring (*d*) milling
[*IES 1994*] (**Ans.** *d*)

63. Match four correct pairs between List I and List II

List I	*List II*
(*a*) Grinding	1. Surface for oil retention
(*b*) Honing	2. Surface for Max. load capacity
(*c*) Super finishing	3. Surface of limiting friction
(*d*) Burnishing	4. Surface of matte finish
	5. Surface for pressure sealing
	6. Surface for interference fit

[*GATE 1997*]

64. Consider the following machine tools:-
1. Hobbing machine 2. Gear shaping machine
3. Broaching machine
The teeth of internal spur gears can be cut in
(*a*) 1, 2 and 3 (*b*) 1 and 2 (*c*) 1 and 3 (*d*) 2 and 3
[*IES 1994*] (**Ans.** *d*)

65. A single point cutting tool made of HSS has the values of constants $C = 80$ and $n = 0.2$ in the basic tool life equation. If the tool cost per regrind is Rs. 2 and the machine hour rate is Rs 30, determine the most economical cutting speed. The tool cost includes the cost of time spent on changing. (*GATE 1994*)

Sol:

Optimum value of V is given as

$$V = C\left[\frac{C_m}{C_t} \cdot \frac{n}{1-n}\right]^n$$

$$C = 80,\ n = 0.2,\ C_m = \text{Rs. } 30,\ C_t = \text{Rs. } 2$$

$$\therefore V = 80\left(\frac{0.2}{0.8}\right)^{0.2} \cdot \left(\frac{30}{2}\right)^{0.2}$$

$$= 104.2 \text{ m/min.}$$

66. A tube of 35 mm O.D. was turned on a lathe and the following data was obtained:-
Rake angle = 35°, Cutting speed = 15 m/min.
Feed rate = 0.1 mm/rev.
Length of continuous chip in one revolution = 60 mm.
Cutting force = 200 N, Feed force)Thrust force) = 800 N
Calculate the chip thickness, shear plane angle, velocity of chip along tool face and co-efficient of friction. *(GATE 2002)*

Solution: $l_c = 60$ mm, $l = \pi D = 35\pi$

$$\therefore r = l_c / l \qquad \text{[eqn. 1.5)}$$

$$= 0.545$$

$$\phi = \tan^{-1} \frac{r \cos \alpha}{1 - r \sin \alpha} \qquad \text{(Eqn. 1.4)}$$

$$= 33°$$

$$V_c = r \times V = 8.175 \, m / \text{min}.$$

$$\mu \text{ (eqn. 14.15)} = 1.528$$

67. During orthogonal cutting of mild steel with a 10° rake angle tool, the chip thickness ratio obtained was 0.4. The shear angle (in degrees) evaluated from this data is
(A) 6.53 (B) 20.22 (C) 22.94 (D) 50.00
(GATE 2001) **(Ans.** C)

68. In a drilling operation under a given condition, the tool life was found to decrease from 20 min to 5 min due to increase in drill speed from 200 rpm to 400 r.p.m. What will be the tool life of that drill under the same condition if the drill speed is 300 r.p.m.? *(IES 2003)*

Solution:- $VT^n = C$

$$\therefore \; 200\,(20)^n = 400\,(5)^n$$

$$\therefore (4)^n = 2 \quad \therefore \; n = 0.5$$

$$\therefore \; C = 200 \times (20)^{0.5} = 894.4272$$

Now, $300 \,.\, (T)^{0.5} = 894.4272$

$\therefore \; T =$ **8.9 min.**

69. Quality screw threads are produced by
(A) thread milling
(B) thread chasing
(C) thread cutting with single point tool
(D) thread casting **(Ans.** A)

70. A cylinder is turned on a lathe with orthogonal machining principle. Spindle rotates at 200 rpm. The axial feed rate is 0.25 mm per revolution. Depth of cut is 0.4 mm. The rake angle is 10°. In the analysis, it is found that the shear angle is 27.75°.
1. The thickness of the produced chip is
(A) 0.51 mm (B) 0.528 mm (C) 0.818 mm (D) 0.846 mm **(Ans.** A)
2. In the above problem, the co-efficient of friction at the chip tool interface obtained using Earnt and Merchant Theory is
(A) 0.18 (B) 0.36 (C) 0.71 (D) 0.98 **(Ans.** D)

71. The effect of rake angle on the mean friction angle in machining can be explained by
(A) Sliding (Coulomb) model of friction
(B) Sticking and then sliding model of friction

(C) sticking friction
(D) Sliding and then sticking model of friction *[GATE 1992]* **(Ans. D)**

72. In horizontal milling process ..... (up/down) milling provides better surface finish and ..... (up/down) milling provides longer tool life. *[GATE 1992]* **(Ans. *down, down*)**

73. Match the following components with the appropriate machining process:

Component	*Process*
(A) Square hole in a high strength alloy	(P) Milling
(B) Square hole in a ceramic component	(Q) Drilling
(C) Blind hole in a die	(R) ECM
(D) Turbine blade profile on high strength alloy	(S) Jig boring
	(T) EDM
	(U) USM

[GATE 1992] **(Ans. A – T, B – U, C – S, D – R)**

74. The cross-feed on a shaper consists of a lead screw having 0.2 threads per mm. A ratchet and pawl on the end of the lead screw is driven from the sharper crank such that the pawl indexes the ratchet by one tooth during each return stroke of the ram. Ratchet has 20 teeth.
(A) Find the feed in mm
(B) If a plate 100 mm wide has to be machined in 10 minutes, find the cutting speed in metres/sec. The ratio of return speed to cutting speed is 2 : 1, and the length of the stroke is 150 mm. *[GATE 1992]*

Sol. Pitch of lead screw $= \dfrac{1}{0.2} = 5$ mm

Now, the pawl indexes $\dfrac{1}{20}$ revolution per stroke

$\therefore$ (A) Cross feed $= \dfrac{5}{20} = 0.25$ mm,

(B) Number of cutting strokes $= \dfrac{\text{width of job}}{\text{cross feed per stroke}}$

$$= \frac{100}{0.25} = 400$$

Refer to Art 3.2.4.

$$V = \frac{LN(1+K)}{1000} \text{ min}$$

L = Length of ram stroke = 150 mm

N = Number of full strokes/min

$$= \frac{400}{\text{Time}} = \frac{400}{10} = 40$$

$$K = \frac{\text{Return time}}{\text{Cutting time}} = \frac{\text{Cutting speed}}{\text{Return speed}} = \frac{1}{2}$$

$$\therefore \quad V = \frac{150 \times 40 \times 1.5}{1000} = 9 \text{ m/min.}$$

$$= 0.15 \text{ m/s}$$

75. A milling cutter having 8 teeth is rotating at 150 rpm. If the feed per tooth is 0.1 mm, the value of the table speed in mm/min is
(A) 120 (B) 187 (C) 125 (D) 70
[GATE 1993] **(Ans. A)**

76. To get good surface finish on a turned job, one should use a sharp tool with a .... feed and ...... speed of rotation of the job. [*GATE1994*] (**Ans.** *Minimum, maximum*)

77. Among the conventional, machining process, maximum specific energy is consumed in
(A) Turning (B) Drilling (C) Planing (D) Grinding
[*GATE 1995*] (**Ans.** D)

78. Cutting power consumption in turning can be significantly reduced by
(A) increasing rake angle of the tool (B) increasing the cutting angle of the tool
(C) widening the nose radius of the tool (D) increasing the clearance angle
[*GATE 1995*] (**Ans.** A)

79. Plain milling of mild steel plate produces
(A) irregular shaped discontinuous chips (B) regular shaped discontinuous chips
(C) continuous chips without built up edge (D) jointed chips [*GATE 1995*] (**Ans.** C)

80. Diamond wheels should not be used for grinding steel components. (True/False)
[*GATE 1995*] (**Ans.** *True*)

81. Match 4 correct pairs

List I	*List II*
(Manufacturing Processes)	(Conditions)
(A) Finish turning	1. Backlash eliminator
(B) Forming	2. Zero rake
(C) Thread cutting	3. Nose radius
(D) Down milling	4. Low speed

[*GATE 1995*] (**Ans.** A – 2, B– 3, C– 4, D– 1).

82. The rake angle in a drill
(A) increases from centre to periphery (B) decreases from centre to periphery
(C) remains constant (D) is irrelevant to the drilling operation
[*GATE 1996*] (**Ans.** D)

83. Helix angle of a fast helix drill is normally
(A) 35° (B) 60° (C) 90° (D) 5°
[*GATE 1997*] (**Ans.** A)

84. In machining using abrasive material, increasing abrasive grain size
(A) increases the material removal rate
(B) decreases the material removal rate
(C) first decreases and then increases the material removal rate
(D) first increases and then decreases the material removal rate [*GATE 1998*] (**Ans.** D)

85. Abrasive material used in grinding wheel selected for grinding ferrous alloys is
(A) Silicon carbide (B) diamond (C) aluminium oxide (D) boron carbide
[*GATE 2000*] (**Ans.** C)

86. Match List I with List II and select the correct answer using the codes given below the lists:

List I	*List II*
(Cutting tool Material)	(Major characteristic constituent)
A. High Speed Steel	1. Carbon
B. Stellite	2. Molybdenum
C. Diamond	3. Nitride
D. Coated Carbide tool	4. Columbuim
	5. Cobalt

Codes:	A	B	C	D		A	B	C	D
(*a*)	2	1	3	5	(*b*)	2	5	1	3
(c)	5	2	4	3	(*d*)	5	4	2	3

[*IES 1993*] (**Ans.** *b*)

87. Consider the following parameters:
1. Grinding wheel diameter
2. Regulating wheel diameter
3. Speed of grinding wheel
4. Speed of the regulating wheel
5. Angle between the axes of grinding and regulating wheels
Among these parameters, those which influence the axial feed rate in centreless grinding would include:
(A) 2, 4 and 5 (B) 1, 2 and 3 (C) 1, 4 and 5 (D) 3, 4 and 5
[*IES 1993*] (**Ans.** A)

88. It is required to cut screw threads of 2 mm pitch on a lathe. The lead screw has a pitch of 6 mm. If the spindle speed is 60 rpm, then the speed of the lead screw will be
(A) 10 rpm (B) 20 rpm (C) 120 rpm (D) 180 rpm
[*IES 1993*] (**Ans.** B)

89. A dynamometer is a device used for the measurement of
(A) chip thickness ratio (B) forces during metal cutting
(C) wear of the cutting tool (D) deflection of the cutting tool
[*IES 1993*] (**Ans.** B)

90. The main purpose of boring operation, as compared to drilling is to:
(A) drill a hole (B) finish the drilled hole
(C) correct the hole (D) enlarge the existing hole
[*IES 1993*] (**Ans.** B, C, D)

91. Climb milling is chosen while machining because
(A) the chip thickness increases gradually
(B) it enables the cutter to dig in and start the cut
(C) the specific power consumption is reduced
(D) better surface finish can be obtained [*IES 1993*] (**Ans.** D)

92. In centreless grinding, the workpiece centre will be
(A) above the line joining the two wheel centres
(B) below the line joining the two wheel centres
(C) on the line joining the two wheel centres
(D) at the intersection of the line joining the wheel centres with the work plate plane
[*IES 1993*] (**Ans.** A)

93. A hole of 30 mm diameter is to be produced by reaming. The minimum diameter permissible is 30.00 mm while the maximum diameter permissible is 30.05 mm. In this regard, consider the following statements about the reamer size:
1. The minimum diameter of the reamer can be less than 30 mm
2. The minimum diameter of the reamer can not be less than 30 mm
3. The maximum diameter of the reamer can be more than 30.05 mm
4. The maximum diameter of the reamer must be less than 30.05 mm
Of these statements
(A) 1 and 4 are correct (B) 1 and 3 are correct
(C) 2 and 3 are correct (D) 2 and 4 are correct [*IES 1993*] (**Ans.** A)

94. A standard dividing head is equipped with the following index plates:
1. Plate with 15, 16, 17, 18, 19, 20 holes circles
2. Plate with 21, 23, 27, 29, 31, 33 holes circles
3. Plate with 37, 39, 41, 43, 47, 49 holes circles
For obtaining 24 divisions on a workpiece by simple indexing
(A) hole plate 2 alone can be used (B) hole plates 1 and 2 can be used
(C) hole plates 1 and 3 can be used (D) any of the three hole plates can be used
[*IAS 1994*] (**Ans.** D)

95. Size of a shaper is given by
(A) stroke length (B) motor power
(C) weight of the machine (D) rate size [*IAS 1995*] (**Ans.** A)

96. Consider the following tool materials:
1. Carbide 2. Cermet 3. Ceramic 4. Borazon
Correct sequence of these tool materials in increasing order of their ability to retain their hot hardness is
(A) 1, 2, 3, 4 (B) 1, 2, 4, 3 (C) 2, 1, 3, 4 (D) 2, 1, 4, 3
[*IES 1994*] (**Ans.** A)

97. Enlarging an existing circular hole with a rotating single point tool is called
(A) boring (B) drilling (C) reaming (D) internal turning
[*IES 1994*] (**Ans.** A)

98. Consider the following statements regarding grinding of high carbon steel
1. Grinding at high speeds results in the reduction of chip thickness and cutting forces per grit.
2. Aluminium oxide wheels are employed.
3. The grinding wheel has be of open structure of these statements
(A) 1, 2 and 3 are correct (B) 1 and 2 are correct
(C) 1 and 3 are correct (D) 2 and 3 are correct [*IES 1994*] (**Ans.** B)

99. Consider the following operations:
1. Cutting keyways on shafts 2. Cutting external screw threads
3. Cutting teeth of spur gears 4. Cutting external splines
Those which can be performed with milling cutters would include:
(A) 1 and 2 (B) 2, 3 and 4 (C) 1 and 3 (D) 1, 2, 3 and 4
[*IES 1994*] (**Ans.** C)

100. In reaming process
(A) metal removal rate is high (B) high surface finish is obtained
(C) high form accuracy is obtained (D) high dimensional accuracy is obtained
[*IES 1994*] (**Ans.** D)

101. Cubic boron nitride is used
(A) as lining material in induction furnaces (B) for making optical quality glass
(C) for heat treatment (D) for none of the above [*IES 1993*] (**Ans.** D)

102. The compositions of some of the alloy steels are as under :
1. 18 W 4 Cr 1 V 2. 12 Mo 1 W 4 Cr 1 V
3. 6 Mo 6W 4 Cr 1 V 4. 18 W 8 Cr 1 V
The compositions of commonly used high speed steels would include :
(A) 1 and 2 (B) 2 and 3
(C) 1 and 4 (D) 1 and 3 (*IES 1995*)(**Ans.** C)

103. The straight grades of cemented carbide cutting tool materials contain
(A) Tungsten carbide only (B) tungsten carbide and titanium carbide
(C) tungsten carbide and cobalt. (D) tungsten carbide and cobalt carbide
(*IES 1995*) (**Ans.** C)

104. In the grinding wheel of A60G7B23, B stands for
(A) resinoid bond (B) rubber bond
(C) shellac bond (D) silicate bond (*IES 1995*) (**Ans.** A)

105. Soft materials cannot be economically ground due to
(A) the high temperature involved (B) frequent wheel clogging
(C) rapid wheel wear (D) low workpiece stiffness.
(*IES 1995*) (**Ans.** B)

106. In a milling operation, two side milling cutters are mounted with a desired distance between them so that both sides of a workprise can be milled simultaneusly. This set up is called
(A) gang milling (B) straddle milling
(C) Side milling *(IES 1995)* [**Ans.** B)

107. In a mechanical shaper, the length of stroke is increased by
(A) increasing the centre distance of bull gear and crank pin
(B) decreasing the centre distance of bull gear and crank pin
(C) increasing the length of ram.
(D) decreasing the length of slot in the slotted lever *(IES 19995)* [**Ans.** A)

108. Cubic Boron Nitride
(A) has a very high hardness which is comparable to that of diamond
(B) has a hardness which is slightly more than that of HSS.
(C) is used for making cylindrical blocks of aircraft engines
(D) is used for making optical glasses *[IES 1996]* [**Ans.** A)

109. The limit to the maximum hardness of a work material which can be machined with HSS tools even at low speeds is set by which of the following tool failure mechanisms?
(A) Attrition (B) Abrasion
(C) Diffusion (D) Plastic deformation under compression
[IES 1996] [**Ans.** A)

110. A machanist desires to turn a round steel block of outside diameter 100 mm at 1000 rpm. The material has tensile strength of 75 kg/mm^2. The depth of cut chosen is 3 mm at a feed rate of 0.3 mm/rev. Which one of the following tool materials will be suitable for machining the component under the specified cutting conditions?
(A) Sintered Carbides (B) Ceramic (C) HSS (D) Diamond
[IES 1996] [**Ans.** B)
[**Hint:** Find cutting speed]

111. In turning of slender rods, it is necessary to keep the transverse force minimum mainly to
(A) improve the surface finish (B) increase productivity
(C) improve cutting efficiency (D) reduce vibrations and chatter.
[IES 1996] [**Ans.** D)

112. A grinding wheel of 150 mm diameter is rotating at 3000 rpm. The grinding speed is
(A) 7.5 π m/s (B) 15 π m/s (C) 45 π m/s (D) 450 π m/s
[IES1996] [**Ans.** A)

113. Which of the following statements are correct?
1. A boring machine is suitable for job shop.
2. A jig boring machine is designed specially for doing more accurate work when compared to a vertical milling machine.
3. A vertical precision boring machine is suitable for boring holes in cylindrical blocks and lines
(A) 1, 2 and 3 (B) 1 and 2 (C) 2 and 3 (D) 1 and 3
[IES 1996] [**Ans.** C)

114. Match List I (Type of drill) with List II (Application) and select the correct answer using the codes given below the lists:

List I	*List II*
A. Straight shank	1. Soft materials
B. Taper Shank	2. Deep holes
C. Single flute	3. General purpose
D. High helix	4. Small hole diameter

Codes:

	A	B	C	D		A	B	C	D
(*a*)	3	4	1	2	(*b*)	3	4	2	1
(*c*)	4	3	2	1	(*d*)	4	3	1	2

[*IES 1997*] (**Ans.** *c*)

115. Which of the following are the advantages of a hydraulic shaper over a mechanically driven shaper?
1. More strokes can be obtained per minute at a given cutting speed.
2. The cutting stroke has a definite stopping point
3. It is simpler in construction
4. Cutting speed is constant throughout most of the cutting stroke

Select the correct answer using the codes given below:
(A) 1 and 2 (B) 1 and 4 (C) 2 and 4 (D) 1, 3 and 4

[*IES 1997*] (**Ans.** B)

116. Consider the following operations:
1. Under cutting 2. Plain turning 3. Taper turning 4. Thread cutting

The correct sequence of these operations in machining a product is
(A) 2, 3, 4, 1 (B) 3, 2, 4, 1 (C) 2, 3, 1, 4 (D) 3, 2, 1, 4

[*IES 1997*] (**Ans.** C)

117. Which one of the following materials is used as the bonding material for grinding wheels?
(A) Silicon carbide (B) Sodium silicate (C) Boron carbide (D) Aluminium oxide

[*IES 1997*] (**Ans.** B)

118. Consider the following statements:
In up-milling process
1. The cutter starts the cut from the machine surface and proceeds upwards.
2. The cutter starts the cut from the top surface and proceeds downwards.
3. The job is fed in a direction opposite to that of cutter rotation.
4. The job is fed in the same direction as that of cutter rotation. Of these statements:

(A) 1 and 3 are correct (B) 1 and 4 are correct
(C) 2 and 3 are correct (D) 2 and 4 are correct [*IES 1997*] (**Ans.** A)

119. The type of quick return mechanism employed mostly in shaping machines is:
(A) DC reversible motor (B) Fast and loose pulleys
(C) Whit worth motion (D) Slotted link mechanism

[*IES 1997*] (**Ans.** D)

120. A 400 mm long shaft has a 100 mm tapered step at the middle with 4° included angle. The tail stock offset required to produce this taper on a lathe would be
(A) 400 sin 4° (B) 400 sin 2° (C) 100 sin 4° (D) 100 sin 2°

[*IES 1998*] (**Ans.** B)

121. A single start thread of pitch 2 mm is to be produced on a lathe having a lead screw with a double start thread of pitch 4 mm. The ratio of speeds between the spindle and lead screw for this operation is
(A) 1 : 2 (B) 2 : 1 (C) 1 : 4 (D) 4 : 1

[*IES 1998*] (**Ans.** D)

122. Match List I with List II and select the correct answer using the codes given below the lists

List I	*List II*
A. Reaming	1. Smoothing and squaring surface around the hole for proper seating
B. Counter-boring	2. Sizing and finishing the hole
C. Counter-sinking	3. Enlarging the end of the hole
D. Spot facing	4. Making a conical enlargement at the end of the hole

Codes:

	A	B	C	D		A	B	C	D
(*a*)	3	2	4	1	(*b*)	2	3	1	4
(*c*)	3	2	1	4	(*d*)	2	3	4	1

[*IES 1998*] (**Ans.** *d*)

123. Which one of the following pairs of parameters and effects is not correctly matched?
(A) Large wheel diameter ...... Reduced wheel wear
(B) Large depth of cut .... Increased wheel wear
(C) Large work diameter .... Increased wheel wear
(D) Large wheel speed ...... Reduced wheel wear [*IES 1998*] (**Ans.** D)

124. A component requires a hole which must be within two limits of 25.03 and 25.04 mm diameter. Which of the following statements about the reamer size are correct?
1. Reamer size cannot be below 25.03 mm
2. Reamer size cannot be above 25.04 mm
3. Reamer size can be 25.04 mm
4. Reamer size can be 25.03 mm
Select the correct answer using the codes given below:
(A) 1 and 3 (B) 1 and 2 (C) 3 and 4 (D) 2 and 4
[*IES 1998*] (**Ans.** D)

125. In metal cutting operation, the approximate ratio of heat distributed among chip, tool and work, in that order is
(A) 80 : 10 : 10 (B) 33 : 33 : 33 (C) 20 : 60 : 10 (D) 10 : 10 : 80
[*IES 1998*] (**Ans.** A)

126. Which one of the following sets of forces are encountered by a lathe parting tool while groove cutting:
(A) Tangential, radial and axial (B) Tangential and radial
(C) Tangential and axial (D) Radial and axial [*IES 1999*] (**Ans.** A)

127. In a single-point turning operation of steel with Cemented carbide tool, Taylor's tool life exponent is 0.25. If the cutting speed is halved, the tool life will increase by
(A) two times (B) four times (C) eight times (D) sixteen times
[*IES 1999*] (**Ans.** D)

128. Match List I (ISO classification of carbide tools) with List II (applications) and select the correct answer using the codes given below the lists :

List I	*List II*
A. P – 10	1. Non-ferrous roughing cut
B. P – 50	2. Non-ferrous, finishing cut
C. K – 10	3. Ferrous materials, roughing cut
D. K – 50	4. Ferrous materials, finishing cut

Codes:

	A	B	C	D		A	B	C	D
(*a*)	4	3	1	2	(*b*)	3	4	2	1
(*c*)	4	3	2	1	(*d*)	3	4	1	2

[*IES 1999*] (**Ans.** *d*)

129. Consider the following statements:
For precision machining of non-ferrous alloys, diamond is preferred because it has
1. low co-efficient of thermal expansion 2. high wear resistance
3. high compression strength 4. low fracture toughness
Which of the following statements are correct?
(A) 1 and 2 (B) 1 and 4 (C) 2 and 3 (D) 3 and 4
[*IES 1999*] (**Ans.** A)

130. Which one of the following processes results in the best accuracy of the hole made?
(A) Drilling (B) Reaming (C) Broaching (D) Boring
[*IES 1999*] (**Ans.** D)

131. Consider the following reasons:
1. Grinding wheel is soft.
2. RPM of grinding wheel is too low.
3. Cut is very fine.
4. An important cutting fluid is used.
A grinding wheel may become loaded due to reasons stated at
(A) 1 and 4 (B) 1 and 3 (C) 2 and 4 (D) 2 and 3
[*IES 1999*] (**Ans.** B)

132. Consider the following statements regarding reaming process:
1. Reaming generally produces a hole larger than its own diameter.
2. Generally rake angles are not provided on reamers.
3. Even numbers of teeth are preferred in reamer design.
Which of these statements are correct?
(A) 1 and 2 (B) 2 and 3 (C) 1 and 3 (D) 1, 2 and 3
[*IES 1999*] (**Ans.** D)

133. Match List-I (Drill bits) with List-II (Applications) and select the correct answer using the codes given below the lists:

List I	*List II*
A. Core drill	1. To enlarge a hole to a certain depth so as to accommodate the bolt head of a screw
B. Reamer	2. To drill and enlarge an already existing hole in a casting
C. Counter bore drill	3. To drill a hole before making internal threads
D. Tap drill	4. To improve the surface finish and dimensional accuracy of the already drilled hole

Codes:

	A	B	C	D		A	B	C	D
(*a*)	1	3	2	4	(*b*)	2	3	1	4
(*c*)	2	4	1	3	(*d*)	3	2	4	1

[*IES 1999*] (**Ans.** c)

134. When milling a slot 20 mm wide × 10 cm long in a rectangular plate 10 cm × 20 cm, the cutting speed = 60 m/min; Diameter of end drill = 20 mm
Number of flutes = 8, Feed = 0.01 mm/flute
Depth of cut = 3 mm
Find the cutting time for the operation to be completed in a single pass. [*GATE 1994*]
Solution. Cutting time is given as,

$$T_m = \frac{l + x + y}{F}, \text{ min}$$

where l = length of job = 100 mm

$$x = \text{Cutter approach} = \frac{D}{2} = 10 \text{ mm}$$

y = Cutter overtravel = 5 mm (say)
F = Table feed, mm/min
$= f_t \times n \times N$
f_t = feed per tooth, per rev. = 0.01 mm
n = number of flutes = 8
N = Spindle rev./min.

$$= \frac{V \times 1000}{\pi D} = \frac{60 \times 1000}{\pi \times 20}$$

$$= 955 \text{ rev./min}$$

$$\therefore \quad T_m = \frac{100 + 10 + 5}{0.01 \times 8 \times 955}$$

$$= \frac{115}{76.4} = 1.505 \text{ min.}$$

135. A grinding wheel A24K7V is specified for finish grinding of a HSS cutting tool. What do you understood about the wheel from the above code? Is this an appropriate choice? *[GATE 1994]*

Solution. Code is A24 K7 V

$A \longrightarrow$ Abrasive type, it is Al_2O_3.

24 $\longrightarrow$ Grain size, it is coarse

$K \longrightarrow$ grade, it is medium grade

7 $\longrightarrow$ Structure, it is medium structure

$V \longrightarrow$ Bond type, it is Vitrified bond

For finish grinding of HSS tool, fine grit is needed. The grit size in the grinding wheel is coarse, so it is not an appropriate choice.

136. Why can we use higher cutting speeds with H.S.S. tools than with carbon steel tools? *[GATE 1994]*

[Hint: Because of the presence of alloying elements]

137. State the Taylor's equation for tool life. *[GATE 1994]*

138. Tool life testing on a lathe under dry cutting conditions gave *n* and *C* of Taylor's tool life equation as 0.12 and 130 m/min respectively. When a coolant was used, C increased by 10%. Find the percentage increase in tool life with the use of coolant at a cutting speed of 90 m/min. *[GATE 2001]*

Solution. $V.T_1^{\,n} = C_1$

$\therefore$ $90.T_1^{\,0.12} = 130$

From here, $T_1 = 21.42$ min.

Now $C_2 = 130 + 13 = 143$

$\therefore$ $90.T_2^{\,0.12} = 143$

$\therefore$ $T_2 = 47.40$ min

$\therefore$ Percentage increase in tool life = 121.29%

139. A lead-screw with half nuts in a lathe, free to rotate in both directions has
(A) V-threads (B) Whitworth threads
(C) Buttress threads (D) ACME threads. (*GATE 2002*) (**Ans.** D)

140. The time taken to drill hole through a 25 mm thick plate with the drill rotating at 300 rpm and moving at a feed rate of 0.25 mm/rev is
(A) 10 s (B) 20 s
(C) 100 s (D) 100 s (*GATE 2002*) (**Ans.** B)

141. The hardness of a grinding whel is determined by the
(A) hardness of abrajuse grains
(B) ability of the bond to retain abrasives
(C) hardness of the bond
(D) ability of grinding wheel to penetrate the workpiece (*GATE 2002*) (**Ans.** B)

142. A built-up-edge is formed while machining
(A) ductile material at high speed
(B) ductile material at low speed
(C) brittle material at high speed
(D) brittle material at low speed
(*GATE 2002*) (**Ans.** B)

143. Trepanning is performed for
(A) finishing a drilled hole
(B) producing a large hole without drilling
(C) truing a hole for alignment
(D) enlarging a drilled hole
(*GATE 2002*) (**Ans.** B)

IES : 2002

144. Consider the following statements :
The strength of a single point cutting tool depends upon
1. Rake angle
2 Clearance angle
3. Lip angle
Which of these statements are correct ?
(*a*) 1 and 3
(*b*) 2 and 3
(*c*) 1 and 2
(*d*) 1, 2 and 3
(**Ans.** *d*)

145. Which of the following materials are used in grinding wheels ?
1. Aluminium oxide
2. Cubic boron nitride
3. Silicon carbide
Select the correct answer using the codes given below :
Codes :
(*a*) 1, 2 and 3
(*b*) 1 and 2
(*c*) 2 and 3
(*d*) 1 and 3
(**Ans.** *a*)

146. Match List I (Machine tools) with List II (Machine tool parts) and select the correct answer using codes given below the lists :

List I	List II
A. Lathe	1. Lead screw
B. Milling machine	2. Rocker arm
C. Shaper	3. Universal indexing
D. Drilling machine	4. Flute

Codes :

	A	B	C	D
(*a*)	4	2	3	1
(*b*)	1	3	2	4
(*c*)	4	3	2	1
(*d*)	1	2	3	4

(**Ans.** *b*)

147. In a machining process, the percentage of heat carried away by the chips is typically
(*a*) 5%
(*b*) 25%
(*c*) 50%
(*d*) 75%
(**Ans.** *d*)

148. Match List I (Machine tools) with List II (Features) and select the correct answer using the codes given below the list :

List I (*Machine Tool*)	List II (*Features*)
A. Lathe	1. Push or Pull tool
B. Drilling machine	2. Ratchet and pawl mechanism
C. Shaper	3. Dividing head
D. Broaching machine	4. Hollow tapered spindle
	5. Face plate

Codes :

	A	B	C	D
(*a*)	2	4	5	1
(*b*)	5	3	2	4
(*c*)	2	3	5	4
(*d*)	5	4	2	1

(Ans. *d*)

149. Which one of the following is the hardest cutting tool material next only to diamond ?
(*a*) Cemented Carbides (*b*) Ceramics
(*c*) Silicon (*d*) Cubic boron nitride **(Ans. *d*)**

150. Crater wear on tools always starts at some distance from the tool tip because at that point
(*a*) cutting fluid does not penerate. (*b*) normal stress on rake face in maximum
(*c*) temperature is maximum (*d*) tool strength is minimum **(Ans. *c*)**

151. A 31.8 mm H.S.S. drill is used to drill a hole in a cast iron block 100 mm thick at a cutting speed of 20 m/min and feed 0.3 mm/rev. If the overtravel of the drill is 4 mm and approach 9 mm, the time required to drill the hole is
(*a*) 1 min 40 s (*b*) 1 min 44 s
(*c*) 1 min 49 s (*d*) 1 min 53 s **(Ans. *d*)**

152. A side and face cutter 125 mm diameter has 10 teeth. It operates at a cutting speed of 14 m/min. with a table traverse 100 mm/min. The feed per tooth of the cutter is
(*a*) 10 mm (*b*) 2.86 mm
(*c*) 0.286 mm (*d*) 0.8 mm **(Ans. *c*)**

153. Why are the lathe-spindles made hollow ?

154. How does application of cutting fluid help in reducing cutting forces and inproving tool life and surface quality ?

155. In which machine tools, teeth of internal spur gears can be cut and which one of those machine tools work fastest ?

156. What are meant by 'A – 60 – K – 8 – V' when used for a grinding wheel specification ?

157. What are the major constituents of high speed steel (HSS) and uncoated single carbides which are used as cutting tool materials ? How are these carbide tool inserts manufactured ?

158. The value of surface roughness '*h*' obtained during the turning operation at a feed '*f*' with a round nose tool having radius '*r*' is given as :

(*a*) $\frac{f}{8r}$ (*b*) $\frac{f^2}{8r}$ (*c*) $\frac{f^3}{8r}$ (*d*) $\frac{f^3}{8r^2}$

[*IES 2003*] **(Ans. *b*)**

159. In economics of machining, which one of the following costs remain constant?
(*a*) Machining cost per piece (*b*) Tool changing cost per piece
(*c*) Tool handling cost per piece (*d*) Tool cost per piece. [*IES 2003*] **(Ans. *c*)**

160. Two tools P and Q have signatures 5°–5°–6°–6°–8°–30°–0 and 5°–5°–7°–7°–8°–15°–0 (both ASA), respectively. They are used to turn components under the same machining conditions. If h_p and h_q denote peak-to-valley heights of surfaces produced by the tools P and Q, the ratio h_p/h_q will be

(*a*) $\frac{\tan 8° + \cot 15°}{\tan 8° + \cot 30°}$ (*b*) $\frac{\tan 15° + \cot 8°}{\tan 30° + \cot 8°}$

(*c*) $\frac{\tan 15° + \cot 7°}{\tan 30° + \cot 7°}$ (*d*) $\frac{\tan 7° + \cot 15°}{\tan 7° + \cot 30°}$ [*GATE 2005*] **(Ans. *b*)**

Hint : See Chapter 1, Art. 1.15.

161. In a machining operation, chip-thickness ratio is 0.3 and the back rake angle of the tool is 10°. What is the value of shear strain?
(*a*) 0.31 (*b*) 0.13 (*c*) 3.00 (*d*) 3.34

[*IES 2004*] **(Ans. *d*)**

162. The rake angle of a cutting tool is 15°, shear angle 45° and cutting velocity 35 m/min. What is the velocity of chip along the tool face?

(*a*) 28.5 m/min. (*b*) 27.3 m/min. (*c*) 25.3 m/min. (*d*) 23.5 m/min.

[*IES 2004*] (**Ans.** *a*)

163. Consider the following statements :

1. As the cutting speed increases, the cost of production initially decreases, then after an optimum cutting speed, it increases.
2. As the cutting speed increases, the cost of production also increases and after a critical value it reduces.
3. Higher feed rate for the same cutting speed reduces cost of production.
4. Higher feed rate for the same cutting speed increases cost of production.

Which of the statements given above is/are correct?

(*a*) 1 and 3 (*b*) 2 and 3 (*c*) 1 and 4 (*d*) 3 only

[*IES 2004*] (**Ans.** *d*)

164. Consider the following statements :

During the third stage of tool wear, rapid deterioration of tool edge takes place because

1. Flank wear is only marginal.
2. Flank wear is large.
3. Temperature of the tool increased gradually.
4. Temperature of the tool increased drastically.

Which of the statements given above are correct ?

(*a*) 1 and 3 (*b*) 2 and 4 (*c*) 1 and 4 (*d*) 2 and 3

[*IES 2004*] (**Ans.** *b*)

165. What are the velocities which come into existance when a metal is cut orthogonally. Show these velocities graphically on a velocity diagram and determine the mathematical relationship in terms of shear and rake angles. [*IES 2004*]

Index